Integrative and Molecular Approach to Brain Function

Integrative and Molecular Approach to Brain Function

Proceedings of the Uehara Memorial Foundation Symposium on Integrative and Molecular Approach to Brain Function, Tokyo, 25–27 June 1996

Editors:

Masao Ito
Chairman of the Organizing Committee
President of the Science Council of Japan
Director-General, Frontier Research Program,
Institute for Physical and Chemical Sciences (RIKEN)

Yasushi Miyashita
Vice-chairman of the Organizing Committee
Professor, the First Department of Physiology
Faculty of Medicine, University of Tokyo

1996

ELSEVIER
Amsterdam – Lausanne – New York – Oxford – Shannon – Tokyo

International Congress Series No. 1116
ISBN 0 444 82380 8

This book is printed on acid-free paper.

Published by:
Elsevier Science B.V.
P.O. Box 211
1000 AE Amsterdam
The Netherlands

In order to ensure rapid publication this volume was prepared using a method of electronic text processing known as Optical Character Recognition (OCR). Scientific accuracy and consistency of style were handled by the author. Time did not allow for the usual extensive editing process of the Publisher.

Printed in the Netherlands

Preface

The present advancement in the brain is characterized by the dichotomy of the molecular approaches to cellular functions in the brain and the integrative approaches to the mechanisms of brain functions. This book contains the proceedings of the Uehara Foundation Symposium held in 1996 in Tokyo, which aimed to synthesize the two lines of approaches.

Our knowledge at the cellular and molecular levels is expanding rapidly, as is represented by the 12 articles from sessions 1 and 2. These articles give detailed descriptions of the chemical mechanisms in neurons and glia which involve numerous messengers, receptors, growth factors and trophic factors, as well as the underlying gene-regulatory processes. One may expect that in the near future we will be able to fully dissect the chemical and gene-regulatory processes which underlie the functions of neurons and glia, and will be able to apply the knowledge to prevention or treatment of neurological and psychiatric diseases, or to the control of aging. Indeed, such medical application will be achieved by means of the creation of new drugs which interact with the chemical processes in neurons and glia, gene therapy which control the gene-regulatory processes, or the transplantation of healthy neurons to lesioned brain areas.

However, the alternate perspective is the view of the brain as a highly complex system. Neurons form sophisticated networks throughout the brain, and a combination of neuronal networks, sensory organs, and effector organs constitutes a neural system which represents a brain function. The brain is the highest level integrator of molecules, cells, tissues, organs, and the body as a whole. The integrative aspect of the brain can be described by using an entirely different set of basic concepts from those which describe molecular and cellular aspects of the brain, as is evident in the 17 articles from sessions 3 to 5.

In order to understand the principles governing the basic functional structure of the central nervous system, it is helpful to follow the evolutionary development of the brain. In lower vertebrates, the brain with the spinal cord attached contains three major executive functions, i.e., reflexes and compound movements (such as locomotion), as well as behavior which is operated by the most rostral part of the brain stem, i.e., hypothalamus. On the other hand, the brainstem, cerebellum, basal ganglia, and limbic system exert regulatory functions upon the five executive systems. The sleep-wakefulness centers in the brainstem switch operational states of the executive functions. The cerebellum provides adaptive mechanisms, and the basal ganglia appears to provide stabilizing mechanisms in the executive functions by selection among competitive activities in the executive functions. The limbic system secures the purposefulness in the operation of the executive functions by means of positive and negative reinforcement. In addition, development of the neocortex generated two additional executive functions, i.e., the sensorimotor cortical function in mammals, and the association cortical functions dominating in primates. These two neocortical executive functions are also under the influences of the four regulatory functions in the same way as the three executive functions in the spinal cord and brainstem.

The combination of the five executive functions and the four regulatory functions explains major functional features of the brain. Our efforts in the area of integration have been successful for the mechanistic mechanisms of reflexes and compound movements, but become more and more difficult as we progress along evolutional development toward the association cortical function. Adaptive mechanisms of the cerebellum have been well dissected, but specific mechanisms of other regulatory systems are much less understood.

The major research strategy employed at the molecular and cellular levels is to pinpoint a molecule, a set of them, a microstructure which is responsible for a cellular function. That undertaken at integrative and functional levels is to comprehend the mechanisms which underlie a system function of the brain. These two strategies will be mutually beneficial, and the need to combine them in brain science will become increasingly important. This monograph thus helps us envision the future of brain science which will be built upon the fusion of the two research strategies.

Masao Ito

Contents

Dynamic organization of cerebral neural network

Mechanisms of memory as the basis of behaviour

Noninvasive measurements in humans

Molecular biology of learning and memory

Integrative and Molecular Approach to Brain Function.
M. Ito and Y. Miyashita, editors.

Synapsins: emerging concepts and unresolved puzzles

Thomas C. Südhof
Howard Hughes Medical Institute and Department of Molecular Genetics, The University of Texas Southwestern Medical School, Dallas, Texas, USA

Introduction

Major progress has been made in recent years in the identification and functional characterization of proteins of presynaptic nerve terminals (reviewed in [1]). A large number of proteins has been described with a diverse range of demonstrated or suspected functions. In addition to the synapse, closely related isoforms of almost all of these proteins or sometimes the proteins themselves have also been found in nonneuronal cells where they are thought to perform functions in membrane traffic that are analogous to their synaptic functions. At present there is only one exception to this observation: Synapsins, abundant phosphoproteins of synaptic vesicles, are currently only known to function at synapses. No ubiquitous isoforms for synapsins have been discovered. This sets synapsins apart and makes them an important tool in studying how nerve terminals differ from other eukaryotic membrane trafficking systems. Although synapsins have been intensely studied, their in vivo functions are still obscure. A large body of confusing, sometimes even contradictory data on synapsins has accumulated over the years that raise a number of puzzling questions. In the present review, I will attempt to summarize these data and to assess their implications.

Historical background

In the early 1970s Paul Greengard, following closely in the footsteps of Earl Sutherland and Edwin Krebs, reasoned that protein phosphorylation was likely to have a major role in the nervous system [2]. In a systematic study of substrates for cAMP-dependent protein kinases (PKA) in brain homogenates, Greengard identified synapsins as among the most abundant substrate proteins for PKA and originally named them proteins I and III in a numerical designation of major substrate proteins [3]. Pioneering studies over the following 15 years in the Greengard laboratory established that proteins I and III, subsequently renamed synapsins I and II [4,5], are major components of synaptic vesicles and that they are physiological substrates for

Address for correspondence: Thomas C. Südhof, Howard Hughes Medical Institute and Department of Molecular Genetics, The University of Texas Southwestern Medical School, Dallas, TX 75235, USA. Tel.: +1-214-648-5022. Fax: +1-214-648-6426.

two Ca^{2+}, calmodulin-dependent protein kinases (CaM Kinases), CaM Kinase I and CaM Kinase II (reviewed in [2]). Together these studies established synapsin I as the most well-published phosphoprotein in brain whose phosphorylation properties and biochemical characteristics are described in an exquisite detail unparalleled by any other protein.

Structure of synapsins: insects and mammals

The full-length structures of all rat synapsins as well as those of bovine and human synapsin I were published in 1989 and 1990 [5,6]. These structures demonstrate that synapsins constitute a family of closely related proteins comprised of four forms, synapsins Ia, Ib, IIa, and IIb (Fig. 1A). The four synapsins are derived by alternative splicing from the primary transcripts of two genes. For both synapsins Ia and Ib and for synapsins IIa and IIb alternative splicing only affects the very C-terminus. Here

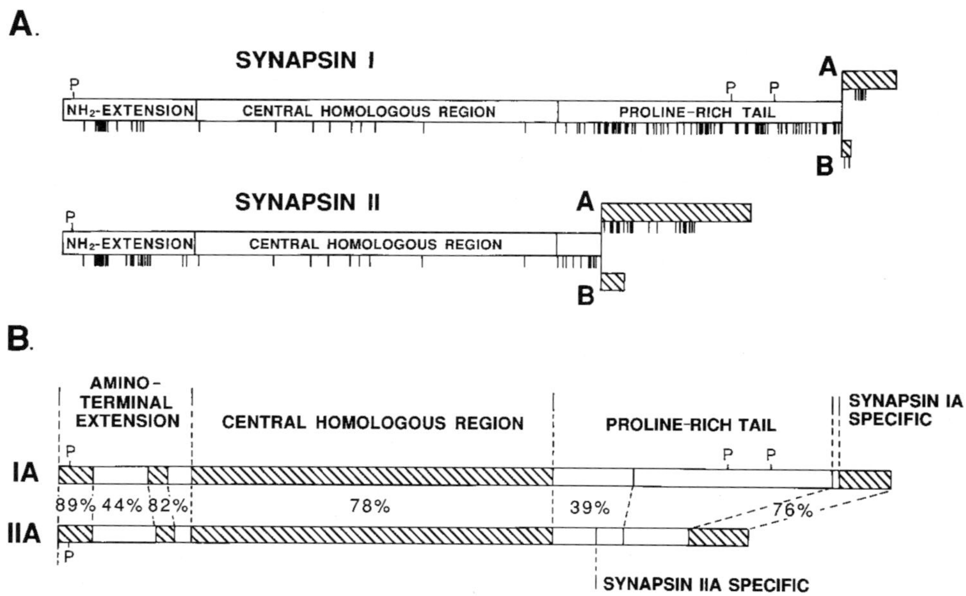

Fig. 1. Structures of synapsins Ia, Ib, IIa, and IIb, and homologies between them. **A**: Diagram of the four synapsins. Synapsins Ia and Ib, and synapsins IIa and IIb are derived from the primary transcripts of the synapsin I and II genes, respectively, by alternative splicing that results in distinct C-termini (hatched). Alternative splicing results in significantly different C-termini for synapsin IIa and IIb but rather small differences for synapsins Ia and Ib. Phosphorylation sites are indicated by a "P" above the bar diagrams. The synapsins have a very nonuniform distribution of amino acids in the sequence as indicated by notches below the bar diagram. These notches indicate positions of proline residues which are clustered in the N-terminal extension and C-terminal proline-rich tail but almost absent from the central homologous region. Note that not only synapsin I but also synapsin II has a proline-rich C-terminus. **B**: Comparison of synapsin Ia with synapsin IIa. Homologies between the synapsins are shown as % identity between the bar diagrams. Highly homologous regions (>70% identity) are hatched. The C-termini of synapsins Ia and IIa that are created by alternative splicing are highly homologous to each other; no such homology is present between the unique C-termini of synapsins Ib and IIb (not shown).

alternative splicing results in a 45-residue C-terminal sequence for synapsin Ia that is replaced by a unique 9 residue sequence in synapsin Ib. Similarly, synapsin IIa contains a 138 residue C-terminal sequence that is absent from synapsin IIb which only exhibits a 21 residue unique C-terminus. Thus synapsin Ia is very similar to Ib, and synapsin IIa to IIb, differing only in their very C-terminal sequences.

Comparison of synapsins I and II revealed that they are highly homologous to each other almost over their entire length (Fig. 1B). The distribution of homology, however, is uneven across the molecules. At the N-terminus, all four synapsins share a highly conserved short domain which includes a phosphorylation site for CaM Kinase I and PKA. This conserved domain is followed by a short rather divergent sequence primarily containing uncharged short-chain amino acids. In the middle of the synapsins, a large central region is present that is highly homologous between different synapsins. Following the central region, all synapsins contain a proline-rich sequence that diverges into the respective synapsin I- and synapsin II-specific domains. Two aspects of the C-terminal structures of synapsins are particularly striking: First, synapsins Ia and Ib contain an extended proline-rich tail that is polybasic and contains two phosphorylation sites for CaM Kinase II. No CaM Kinase II phosphorylation sites are observed in synapsins IIa and IIb. Second, the very C-termini of synapsins Ia and IIa created by alternative splicing are highly homologous to each other (Fig. 1B) whereas no such homology is observed between the unique C-termini of synapsins Ib and IIb. Comparison of the rat, bovine, and human synapsin I sequences demonstrated that they are highly conserved with 95% invariant residues [6].

Taken together, the sequence analyses characterize synapsins as conserved mammalian proteins that are composed of a highly homologous central domain flanked by an N-terminal phosphorylation sequence and C-terminal variable domains. Because of their abundance and conservation in mammals, it was long suspected that synapsins should also be present in invertebrates. However, only recently has a *Drosophila* synapsin homologue been identified [7]. Only a single synapsin gene in *Drosophila* was found that specifies production of at least two proteins which differ in the C-terminus similar to mammalian synapsins. However, the two *Drosophila* synapsin forms are not created by alternative splicing as in mammals but instead by read-through of an amber stop codon, resulting in a smaller, more abundant protein when the stop codon is used and a larger, less abundant form when the stop codon is suppressed [6]. Sequence comparisons revealed that *Drosophila* synapsin is 50% identical with rat synapsins I and II in their entire central homologous domain extending over 312 amino acids. Strikingly, however, the N- and C-terminal domains of mammalian synapsins are not conserved in *Drosophila*. Although mammalian and *Drosophila* synapsins have the same general architecture containing a shared central homologous region flanked by a shorter N-terminal extension and a variable C-terminal extension, there is no sequence homology between the N- and C-terminal domains of mammalian and *Drosophila* synapsins, and the phosphorylation sites of mammalian synapsins are not conserved. The only homology detected was between the very C-terminus of the shorter *Drosophila* synapsin and of mammalian synapsins

Ia and IIa which exhibit 31.3% identity (Fig. 2).

Together these results suggest that the central homologous regions of synapsins are their functionally most important region, and that the phosphorylation sites may have important modulatory roles but are not necessarily central to the function of synapsins. Furthermore, the results point to a possible important role for the C-termini of synapsins Ia and IIa.

Expression of synapsins

Outside of brain, synapsins have only been detected in neuroendocrine cells. In brain, synapsins are present in almost all neurons tested where they invariably localize to presynaptic nerve terminals [4]. Only a subclass of neurons does not express synapsins. This subclass includes photoreceptor cells and hair cells and is characterized by forming atypical synapses called ribbon synapses (reviewed in [1]). The hallmark of ribbon synapses consists in the presence of a cytoplasmic dense body close to the active zone which appears to collect synaptic vesicles in these neurons and to guide them to the active zone. Ribbon synapses contain members of all other synaptic vesicle protein families except for synapsins [8]. Physiologically, they are characterized by a high rate of tonic neurotransmitter release during rest that is inhibited by excitation of the cells. The fact that synapsins are selectively absent from ribbon synapses could be explained by at least three hypotheses:

1. Synapsins have an inhibitory regulatory function that would interfere with the tonic release process at ribbon synapses.
2. Synapsins have an essential function at synapses that is replaced in ribbon synapses by a distinct ribbon-specific component.
3. Synapsins have a function that is not needed for the tonic release mode of ribbon synapses, such as increasing rapid turnover rates during repetitive firing.

These hypotheses were partially tested by expression of synapsin Ib in photoreceptor cells in transgenic mice [9]. In these mice, synapsin Ib was correctly targeted to synaptic vesicles but had no measurable effect on photoreceptor function, ruling out hypothesis 1 but still allowing hypotheses 2 and 3.

In brain, the four isoforms of synapsin are distributed rather homogeneously. For almost all synaptic vesicle proteins, multiple isoforms have been described that are almost always distributed in a very heterogeneous manner. Again synapsins form an exception in that in most neurons at least three isoforms appear to be coexpressed

```
                 *** * : :    :  *:   : ::   ** *   *** **  * *
Dros.          QTSQSSTISSSVSR AGQRPPQTQNSVVEDAEDTMKNLRKTFAGIFGDM    596
Rat Ia         NKSQSLTNAFNLPEPAPPRPSLSQDEV  KAE TIRSLRKSFASLFSD     704
Rat IIa        NKSQSLTNAFSFSESSFFRSSANEDEA  KAE TIRSLRKSFASLFSD     586
```

Fig. 2. Alignment of the C-terminal sequences of the shorter form of *Drosophila* synapsin (Dros.) [7] with those from rat synapsins Ia and IIa (Rat Ia and IIa, respectively) [5]. Sequences are shown in single letter amino acid code and numbered on the right. Residues that are identical in all three sequences are marked by an asterisk, and residues shared between the *Drosophila* sequence and at least one of the rat sequences by a colon. Overall identity is 15 out of 48 residues.

[10]. Detailed studies identified classes of synapses in brain in which one or two isoforms of synapsins are lacking [5] but the absence of more than one isoform from a synapse appears to be more of an exception than a rule. This implies that the four synapsins probably do not have alternative functions but perform functions that are unique to individual isoforms and coexecuted in most neurons. This implication was partially confirmed in the knockout studies (see below).

Cell biology of synapsins

In brain or cultured neurons, synapsins are one of the best markers for synaptic vesicles, suggesting that they are targeted efficiently to these organelles. However, they do not contain a classical transmembrane region, and it is unclear how they are attached to synaptic vesicles. Since they are peripheral membrane proteins, it is possible that they could dissociate during membrane depolarization. Indeed, it has been reported that a low percentage of synapsin I becomes soluble after exocytosis [11] although a separate study from the same laboratory concluded that synapsin remains attached to synaptic vesicles after exocytosis [12]. Future experiments will have to investigate if synapsins dissociate from synaptic vesicles after exocytosis, maybe by studying if clathrin-coated vesicles isolated in low ionic strength conditions contain synapsins.

Since synapsins I and II are both specific for synaptic vesicles and since they are highly homologous to each other, it seems likely that they are associated with vesicles via similar domains. Biochemical studies have demonstrated that the N-terminal domain of synapsin I spontaneously associates with phospholipids and contacts the hydrophobic interior of the bilayer, suggesting that this may bind synapsins to vesicles [13]. This suggestion is compatible with the observation that in transgenic mice, the N-terminal but not the C-terminal domain of synapsin I is correctly targeted to synaptic vesicles [9]. Elucidation of the targeting domain and of the mechanism by which synapsins are targeted would be very interesting and could yield further insights into its function.

Synapsins are essential regulators of neurotransmitter release

Mice that do not express proteins derived from one or both synapsin genes are viable and fertile [14,15]. This establishes the baseline: synapsins are not essential for life in a cage, they are not indispensable for the mechanics of the release apparatus. However, mice lacking synapsins have a series of symptoms that demonstrate an essential and partially redundant function for these proteins in regulating neurotransmitter release.

First, they suffer from spontaneous seizures whose frequency is inversely proportional to the number of healthy synapsin I and II alleles present [15].

Second, although they form normal synapses and exhibit no major changes in synaptogenesis, synapse numbers, or nerve terminal size, the synapses in synapsin-deficient mice contain fewer synaptic vesicles [15]. Again, the decrease in synaptic vesicle proteins is inversely proportional to the number of healthy synapsin I and II alleles present. However, the vesicles are still clustered close to the active zone, indicating that the clustering mechanism is distinct from synapsin function.

Third, mice lacking synapsins exhibit changes in synapse function that are different between synapsin I knockouts, synapsin II knockouts, or double knockouts [14,15]. In synapsin I knockouts, paired pulse facilitation, a form of short-term synaptic plasticity, is increased. Furthermore, the response to repetitive stimulation is decreased. By contrast, in synapsin II knockouts posttetanic potentiation, another form of short-term synaptic plasticity, is decreased. In addition, the response to repetitive stimulation is also decreased in synapsin II knockouts. Double knockouts resemble in phenotype synapsin II knockouts except that the phenotype is much more severe.

There is currently no single concept for synapsin function that would satisfactorily explain the phenotype of the knockout mice. However, a few conclusions can be drawn beyond the fundamental insight that synapsins are primarily regulatory proteins and are not involved in the mechanics of release. The presence of distinct unique phenotypes in the synapsin I and the synapsin II knockouts in addition to shared symptoms indicates that the two principal synapsin genes exert common and unique functions. Shared functions are manifested by the observation that epileptic seizures, synaptic depression upon repetitive stimulation, and the decrease in synaptic vesicle levels are substantially increased in the double knockouts over each of the single knockouts. The fact that the electrophysiological phenotype of the synapsin I knockout is much milder than that of the synapsin II knockout in spite of its greater abundance suggests that synapsin I may have both positive and negative functions that are balanced in the presence of synapsin II. This would also explain the major increase in the phenotype of the double knockout over either the synapsin I or the synapsin II knockout. However, although the synapsin knockouts rule out some of the essential functions proposed for synapsins, such as a role in synaptogenesis or membrane fusion, they do not themselves immediately suggest an obvious function for synapsins except for that in the rather large realm of regulation at the active zone.

A puzzling result was recently reported with a synapsin I knockout which revealed that in these knockouts the cross-bridges between vesicles were absent [16]. This is an amazing result because synaptic vesicles are still clustered and still contain the highly homologous synapsin II proteins in the knockouts. Future experiments will have to test if this observation corresponds to a true change in synaptic vesicle morphology.

Synapsins bind to multiple proteins in vitro

Another approach to study synapsin function is to investigate its properties and binding proteins. Since synapsin I is easy to purify but synapsin II is relatively resistant to purification, almost all of these studies were performed on synapsin I. These studies demonstrated that synapsin I is a highly surface active molecule which forms a stable monolayer at an air-water interface and has one of the highest collapse pressures of all surface-active proteins [17]. Synapsin I spontaneously binds to acidic phospholipid vesicles, probably via the N-terminal half that is homologous to that of synapsin II, and can be labeled with hydrophobic probes, indicating it may actually insert partially into the bilayer [13]. C-terminal phosphorylation by CaM Kinase II but not N-terminal phosphorylation by CaM Kinase I or PKA cause a major conformational change in synapsin I suggestive of a phosphorylation-dependent folding of synapsin I into a more globular protein [18]. Together these studies indicate that synapsin I is a peripheral membrane protein, maybe even a monotopic membrane protein. It seems likely that the interactions of synapsins I and II with phospholipids and synaptic vesicles are mediated by sequences that are shared between them.

Many groups have attempted to identify binding proteins for synapsins as an approach to gain insight into its function. A bewildering number of biochemically interacting proteins was identified by this approach. These include the major elements of the cytoskeleton, actin microfilaments, neurofilaments, microtubuli, and spectrin [19–23]. The interaction of synapsin with actin microfilaments and with neurofilaments is modulated by the C-terminal phosphorylation of synapsin I [19,20,23]. All of the binding reactions appear to be of high affinity to occur in vivo and could potentially be physiological, although none has been demonstrated in vivo. In addition, synapsin I has been shown to bind to calmodulin and annexin VI in a Ca^{2+}-dependent manner [24,25]. Furthermore, synapsins binds to the MARCKS protein and to Grb2 [26,27]. Although models for synapsin function have been elaborated based on these in vitro interactions, it is unclear which of these interactions are physiological. Because of the large number of interacting proteins, no unequivocal model for synapsin function can be based on these interactions. Since synapsin I is highly charged and amphipathic, it has a tendency to stick to polyanionic surfaces such as those formed by cytoskeletal elements, and this could explain the actin bundling activity of synapsin I [28]. The active zone is relatively poor in cytoskeletal proteins which appear to be much more abundant in surrounding regions of the presynaptic plasma membrane [29]. In view of the regulatory function of synapsins at the active zone, the interaction of synapsins with the cytoskeleton seems to be an unlikely candidate for important interactions at this point.

What we have always wanted to know about synapsins

Clearly synapsins are important regulatory proteins with a function in mature synapses. The somewhat contradictory results in the literature pose a number of questions which will have to be resolved in future studies:

1. What do synapsins interact with physiologically? In view of their stickiness, new techniques are needed to address this question. Particularly useful would be assays to test in vivo interactions.
2. Massive depolarization of nerve terminals leads to quantitative phosphorylation of synapsins. However, what happens during a single action potential? Is there a subpopulation of synapsins that is preferentially phosphorylated, maybe synapsins

at the active zone? Experiments with synaptosomes suggest that Ca^{2+} influx at the active zone selectively increases synapsin phosphorylation, supporting this hypothesis [30].

3. What functional properties of synapsins are regulated by phosphorylation? The fact that the phosphorylation sites are not conserved in *Drosophila* synapsin suggests that the phosphorylation events may modulate a central function of synapsins which is shared between vertebrates and invertebrates but only regulated by this mechanism in vertebrates.
4. The presence of two sequence regions in synapsins that are conserved between vertebrates and invertebrates indicates that these sequences are responsible for the physiological actions of synapsins. The central homologous domain is a large region with a very high degree of conservation (Fig. 1); this region probably mediates the central functions of synapsins. In addition, there is weak homology between the C-termini of synapsins Ia and IIa and the C-terminus of the shorter form of *Drosophila* synapsin, indicative of a second possibly important functional region. Interacting partners for these regions would be particularly valuable.
5. Are there nonneuronal synapsins? Until the mammalian genome has been completely sequenced, this question will not be answered. However, it seems evolutionarily unlikely that synapsins will not have a distant relative in non-neuronal cells, although current results indicate that different from other synaptic proteins, synapsins do not have close relatives.

References

1. Südhof TC. The synaptic vesicle cycle: A cascade of protein-protein interactions. Nature 1995; 375:645–653.
2. Greengard P. Neuronal phosphoproteins. Molec Neurobiol 1987;1:81–119.
3. Johnson EM, Ueda T, Maeno H, Greengard P. Adenosin 3′:5′-monophosphate-dependent phosphorylation of a specific protein in synaptic membrane fractions from rat cerebrum. J Biol Chem 1972; 247:5650–5652.
4. De Camilli P, Cameron R, Greengard P. Synapsin I (protein I), a nerve terminal-specific phosphoprotein. I. Its general distribution in synapses of the central peripheral nervous system demonstrated by immunofluorescence in frozen plastic sections. J Cell Biol 1983a;96:1337–1354.
5. Südhof TC, Czernik AJ, Kao H, Takei K, Johnston PA, Horiuchi A, Wagner M, Kanazir SD, Perin MS, DeCamilli P and Greengard P. Synapsins: mosaics of shared and individual domains in a family of synaptic vesicle phosphoproteins. Science 1989;245:1474–1480.
6. Südhof TC. The structure of the human synapsin I gene and protein. J Biol Chem 1990;265: 7849–7852.
7. Klagges BRE, Heimbeck G, Godenschwege T, Hobauer A, Pflugfelder GO, Reifegerste R, Reisch D, Schaupp M, Buchner S, Buchner E. Invertebrate synapsins: A single gene codes for several isoforms in *Drosophila*. J Neurosci 1996;(In press).
8. Ullrich B, Südhof TC. Distribution of synaptic markers in the retina: Implications for synaptic vesicle traffic in ribbon synapses. J Physiol (Paris) 1994;88:249–257.
9. Geppert M, Ullrich B, Green DG, Takei K, Daniels L, De Camilli P, Südhof TC, Hammer RE. Synaptic targeting domains of synapsin I revealed by transgenic expression in photoreceptor cells. EMBO J 1994;13:3720–3727.
10. Ullrich B, Südhof TC. Differential distribution of novel synaptotagmins: Comparison to synapsins. Neuropharmacology 1995;34:1371–1377.

11. Sihra TS, Wang JKT, Gorelick FS, Greengard P. Translocation of synapsin I in response to depolarization of isolated nerve terminals. Proc Natl Acad Sci USA 1989;86:8108–8112.
12. Torri-Tarelli F, Villa A, Valtorta F, De Camilli P, Greengard P, Ceccarelli B. Redistribution of synaptophysin and synapsin I during α-latrotoxin-induced release of neurotransmitter at the neuromuscular junction. J Cell Biol 1990;110:449–459.
13. Benfenati F, Greengard P, Brunner J, Bahler M. Electrostatic and hydrophobic interactions of synapsin I and synapsin I fragments with phospholipid bilayers. J Cell Biol 1989;108:1851–1862.
14. Rosahl TW, Geppert M, Spillane D, Herz J, Hammer RE, Malenka RC, Südhof TC. Short term synaptic plasticity is altered in mice lacking synapsin I. Cell 1993;75:661–670.
15. Rosahl TW, Spillane D, Missler M, Herz J, Selig DK, Wolff JR, Hammer RE, Malenka RC, Südhof TC. Essential functions of synapsins I and II in synaptic vesicle regulation. Nature 1995;375: 488–493.
16. Takei Y, Harada A, Takeda S, Kobayashi K, Terada S, Noda T, Takahashi T, Hirokawa N. Synapsin I deficiency results in the structural change in the presynaptic nerve terminals in the murine nervous system. J Cell Biol 1995;131:1789–1800.
17. Ho M, Bahler M, Czernik AJ, Schiebler W, Kezdy FJ, Kaiser ET, Greengard P. Synapsin I is a highly surface-active molecule. J Biol Chem 1991;266:5600–5607.
18. Benfenati F, Neyroz P, Bahler M, Masotti L, Greengard P. Time-resolved fluorescence study of the neuron-specific phosphoprotein synapsin I. J Biol Chem 1990;265:12584–12595.
19. Petrucci TC, Morrow J. Synapsin I: An actin-bundling protein under phosphorylation control. J Cell Biol 1987;105:1355–1363.
20. Bähler MJ, Greengard P. Synapsin I bundles F-actin in a phosphorylation-dependent manner Nature 1987;326:704–707.
21. Baines AJ, Bennett V. Synapsin I is a spectrin-binding protein immunologically related to erythrocyte protein 4.1. Nature 1985;315:410–413.
22. Baines AJ, Bennett V. Synapsin I is a microtubule-bundling protein. Nature 1986;319:145–147.
23. Steiner JP, Ling E, Bennett V. Nearest neighbor analysis for brain synapsin I. Evidence for in vitro reassociation assays for association with membrane protein(s) and the M_r = 68,000 neurofilament subunit. J Biol Chem 1987b;262:905–914.
24. Hayes NV, Bennett AF, Baines AJ. Selective Ca^{2+}-dependent interaction of calmodulin with the head domain of synapsin I. Biochem J 1991;275:93–97.
25. Inui M, Watanabe T, Sobue K. Annexin VI binds to a synaptic vesicle protein, synapsin I. J Neurochem 1994;63:1917–1923.
26. Mizutani A, Tokumitsu H, Hidaka H. Acidic calmodulin-binding protein, ACAMP-81, is MARCKS protein interacting with synapsin I. Biophys Biochem Res Commun 1992;182:1395–1401.
27. McPherson, Czernik AJ, Chilcote TJ, Onofri F, Benfenati F, Greengard P, Schlessinger J, De Camilli P. Interaction of Grb2 via its Src homology 3 domains with synaptic proteins including synapsin I. Proc Natl Acad Sci USA 1994;91:6486–6490.
28. Tang JX, Janmey PA. The polyelectrolyte nature of F-actin and the mechanism of actin bundle formation. J Biol Chem 1996;271:8556–8563.
29. Peters A, Palay SL, Webster H deF. The Fine Structure of the Nervous System: Neurons Their Supporting Cells, 3rd edn. New York, Oxford: Oxford University Press, 1991.
30. Sihra TS, Bogonez E, Nicholls DG. Localized Ca^{2+} entry preferentially effects protein dephosphorylation, phosphorylation, and glutamate release. J Biol Chem 1992;267:1983–1989.

Integrative and Molecular Approach to Brain Function.
M. Ito and Y. Miyashita, editors.

Mode of action of Rab3A system in neurotransmitter release

Hiromichi Shirataki, Takuya Sasaki, and Yoshimi Takai
Department of Molecular Biology and Biochemistry, Osaka University Medical School, Suita, Japan

Abstract. Evidence is accumulating that Rab3A small G protein and its downstream target, named rabphilin3, both of which have been discovered in our laboratory, are involved in Ca^{2+}-dependent exocytosis, particularly in Ca^{2+}-dependent neurotransmitter release. Moreover, rabphilin3 interacts with Ca^{2+} and may serve as a Ca^{2+} sensor for neurotransmitter release.

Key words: Ca^{2+} sensor, rabphilin3, small G protein, vesicle transport.

Introduction

The Rab family small G proteins, consisting of more than 30 members, regulate intracellular vesicle trafficking. Of the Rab family members, Rab3A, a member of the Rab3 subfamily, has been most extensively investigated, and evidence has accumulated that Rab3A regulates Ca^{2+}-dependent exocytosis, particularly neurotransmitter release. Regulators of Rab3A, named Rab GDP dissociation inhibitor (Rab GDI), Rab3 GDP/GTP exchange protein (Rab3 GEP), and Rab3 GTPase activating protein (Rab3 GAP), and a downstream target of Rab3A, named rabphilin3, have been isolated and characterized. Our current model for the mode of action of Rab3A in neurotransmitter release is schematically shown in Fig. 1. The GDP-bound form of Rab3A is complexed with Rab GDI and stays in the cytosol. After it is released from Rab GDI by the action of a putative factor, named Rab3 GDI displacement factor (Rab3 GDF), and is converted to the GTP-bound form by the action of Rab3 GEP, the GTP-bound form is associated with rabphilin3 on synaptic vesicles, which are consequently transported to the presynaptic plasma membrane. Rabphilin3 complexed with the GTP-bound form of Rab3A interacts with an acceptor protein on the membrane and the vesicles are docked with the membrane. Before the fusion of the vesicles with the membrane, the GTP-bound form is converted to the GDP-bound form by the action of Rab3 GAP. Once the GDP-bound form is produced on the membrane, it is complexed with Rab GDI and is translocated from the membrane to the cytosol. Of these Rab3A-interacting molecules, rabphilin3 interacts with Ca^{2+} and may serve as a Ca^{2+} sensor for neurotransmitter release. In this review article, we focus on rabphilin3 and describe its function and mode of action in neurotransmitter release.

Address for correspondence: Yoshimi Takai, Department of Molecular Biology and Biochemistry, Osaka University Medical School, Suita 565, Japan. Tel.: +81-6-879-3410. Fax.: +81-6-879-3419. E-mail: ytakai@molbio.med.osaka-u.ac.jp

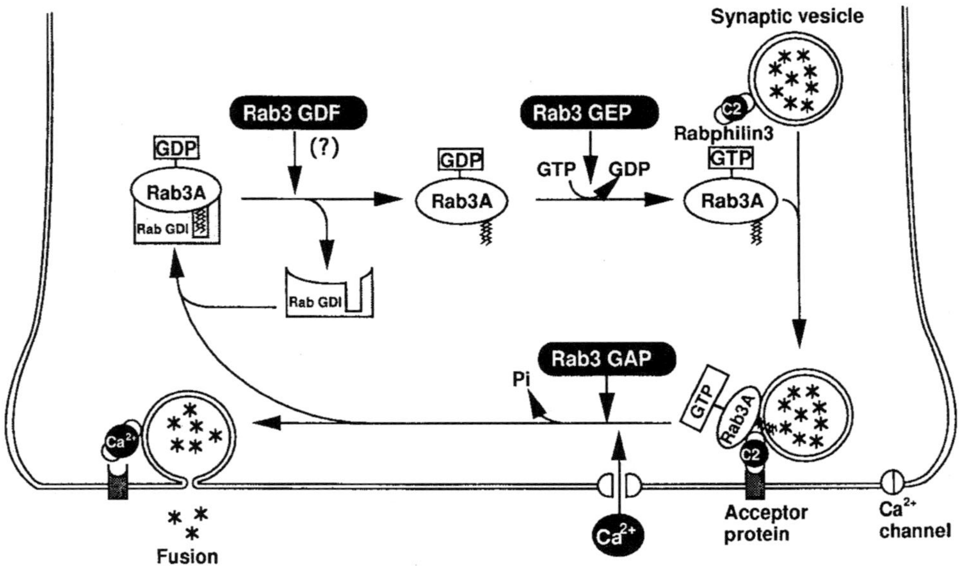

Fig. 1. A model for the mode of action of Rab3A in neurotransmitter release.

Isolation of rabphilin3

We attempted to isolate a downstream target of Rab3A from bovine brain by use of a cross-linking technique [1]. One protein, which was specifically cross-linked with the GTP-bound form of radioiodinated Rab3A, was detected in the bovine brain crude membrane fraction. This cross-linking was competitively inhibited by the GTP-bound form of nonradioiodinated Rab3A, but not by the GDP-bound form. By use of this assay method, this protein was purified and its partial amino acid sequence was determined. On the basis of this information, its cDNA was cloned from a bovine brain cDNA library, and its primary structure was determined [2]. We named this protein rabphilin-3A. Recombinant rabphilin-3A indeed prefers the GTP-bound form of Rab3A to the GDP-bound form. It also interacts with the GTP-bound form of Rab3B, Rab3C, and Rab3D, but does not interact with other Rab family members, including at least Rab1B, Rab5, and Rab11 [3,4] (our unpublished data). It may be therefore better to change the name of rabphilin-3A to rabphilin3.

Structure of rabphilin3

Rabphilin3 shows a calculated Mr of about 78,000. Structural analysis indicates that rabphilin3 has no transmembrane segment but has two C2-like domains (C2A domain and C2B domain) at its C-terminal region as schematically shown in Fig. 2 [2]. The C2 domain was originally found in protein kinase C, which is activated by Ca^{2+} and phospholipid, particularly phosphatidylserine [5]. Protein kinase C has one C2

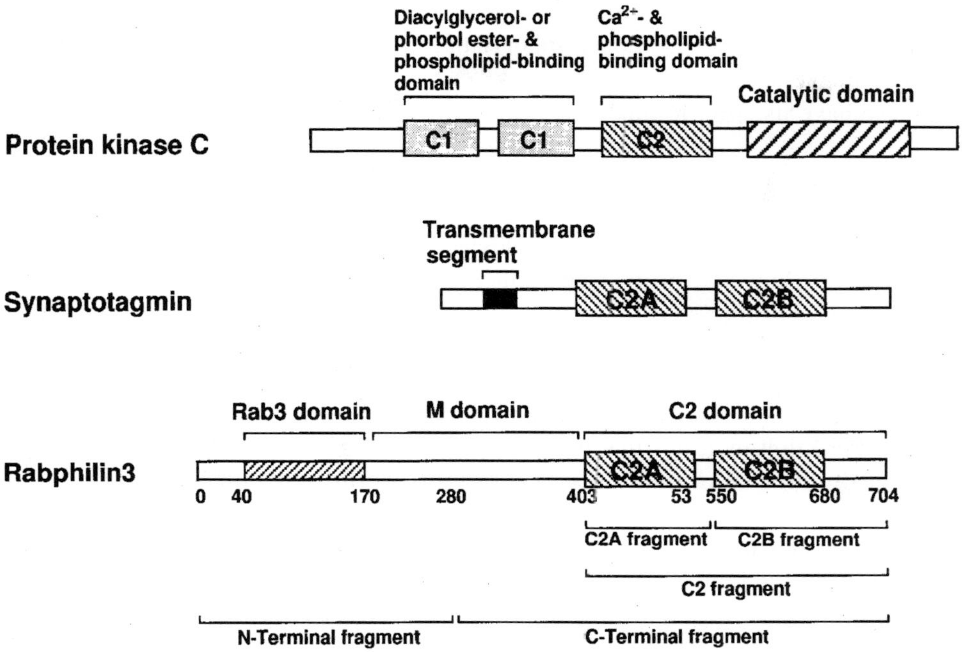

Fig. 2. Structures of rabphilin3, protein kinase C, and synaptctagmin.

domain, but synaptotagmin has been shown to have two C2-like domains (C2A domain and C2B domain) and to interact with Ca^{2+} and phospholipid, particularly phosphatidylserine [6]. This protein constitutes a family consisting of at least eight members [7]. The originally isolated one, synaptotagmin I, is specifically localized on synaptic vesicles through its one transmembrane segment and has been suggested to serve as a Ca^{2+} sensor for neurotransmitter release [6,8,9]. Recombinant rabphilin3 actually interacts with Ca^{2+} and phospholipid, particularly phosphatidylserine, at the C2-like domains, whereas it interacts with Rab3A at its N-terminal region [10]. Thus, rabphilin3 has at least two functionally different domains, the N-terminal Rab3A-binding domain and the C-terminal Ca^{2+}- and phospholipid-binding C2-like domain. We name these domains Rab3 domain and C2 domain, respectively. Recent analysis indicates that Rab3A interacts with amino acid residues 40–170 of rabphilin3 [11]. This region has two repeats of a Cys-X2-Cys (X: any amino acid) structure which is also found in the region of Raf protein kinase which interacts with Ras and Rap1.

In addition to these two domains of rabphilin3, the middle region shown in Fig. 2 is rich in prolines and serine/threonine phosphorylation sites and we name it M domain. Indeed, rabphilin3 is phosphorylated by protein kinase A and calmodulin-dependent protein kinase II (CaMKII) in a cell-free system [12–14]. We have shown that Ser^{34}, Thr^{205}, Thr^{209}, and Thr^{537} are phosphorylated by CaMKII [12], whereas another group has shown that Ser^{234} and Ser^{274} are phosphorylated [14]. The reason for these different results is unknown, but may be due to different experimental conditions. The function of the phosphorylation of rabphilin3 remains to be clarified, but four of the six phosphorylation sites are located at M domain and this domain may also have its specific function.

Tissue and subcellular distributions of rabphilin3

Northern and Western blot analyses indicate that rabphilin3 is expressed only in neuronal cells, but not in other cells, including exocrine cells and endocrine cells where Rab3A is expressed [2]. In the cells where Rab3A is expressed but rabphilin3 is not, a rabphilin3 isoform may be present. In brain, rabphilin3 is highly concentrated on synaptic vesicles in the presynaptic nerve terminal [15]. Immunoelectronmicroscopic analysis on the subcellular distribution of rabphilin3 in neuromuscular junction indicates that it is uniformly distributed to most synaptic vesicles in the nerve terminal of motor neurons, and that its distribution is similar to that of Rab3A [15].

We have shown that rabphilin3 interacts with isolated synaptic vesicles in dose-dependent and saturable manners [16]. This interaction is observed with the synaptic vesicles beforehand deprived of Rab3A by use of Rab GDI or associated with large amounts of exogenous Rab3A, but is abolished by prior digestion of the vesicles with trypsin. These results suggest that rabphilin3 is associated with the vesicles through its anchoring protein, irrespective of the presence and absence of Rab3A. This result is consistent with the structural property of rabphilin3 that it has no transmembrane segment. Therefore, it is likely that the specific localization of rabphilin3 is determined by the specific localization of its anchoring protein. In contrast to our results, however, another group suggests that rabphilin3 is associated with synaptic vesicles through Rab3A, on the basis of the observations that, in Rab3A-deficient mice, rabphilin3 was decreased in synapses [3], and that Rab GDI removed both Rab3A and rabphilin3 from isolated synaptic vesicles [11]. However, these studies have not directly shown that rabphilin3 interacts with the vesicles in a Rab3A-dependent manner, whereas we have directly shown that rabphilin3 interacts with the vesicles in a Rab3A-independent manner.

Evidence for involvement of rabphilin3 in Ca^{2+}-dependent neurotransmitter release

Because rabphilin3 interacts with the GTP-bound form of Rab3A, it is specifically localized on synaptic vesicles, and it interacts with Ca^{2+}, it is likely that rabphilin3 as well as synaptotagmin plays a role in Ca^{2+}-dependent neurotransmitter release. This assumption has recently been proved by use of human growth hormone (GH) co-expression assay system [17,18]. In this system, GH is overexpressed in bovine adrenal chromaffin cells or PC12 cells. Exogenous GH is stored in dense core vesicles and released in response to an agonist in an extracellular Ca^{2+}-dependent manner. When the N-terminal fragment of rabphilin3 shown in Fig. 2 is co-overexpressed with human GH in bovine adrenal chromaffin cells, the Ca^{2+}-dependent, agonist-induced GH release is markedly inhibited. In contrast, overexpression of the C-terminal fragment, which contains both a part of the M domain and the C2 domain, has no effect on the GH release. The similar results have also been obtained by use of PC12 cells [18]. However, in the PC12 cell experiment, overexpression of the C2 fragment inhibits the Ca^{2+}-dependent, high K^+-induced GH release. The exact reason for the different effects of these two fragments is not known, but it may be due to the inhibitory effect of the M domain on the C2 domain. It has furthermore been shown that overexpression of the C2B fragment, but not the C2A fragment, inhibits the secretion in the PC12 cell experiment.

Evidence for the involvement of rabphilin3 in Ca^{2+}-dependent exocytosis has been also obtained by use of the squid giant axon system in which the presynaptic nerve terminal is large enough to be microinjected with protein samples to be tested (Burns et al. manuscript in preparation). The N-terminal fragment or the C2 fragment is microinjected into the nerve terminal while electrically stimulating this terminal. Postsynaptic electrical responses are recorded simultaneously to monitor neurotransmitter release. Microinjection of the N-terminal fragment or the C2 fragment inhibits Ca^{2+}-dependent neurotransmitter release. Electron microscopy has revealed that injections cause a reduction in the number of synaptic vesicles within 0.5 μm of the plasma membrane and a relative accumulation of predocked vesicles 50–100 nm away from the plasma membrane. In contrast, while microinjection of the peptides from the C2-like domains of synaptotagmin also inhibits Ca^{2+}-dependent neurotransmitter release, this perturbation causes an accumulation of docked vesicles within 50 nm from the plasma membrane [19]. These ultrastructural data suggest that rabphilin3 is involved in the predocking process or the docking process, whereas synaptotagmin is involved in the fusion process that follows the docking process. These two lines of evidence indicate that the Rab3A-rabphilin3 system indeed functions in Ca^{2+}-dependent exocytosis in intact cells.

Possible function of rabphilin3 as a linker for the Rab3A system, the NSF/SNAP and SNARE systems, and the Ca^{2+} system

The exact mode of action of synaptotagmin in Ca^{2+}-dependent neurotransmitter release has not been established, but one fascinating function proposed is that synaptotagmin serves as a Ca^{2+}-controlled negative regulator as schematically shown in Fig. 3. Namely, the docking and fusion of synaptic vesicles with the presynaptic plasma membranes are mediated by the NSF/SNAP and SNARE systems. NSF indicates *N*-ethylmaleimide-sensitive fusion protein; SNAP indicates soluble NSF attachment protein; and SNARE indicates SNAP receptor. The NSF and SNAP system is considered to be the general fusion machinery, which induces the fusion of the vesicles with the acceptor membrane. The SNARE system is considered to regulate the docking of the vesicles with the acceptor membrane. In the SNARE system, the component located on the vesicles is vesicle-associated membrane protein (VAMP) and the components located on the plasma membrane are syntaxin and the

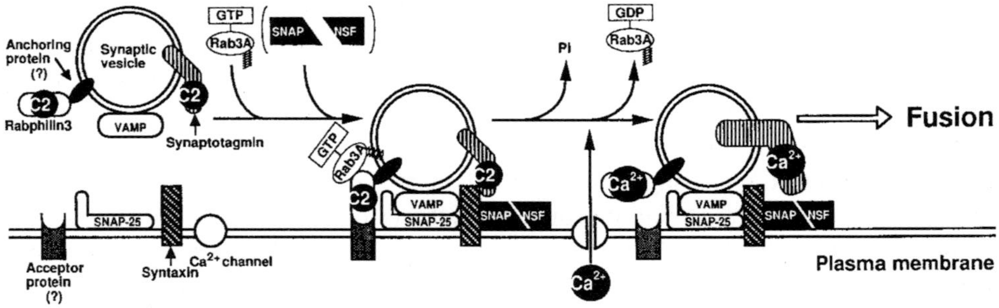

Fig. 3. A model for the modes of action of the Rab3A system, the synaptotagmin system, and the NSF/SNAP and SNARE systems in neurotransmitter release.

25-kDa synaptosomal-associated protein (SNAP-25). Before Ca^{2+} influx, synaptotagmin interacts with syntaxin and inhibits the NSF/SNAP and SNARE systems to operate. When Ca^{2+} influxes into the cytoplasm through Ca^{2+} channel, it binds to synaptotagmin, causing the release of this inhibitory action to operate the NSF/SNAP and SNARE systems. A recent biochemical observation, however, shows that the C2A domain of synaptotagmin interacts with syntaxin in a Ca^{2+}-dependent manner [7,20]. It is, therefore, likely that synaptotagmin interacts with syntaxin after Ca^{2+} influx and then operates the NSF/SNAP and SNARE systems. In this case, synaptotagmin serves as a positive regulator to stimulate Ca^{2+}-dependent exocytosis.

It is not known either how rabphilin3 regulates synaptic vesicle trafficking, but evidence is accumulating that the Rab system is an upstream regulator of the NSF/SNAP and SNARE systems: 1) Genetic analysis using the budding yeast has shown that Ypt1 induces the assembly of the SNARE components directly or indirectly [21,22]; 2) Inhibition of the rabphilin3 function by microinjecting the N-terminal fragment or the C2 fragment into squid giant axon causes an accumulation of predocked vesicles 50–100 nm away from the plasma membrane as described above; and 3) A dominant active mutant of Rab3A inhibits Ca^{2+}-dependent exocytosis from chromaffin cells and PC12 cells [23,24]. These results suggest that the Rab3A-rabphilin3 system regulates the NSF/SNAP and SNARE systems also in neurotransmitter release. After the GTP-bound form of Rab3A is converted to the GDP-bound form by the action of Rab3 GAP, the Rab3A-rabphilin3 complex may dissociate and the fusion is induced by the action of the NSF/SNAP and SNARE systems. This model for the mode of action of Rab3A seems to be consistent with the phenotype of the Rab3A-deficient mice [25]. These mice show prominent decrease in synaptic transmission only with repetitive stimulation, for which sequential synaptic vesicle recruitment to the presynaptic membrane is essential.

In addition to this function of rabphilin3, it may also serve as a Ca^{2+} sensor, because rabphilin3 has two C2-like domains like synaptotagmin. Our working model for the mode of action of the Rab3A-rabphilin3 system in neurotransmitter release is schematically shown in Fig. 3. There is an acceptor protein directly or indirectly recognized by rabphilin3 complexed with the GTP-bound form of Rab3A on the presynaptic plasma membrane. When rabphilin3 is complexed with the GTP-bound form of Rab3A, it interacts with this acceptor protein and thereby causes the docking of the vesicles with the presynaptic plasma membrane. Thus, rabphilin3 may have at least three functional domains: the domains interacting with Rab3A, an anchoring protein on synaptic vesicles, and an acceptor protein on the plasma membrane. This assumption is consistent with the structural properties of rabphilin3 that it has at least three domains, and also with the results obtained by the GH overexpression experiment and the squid giant axon experiment described above that both the N-terminal fragment and the C2 fragment inhibit the Ca^{2+}-dependent exocytosis. The Rab3A-dependent interaction of rabphilin3 with the acceptor protein may be followed by the assembly of synaptotagmin and the SNARE components. Binding of influxed Ca^{2+} to rabphilin3 as well as to synaptotagmin may cause either dissociation from or tighter interactions with their respective interacting proteins, and thereby lead to the operation of the NSF/SNAP and SNARE systems. The functional relationships between rabphilin3 and synaptotagmin and between rabphilin3 and the NSF/SNAP and SNARE systems is unknown at present, but rabphilin3 appears to be an important linker for the Rab3A system, the NSF/SNAP and SNARE systems, and the Ca^{2+} systems.

References

1. Shirataki H, Kaibuchi K, Yamaguchi T, Wada K, Horiuchi H, Takai Y. A possible target protein for smg 25A/rab3A small GTP-binding protein. J Biol Chem 1992;267:10946–10949.
2. Shirataki H, Kaibuchi K, Sakoda T, Kishida S, Yamaguchi T, Wada K, Miyazaki M, Takai Y. Rabphilin-3A, a putative target protein for smg p25A/rab3A p25 small GTP-binding protein related to synaptotagmin. Mol Cell Biol 1993;13:2061–2068.
3. Li C, Takei K, Geppert M, Daniell L, Stenius K, Chapman ER, Jahn R, De Camilli P, Südhof TC. Synaptic targeting of rabphilin-3A, a synaptic vesicle Ca^{2+}/phospholipid-binding protein, depends on rab3A/3C. Neuron 1994;13:885–898.
4. Weber E, Jilling T, Kirk KL. Distinct functional properties of Rab3A and Rab3B in PC12 cells. J Biol Chem 1996;271:6963–6971.
5. Nishizuka Y. The molecular heterogeneity of protein kinase C and its implications for cellular regulation. Nature 1998;334:661–665.
6. Perin MS, Fried VA, Mignery GA, Jahn R, Südhof TC. Phospholipid binding by a synaptic vesicle protein homologous to the regulatory region of protein kinase C. Nature 1990;345:260–263.
7. Li C, Ullrich B, Zhang JZ, Anderson RG, Brose N, Südhof TC. Ca^{2+}-dependent and -independent activities of neural and non-neural synaptotagmins. Nature 1995;375:594–599.
8. Brose N, Petrenko AG, Südhof TC, Jahn R. Synaptotagmin: a calcium sensor on the synaptic vesicle surface. Science 1992;256:1021–1025.
9. Geppert M, Goda Y, Hammer RE, Li C, Rosahl TW, Stevens CF, Südhof TC. Synaptotagmin I: a major Ca^{2+} sensor for transmitter release at a central synapse. Cell 1994;79:717–727.
10. Yamaguchi T, Shirataki H, Kishida S, Miyazaki M, Nishikawa J, Wada K, Numata S, Kaibuchi K, Takai Y. Two functionally different domains of rabphilin-3A, Rab3A p25/smg p25A-binding and phospholipid- and Ca^{2+}-binding domains. J Biol Chem 1993;268:27164–27170.
11. Stahl B, Chou JH, Li C, Südhof TC, Jahn R. Rab3A reversibly recruits rabphilin to synaptic vesicles by a mechanism analogues to raf recruitment by ras. EMBO J 1996;15:1799–1809.
12. Kato M, Sasaki T, Imazumi K, Takahashi K, Araki K, Shirataki H, Matsuura Y, Ishida A, Fujisawa

H, Takai Y. Phosphorylation of Rabphilin-3A by calmodulin-dependent protein kinase II. Biochem Biophys Res Commun 1994;205:1776–1784.
13. Numata S, Shirataki H, Hagi S, Yamamoto T, Takai Y. Phosphorylation of Rabphilin-3A, a putative target protein for Rab3A, by cyclic AMP-dependent protein kinase. Biochem Biophys Res Commun 1994;203:1927–1934.
14. Fykse EM, Li C, Südhof TC. Phosphorylation of rabphilin-3A by Ca^{2+}/calmodulin- and cAMP-dependent protein kinase in vitro. J Neurosci 1995;15:2385–2395.
15. Mizoguchi A, Yano Y, Hamaguchi H, Yanagida H, Ide C, Zahraoui A, Shirataki H, Sasaki T, Takai Y. Localization of Rabphilin-3A on the synaptic vesicle. Biochem Biophys Res Commun 1994;202:1235–1243.
16. Shirataki H, Yamamoto T, Hagi S, Miura H, Oishi H, Jinno Y, Senbonmatsu T, Takai Y. Rabphilin-3A is associated with synaptic vesicles through a vesicle protein in a manner independent of Rab3A. J Biol Chem 1994;269:32717–32720.
17. Chung SH, Takai Y, Holz RW. Evidence that the Rab3a-binding protein, rabphilin3a, enhances regulated secretion. Studies in adrenal chromaffin cells. J Biol Chem 1995;270:16714–16718.
18. Komuro R, Sasaki T, Orita S, Maeda M, Takai Y. Involvement of rabphilin-3A in Ca^{2+}-dependent exocytosis from PC12 cells. Biochem Biophys Res Commun 1996;219:435–440.
19. Bommert K, Charlton MP, DeBello WM, Chin GJ, Betz H, Augustine GJ. Inhibition of neurotransmitter release by C2-domain peptides implicates synaptotagmin in exocytosis. Nature 1993;363:163–165.
20. Chapman ER, Hanson PI, An S, Jahn R. Ca^{2+} regulates the interaction between synaptotagmin and syntaxin 1. J Biol Chem 1995;270:23667–23671.
21. Lian JP, Stone S, Jiang Y, Lyons P, Ferro-Novick S. Ypt1p implicated in v-SNARE activation. Nature 1994;372:698–701.
22. Sogaard M, Tani K, Ye RR, Geromanos S, Tempst P, Kirchhausen T, Rothman JE, Dollner T. A rab protein is required for the assembly of SNARE complexes in the docking of transport vesicles. Cell 1994;78:937–948.
23. Holz RW, Brondyk WH, Senter RA, Kuizon L, Macara IG. Evidence for the involvement of Rab3A in Ca^{2+}-dependent exocytosis from adrenal chromaffin cells. J Biol Chem 1994;269: 10229–10234.
24. Johannes L, Lledo PM, Roa M, Vincent JD, Henry JP, Darchen F. The GTPase Rab3a negatively controls calcium-dependent exocytosis in neuroendocrine cells. EMBO J 1994;13:2029–2037.
25. Geppert M, Bolshakov VY, Siegelbaum SA, Takei K, De Camilli P, Hammer RE, Südhof TC. The role of Rab3A in neurotransmitter release. Nature 1994;369:493–497.

1996 Elsevier Science B.V.
Integrative and Molecular Approach to Brain Function.
M. Ito and Y. Miyashita, editors.

Roles of glutamate receptors in synaptic transmission and plasticity

Shigetada Nakanishi

Department of Biological Sciences, Kyoto University Faculty of Medicine, Yoshida, Sakyo-ku, Kyoto, Japan

Abstract. Glutamate is a major excitatory neurotransmitter in the central nervous system. Our molecular cloning study has disclosed the existence of diverse members of both NMDA receptors and metabotropic glutamate receptors (mGluRs). Functional analysis of mGluR subtypes has demonstrated that the specific subtypes of the mGluR family play critical roles in information processing in the retina and synaptic plasticity in the hippocampus.

Key words: gene targeting, glutamate transmission, hippocampal synaptic plasticity, visual transmission.

Background

Glutamate is a major excitatory neurotransmitter in the central nervous system and plays a critical role in neural plasticity, neural development and some neurodegenerative disorders [1–3]. Glutamate receptors are categorized into ionotropic glutamate receptors and metabotropic glutamate receptors (mGluRs) [2,3]. The ionotropic glutamate receptors comprise ligand-gated, cation-specific ion channels and are subdivided into AMPA/kainate receptors and NMDA receptors. mGluRs are linked to the G protein-signaling cascades and modulate the production of intracellular second messengers. We cloned and characterized the NMDA receptors and mGluRs with the aid of our functional cloning strategy that combines electrophysiology and a *Xenopus* oocyte expression system [4,5]. Both receptors exist as a family of the diverse members of receptor subunits and subtypes (Fig. 1). NMDAR1 is a key subunit that possesses all properties of the NMDA receptor-channel complex, whereas four NMDAR2 subunits (NMDAR2A-2D) serve as modulatory subunits that potentiate and differentiate the NMDA receptor activity by heteromeric formation with NMDAR1 [6]. The mGluRs consist of at least eight different subtypes (mGluR1-mGluR8) that show the diversity in signal transduction, expression patterns and subcellular localization. This article deals with our recent study that indicates the specialized functions of the diverse members of mGluRs in the central nervous system.

Address for correspondence: Shigetada Nakanishi, Department of Biological Sciences, Kyoto University Faculty of Medicine, Yoshida, Sakyo-ku, Kyoto 606, Japan.

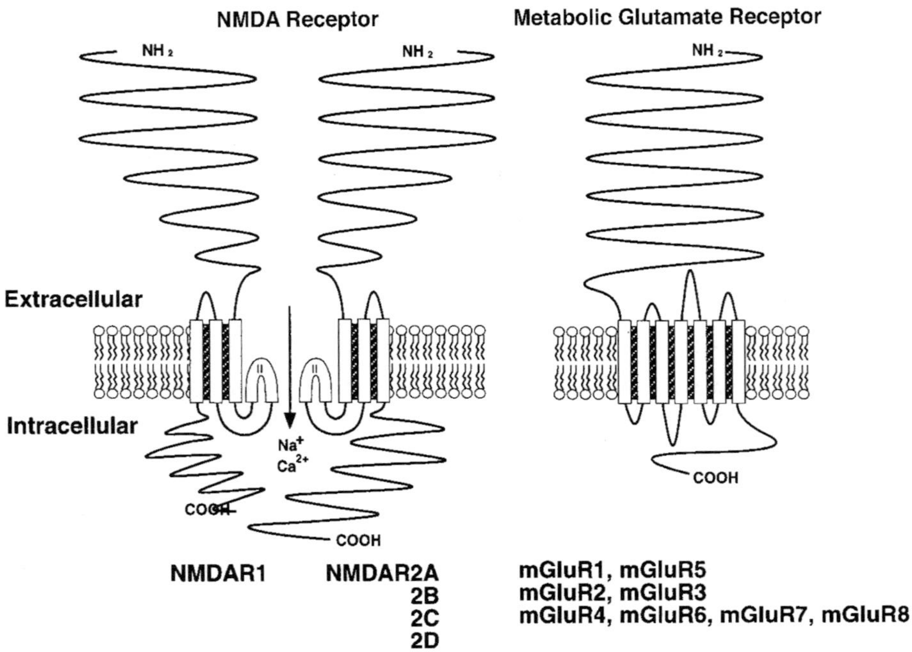

Fig. 1. Structural models of NMDA receptors and mGluRs.

Experimental procedures

Targeted disruption of the mGluR2 and mGluR6 genes was carried out with use of an embryonic cell line [7,8]. ON and OFF responses were recorded from the superior colliculus with extracellular recording techniques [7]. Basal synaptic transmission and paired-pulse facilitation at the mossy fiber-CA3 synapses were determined in hippocampal slice preparations with extracellular recording techniques [8,9]. Long-term potentiation (LTP) and long-term depression (LTD) at the mossy fiber-CA3 synapses were induced by tetanic stimulation (100 Hz, 1 s) and low-frequency stimulation (1 Hz, 15 min), respectively [8]. Immunostaining of mGluR2 and mGluR6 was carried out with use of the mGluR2/mGluR3 antibody and the mGluR6 antibody, respectively [8,10]. The ability of wild-type and mutant mice to respond to visual stimulation was assessed by measurements of shuttle box avoidance behavior in conjunction with light exposure, a daily locomotor activity and Fos inducibility in the suprachiasmatic nuclei [7,11]. Optokinetic nystagmus was measured by rotating a striped drum with different contrasts [11].

Results and Discussion

Molecular diversity of the mGluR family

The mGluR subtypes are subdivided into three groups according to their sequence similarities, signal transduction mechanisms and agonist selectivities [3] (Fig. 1). mGluR1 and mGluR5 are coupled to IP_3/Ca^{2+} signal transduction and respond strongly to quisqualate. The other six subtypes are linked to the inhibitory cascade of cAMP formation and can be subdivided into two groups. mGluR2 and mGluR3 react effectively with (2S, 1′R, 2′R, 3′R)-2-(2,3-dicarboxycyclopropyl) glycine (DCG-IV) [3,12], whereas mGluR4, mGluR6, mGluR7 and mGluR8 respond effectively to L-2-amino-4-phosphonobutyrate (L-AP4) [3]. Different mGluR subtypes show different expression patterns and distinct subcellular localizations in neuronal and glial cells. Thus, different mGluR subtypes are thought to have specialized functions in a variety of nerve cells.

mGluR6 in visual transmission

On the basis of our knowledge of the molecular diversity of the glutamate receptors, it is important to explore physiological roles of individual receptor subtypes or subunits in the brain function. We have investigated roles of specific mGluR subtypes in synaptic transmission in the retina and the hippocampus.

Visual signals are transmitted from photoreceptors to bipolar cells. A key processing of visual transmission is detecting visual contrasts at the level of bipolar cells, and this processing occurs by segregating the visual signals into the ON-center and OFF-center pathways [13]. Electrophysiological evidence has suggested that a putative mGluR that is sensitive to L-AP4 mediates postsynaptic responses of ON-bipolar cells by enhancing cGMP hydrolysis similar to that of signal transduction in photoreceptors [14]. We cloned mGluR6 from a retinal cDNA library and indicated that this mGluR subtype selectively responds to L-AP4 after DNA transfection in CHO cells and is exclusively expressed in the bipolar cell layer [15].

Immunohistochemical and immuno-electron-microscopic analysis disclosed that mGluR6 is restrictedly located at the postsynaptic site of rod (ON-type) bipolar cells in the adult rat retina [10]. In addition, mGluR6 is localized from cell bodies to the postsynaptic site during retinal development, and this developmental change in mGluR6 distribution agrees well with the synaptic formation between photoreceptors and bipolar cells [10]. These findings strongly suggest that mGluR6 mediates glutamate transmission in ON responses. Direct evidence for the role of mGluR6 in ON responses was indicated by targeted disruption of the mGluR6 gene [7]. The homozygous mutant mice lacking mGluR6 showed no apparent behavioral abnormality, nor any overall anatomical changes in the retinal cell organization. These mutant mice showed an unchanged OFF response but displayed a complete loss of ON responses to light stimulus. The mGluR6-deficient mice retained the ability to respond to light stimulation as examined by shuttle box avoidance behavioral analysis in conjunction with light exposure. Furthermore, the mGluR6-deficient mice showed an unaltered light-dark cycle of daily locomotor activity and exhibited light-simulated induction of Fos immunoreactivity in the suprachiasmatic nucleus [11]. The mGluR6 deficiency, however, markedly reduced the sensitivity of pupillary responses to light stimulus and severely impaired the ability to drive optokinetic nystagmus in response to visual contrasts [11]. These observations demonstrate that mGluR6 is essential in synaptic transmission to ON bipolar cells and plays a critical role in discrimination of visual contrasts.

On the basis of the above characterization of mGluR6, together with the electrophysiological study of the mGluR6-mediated synaptic transmission [14], the synaptic mechanism in visual transmission from photoreceptors to ON-bipolar cells can be summarized [3,13] as illustrated in Fig. 2. In response to light exposure, photoreceptors lower cGMP concentrations through the stimulation of phosphodiesterase via transducin. This decrease in cGMP hyperpolarizes photoreceptors by closing the cGMP-gated ion channel and reduces glutamate release. The mGluR6-G protein-phosphodiesterase system in ON-bipolar cells becomes inactive, and a high concentration of cGMP stimulates the cGMP-gated ion channel of ON-bipolar cells and depolarizes these cells. Glutamate release is then augmented in ON-bipolar cells and in turn excites the subsequent ON pathway.

The key role of mGluR6 in ON-bipolar cells is thus to mediate sign inversion between photoreceptors and bipolar cells, so that the presynaptic membrane polarization is converted to the opposite polarization at the postsynaptic bipolar cells. In contrast, OFF-bipolar cells use AMPA receptors and thus preserve the sign of photoreceptors. Two distinct types of glutamate receptors thus effectively serve for the ON and OFF segregation in response to the common glutamate transmitter. Our investigation has provided the compelling evidence that a specific mGluR subtype plays an essential role in synaptic transmission.

The two G protein-coupled receptors, rhodopsin and mGluR6, at the initial steps of visual transmission are thought to be indispensable for the amplification of visual inputs through a second messenger system. The peripheral sensory organs receive

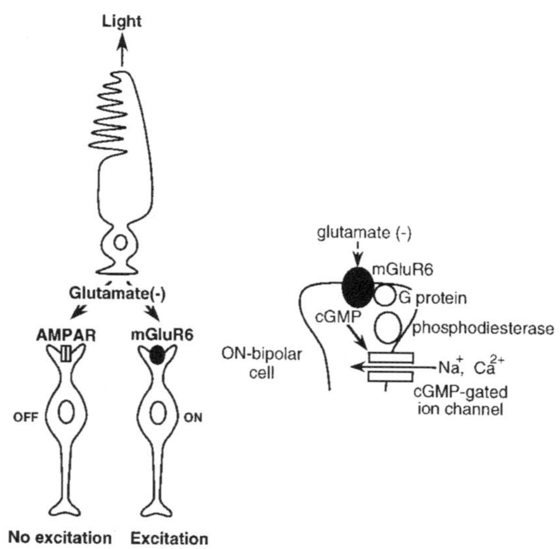

Fig. 2. A model of the function of mGluR6 in the visual ON pathway. See the text for the explanation of this model.

graded external signals, and these graded signals must be converted to action potentials in an early step of information processing. Thus, the fundamental function of the early step of synaptic transmission is to discriminate and amplify the signal-to-noise ratio of external stimuli prior to transmission of these signals to higher centers of the brain. The coupling of mGluR6 to the intracellular signaling cascade is thus critical for enhancing external signals in the sensory system.

mGluR2 in hippocampal synaptic plasticity

Long-lasting modifications in synaptic efficacy, both long-term potentiation (LTP) and long-term depression (LTD), at the CA1 region of the hippocampus are mediated by the activation of NMDA receptors at the postsynaptic cell [16]. It is generally accepted that these modifications in synaptic efficacy are of fundamental importance for memory acquisition and learning [16]. Recent evidence, however, has also demonstrated that both LTP and LTD in hippocampal mossy fiber-CA3 synapses result entirely from changes in the presynaptic cells [9,16]. Furthermore, these changes are independent of NMDA receptor activation [9,16]. In situ hybridization study and immunostaining in combination with lesion analysis indicated that mGluR2 is predominantly expressed in granule cells in the hippocampal dentate gyrus and is distributed to the presynaptic site of mossy fibers [17]. Moreover, immuno-electron-microscopic techniques showed a unique localization of mGluR2 on the preterminal zone of the presynaptic site of mossy fibers [8]. We investigated the role of the presynaptic mGluR2 in synaptic plasticity in the mossy fiber-CA3 synapses by generating mutant mice deficient in mGluR2 [8].

The mGluR2-deficient mice grew normally and showed no apparent behavioral abnormality, nor any gross anatomical changes in the brain. Wild-type and mutant mice showed no difference in the field excitatory postsynaptic potential (EPSP) evoked by stimulation of mossy fibers. Paired-pulse facilitation was also indistinguishable between the two genotypes. Bath application of the mGluR2/mGluR3-selective agonist, DCG-IV, markedly and reversibly depressed EPSPs at the mossy fiber-CA3 synapses in wild-type mice, but this depression was greatly reduced in mutant mice. This analysis of the mGluR2 deficiency indicates that the presynaptic mGluR2 plays an important role in presynaptic inhibition. We investigated effects of mGluR2 deficiency on LTP and LTD at the mossy fiber-CA3 synapse. mGluR2-deficient mice showed normal LTP. In contrast, LTD at the mossy fiber-CA3 synapses is significantly impaired in mutant mice. Therefore, the presynaptic mGluR2 is not required for mossy fiber LTP but is essential for mossy fiber LTD.

We tested the involvement of mossy fiber LTD in spatial learning by performing the Morris water-maze tasks [8]. The wild-type and mutant mice showed no differences in their ability to perform either the visible or hidden platform tasks. We also evaluated spatial learning flexibility by examining the ability of trained mice to adapt to a new platform location. Both wild-type and mutant mice capably adapted to a new platform location with no significant difference. These observations demonstrate that impairment of the CA3 LTD due to mGluR2 deficiency does not

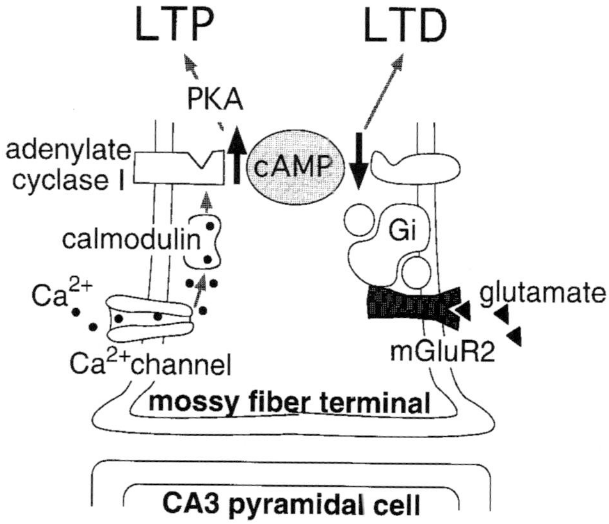

Fig. 3. A model for induction of LTP and LTD at the presynaptic site of mossy fiber-CA3 synapses. This model is discussed in the text by taking together the mechanism reported for LTP [16].

hinder spatial learning in a water-maze.

On the basis of the above findings, we propose a model in which the cAMP cascade in the presynaptic mossy fibers regulates synaptic plasticity at the mossy fiber-CA3 synapses [8,16] (Fig. 3). Recent evidence has suggested that tetanic stimulation enhances calcium entry into the presynaptic terminal, and that this enhancement activates Ca^{2+}/calmodulin-sensitive adenylate cyclase 1, which would in turn produce LTP through protein kinase A [16]. mGluR2, when expressed in CHO cells, is coupled to inhibition of the cAMP cascade [3]. Thus, slowly accumulating glutamate during low-frequency stimulation would activate presynaptic mGluR2 and this activation of mGluR2 could inhibit the cAMP cascade, thus resulting in induction of LTD at the mossy fiber-CA3 synapses. It is thus conceivable that bidirectional modifications of synaptic efficacy, LTP and LTD, can occur by up- and downregulation of the cAMP cascade at the presynaptic site of the mossy fiber-CA3 synapses.

Although impairment of mossy fiber LTD does not hinder water-maze learning, LTD may have a variety of physiological roles; these include, for example, a dynamic modulation of information processing; an increase in the flexibility of circuit formation; protection of postsynaptic cells from overexcitation; or cancellation of acquired memory. Thus, the mGluR2-deficient mouse is interesting for studying the LTD-relevant brain function.

Acknowledgements

I thank Drs Masayuki Masu, Hideki Iwakabe, Akinori Nomura and Mineto Yokoi in my laboratory for study presented in this article and Drs Tomomitsu Miyoshi, Masayuki Yamashita, Yutaka Fukuda, Hitoshi Sasaki, Kano Hiroi, Yasuhisa Nakamura, Katsunori Kobayshi, Toshiya Manabe, Tomoyuki Takahash, Isao Sakaguchi, Goro Katsuura, Ryuuichi Shigemoto, Hitoshi Oishi, Sakashi Nomura, Kenji Nakamura, Kazuki Nakao and Motoya Katsuki for their collaboration work.

References

1. Nakanishi S. Molecular diversity of glutamate receptors and implications for brain function. Science 1992;258:597–603.
2. Nakanishi S, Masu M. Molecular diversity and functions of glutamate receptors. Ann Rev Biophys Biomol Struct 1994;23:319–348.
3. Nakanishi S. Metabotropic glutamate receptors: synaptic transmission, modulation, and plasticity. Neuron 1994;13:1031–1037.
4. Moriyoshi K et al. Molecular cloning and characterization of the rat NMDA receptor. Nature 1991; 354:31–37.
5. Masu M et al. Sequence and expression of a metabotropic glutamate receptor. Nature 1991;349: 760–765.
6. Ishii T et al. Molecular characterization of the family of the N-methyl-D-aspartate receptor subunits. J Biol Chem 1993;268:2836–2843.
7. Masu M et al. Specific deficit of the ON response in visual transmission by targeted disruption of the mGluR6 gene. Cell 1995;80:757–765.
8. Yokoi M et al. Impairment of hippocampal mossy fiber LTD in mice lacking metabotropic glutamate receptor subtype 2. Science 1996;273:645–647.
9. Kobayashi K et al. Presynaptic long-term depression at the hippocampal mossy fiber-CA3 synapse. (Submitted).
10. Nomura A et al. Developmentally regulated postsynaptic localization of a metabotropic glutamate receptor in rat rod bipolar cells. Cell 1994;77:361–369.
11. Iwakabe H et al. Impairment of pupillary responses and optokinetic nystagmus in the mGluR6-deficient mouse. Science 1996;273:648–650.
12. Hayashi Y et al. Role of a metabotropic glutamate receptor in synaptic modulation in the accessory olfactory bulb. Nature 1993;366:687–690.
13. Nakanishi S. Second-order neurons and receptor mechanisms in visual and olfactory information processing. Trends Neurosci 1995;18:359–364.
14. Shiells R. Glutamate receptors for signal amplification. Current Biology 1994;4:917–918.
15. Nakajima Y et al. Molecular characterization of a novel retinal metabotropic glutamate receptor mGluR6 with a high agonist selectivity for L-2-amino-4-phosphonobutyrate. J Biol Chem 1993;268: 11868–11873.
16. Nicoll RA, Malenka RC. Contrasting properties of two forms of long-term potentiation in the hippocampus. Nature 1995;377:115–118.
17. Ohishi H et al. Distribution of the messenger RNA for a metabotropic glutamate receptor, mGluR2, in the central nervous system of the rat. Neuroscience 1993;53:1009–1018.

Integrative and Molecular Approach to Brain Function.
M. Ito and Y. Miyashita, editors.

Are experience-induced morphological changes reflected in brain physiology?

William T. Greenough[1,2,3,4], Rodney A. Swain[1,4], Jeffrey A. Kleim[1,4] and Ivan Jeanne Weiler[1,4]

Departments of [1]Psychology and [2]Cell and Structural Biology; [3]Neuroscience Program; and [4]Beckman Institute, University of Illinois, Urbana-Champaign, Illinois, USA

Abstract. Growing evidence indicates that experience-triggered synaptogenesis plays a role in memory. This argument is based both upon causation of morphological change, as synaptic changes are brought about by learning, and not by mere activity, and upon its functional correlates indicating that the organization of the nervous system is altered in a manner compatible with involvement of the anatomical changes in memory. A surprising finding is that multiple postsynaptic contacts with varicosities are more frequent following training or complex experience, suggesting re-use of existing components and strengthening of pre-existing circuits in the cerebellar cortex. Finally, it is of note that morphological correlates of learning appear to be distributed in the central nervous system. While widespread effects might be expected from an experience as diverse as that of a complex environment, a distributed representation is suggested by the fact that both cerebellar cortex and motor cerebral cortex are altered by motor skill training.

Key words: memory, plasticity, synaptogenesis.

There have been many reports of alteration of brain morphology in response to behavioral experiences involving brain information storage. Many of these reports have suggested that formation of new synapses, or in some cases modification of existing synapses, may play a role in memory. There have also been a substantial number of reports of behaviorally induced alterations in brain physiology that outlast the behavioral experience by a significant amount of time, as expected of a memory. If the morphological changes are, in fact, involved in memory arising from these experiences, this should be reflected in the physiology of the pathways they mediate. This paper describes morphological changes brought about by experience and addresses the issue of whether experience-induced morphological changes are accompanied by physiological changes of a sort that indicate that the morphological changes alter the functional organization of the brain.

The changes in brain morphology that historically have been viewed as most relevant to memory are those involving changes in synaptic numbers. Synapse numbers or equivalent dendritic measures have been shown to increase following

---

Address for correspondence: William T. Greenough, Beckman Institute, University of Illinois, 405 N. Mathews, Urbana, IL 61801, USA. Tel.: +01-217-333-4472. Fax: +01-217-244-5180. E-mail: wgreenou@s.psych.uiuc.edu housing in a complex environment (EC) vs. individual cages (IC) [1], maze training [2] and other forms of learning [3].

Physiological correlates of experience-induced changes in brain morphology

To assess the physiological consequences of these alterations, our laboratory has recently characterized the rat layer II/III visual cortical response to probe stimulation of lateral white matter (containing thalamic afferents to visual cortex) following at least 30 days of differential rearing in the EC and IC environments [4]. Our results indicated that fEPSP amplitude was constant across groups but that among EC rats PS amplitude increased while latency and the threshold to elicit the PS decreased. These results are in accord with the anatomical results and suggest that one effect of the morphological adjustments (25% more synapses per neuron) to the EC environment is to enhance cortical responsivity (via PS thresholds) and to augment the strength of the cortical response (via synchrony of synapses) to visual stimuli. The absence of a difference in fEPSP amplitude reflects, we believe, the fact that synaptic density is identical in EC and IC rats [1].

Learning vs. neural activity as causes of morphological change: the acrobat paradigm

To determine whether learning or mere neural activity caused synapse change, we [5] attempted to separate effects of activity from learning by creating groups of rats with relatively large amounts of one and relatively little of the other. Rats that learned acrobatic motor skills in an elevated obstacle course (AC) were compared with rats forced to exercise (FX) on a treadmill, rats allowed voluntary exercise (VX) in activity wheels accessible from their cages and inactive rats (IC) kept in individual cages except for brief periods of daily handling to simulate the treatment of the other groups. After 30 days in these conditions, the density of Purkinje neurons, synapses, and capillaries was investigated in the cerebellar paramedian lobule. In the activity groups (FX, VX), the density of blood vessels was elevated compared to that of both the inactive controls (IC) and the acrobatic motor learners, while the number of synapses per Purkinje cell did not differ between the FX and VX rats and the IC group. By contrast AC rats did not differ from ICs in vascular density but had a dramatically greater number of synapses per Purkinje cell. Thus synapse number changes were associated with learning whereas vascular changes were associated with high levels of activity.

We have obtained very similar results in a study of layer II-III of motor cerebral cortex [6]. Here groups of animals were examined after 1, 2, 5, 10 and 20 days of training. A motor control (MC) group was substituted for the VX and FX groups, to address specifically the effects of prodding and consequent arousal associated with the AC training – MC rats were run in a closed alley and were prodded each time that it was necessary to prod AC rats to encourage them to complete the obstacle course. Area M1 of one hemisphere was processed for electron microscopy while the other was immunostained for the early response transcription factor Fos, to assess its possible role in the synaptic response. Synapse number increased gradually across the period of training, reaching apparently asymptotic levels by day 5. This indicates that synapse formation in motor cortex occurs as a consequence of motor learning and suggests that plasticity associated with learned motor skills is distributed across brain regions. Expression of Fos was selectively elevated in the AC rats in the early stages of training, after which it returned towards baseline levels. This is compatible with a role for Fos in orchestrating early aspects of the cellular response to training.

Having seen that the synaptic response develops gradually with motor skill training in the motor cortex, it is of interest to examine the degree to which the effects persist in the absence of further training. Taking advantage of the fact that the cerebellar cortex exhibits a robust and rapid synaptic response to training, Kleim et al. [7] compared rats trained to a high level of behavioral performance for 10 days and examined immediately (10T/0D), rats similarly trained but examined after an additional 4 weeks without further training (10T/28D) and rats continuously trained for the entire period prior to examination (38T/0D). Behavioral results indicated that, at least in terms of latency to complete the tasks, there was very little decrement in performance across a 4-week period without practice. It was also clear that the vast majority of the synaptic change induced by training was stable: there was no statistically reliable difference among the three groups in the number of synapses per Purkinje cell.

Several additional studies have further characterized the types of alteration brought in the cerebellar cortical network by motor skill learning. Anderson et al. [8] first noted that the primary source of the synaptic number change was an increase in the number of parallel fiber to Purkinje cell synapses, the synapse most commonly proposed to exhibit plasticity by cerebellar theorists (some of whom, however, predicted synaptic weakening [9]). Anderson et al. [8] did, however, also find evidence for an increase in climbing fiber to Purkinje cell synapses in the motor skill trained rats. Kleim et al. [10] confirmed the parallel fiber synapse increase and also noted, using quantitation of Golgi-impregnated tissue, that stellate inhibitory interneurons in the molecular layer had larger dendritic fields in the motor skill trained rats [11].

A particularly interesting finding with potential implications for modelling of changes in cerebellar circuitry with learning emerged from a study by Federmeier et al. [12]. The first experiment quantified the distance between parallel fiber varicosities in the paramedian lobule in an attempt to determine whether synapses were being made with new Purkinje neurons (which should result in reduced overall intervaricosity distance). Surprisingly, AC, VX and IC rats had equivalent inter-varicosity distance distributions. While it is possible that parallel fibers were adding synapses by growing longer, an additional possibility was that they were forming additional synapses on existing varicosities. Such multiple synapses are common. When this was examined using electron microscopy, AC rats had about twice as many multiple synapses on a varicosity as VX and IC rats. Assuming that the report [13] that at least 70% of postsynaptic contacts on the same varicosity arise from the same postsynaptic neuron also holds for these newly generated multiple synapses, an implication of this finding, is that motor skill learning is associated largely with a strengthening of pre-existing pathways, rather than with the incorporation of new elements into the circuitry. One could imagine that this synaptogenic strengthening might be combined with precise synaptic strength adjustment, using mechanisms such as LTD, to optimize cerebellar output. Advantages of bidirectional modifiability of parallel fiber synapses were noted in the modelling of Fujita [14]. Alternatively, multiple synapses could possess unpredictable nonlinear properties. DeSchutter and Bower [15] have proposed that parallel fiber input to Purkinje cells may be substantially enhanced by effects of voltage activated channels. If so, overlapping regions of suprathreshold activation of channels by adjacent synapses at appropriate distances could yield greater effects than predicted from activation of single synapses.

Multiple postsynaptic contacts with boutons are quite common across many brain regions (e.g., [16]). In addition, they are increased in relative numbers following a variety of manipulations such as exposure to a complex environment (Jones, Klintsova, and Greenough, in preparation), synapses associated with the open eye in monocular deprivation [17], compensatory change following distal brain damage [18], and dehydration [19].

These anatomical findings suggest that cerebellar cortical plasticity may be considerably more complicated than predicted by models that emphasize unidirectional changes in synaptic strength and changes restricted to a particular class of synapses. The parallel fiber synapse increase, taken with evidence for LTD in this pathway, suggests the capacity for plastic change in either direction. Such an increase in range could increase the computational power of this system by allowing for a greater range of tuning. Moreover, the "positive" change in excitatory synapses suggests that a physiological change in the opposite direction from LTD might be expected in the parallel fiber system.

Physiological changes in the acrobat paradigm: a novel form of cerebellar cortical plasticity?

We are examining the impact of this motor learning-induced synaptogenesis on the physiology of the rat paramedian lobule. Purkinje cell extracellular field potential responses to afferent stimulation of the lobule (via middle cerebellar peduncle or intracortical stimulation) have been recorded following 10 days AC training, or MC training on an elevated alleyway with no obstacles, compared to VX and IC [20]. Stimulus-evoked responses, a series of oscillations, are evident in most rats following stimulation with currents exceeding 200 μA. The earliest oscillation (N1) occurs within 2 ms of stimulus onset and is elicited at lower current levels than the other oscillations. Similar patterns have been observed by others and it is believed that this N1 component reflects the presynaptic volley along the parallel fiber axon [21]. Our results indicate that the amplitude of the N1 component increases across current levels in the AC rats alone.

Our most provocative finding has been that elicitation of additional components

(N2-N4) is almost universally present above a varying threshold stimulus intensity in AC animals but much less frequently evident in MC, VX or IC animals. Comparison of high frequency stimulation trials (at levels exceeding the capacity for synaptic transmission) with low frequency stimulation trials indicates that both the N1 and N2 signals reflect to a large extent presynaptic activity along the parallel fiber axon. Difference scores derived by subtracting high frequency trials from low frequency trials consistently indicate that the postsynaptic elements of these components (N1 and N2) are approximately 10% larger in AC rats vs. control animals.

These results suggest a behaviorally induced change, consistent with the anatomical changes, that is maintained outside the learning context. We hypothesize that the deviation of our results from those obtained, for example, in classical conditioning experiments (i.e., facilitation vs. depression) may reflect task-specific plasticity of the cerebellum. Classical conditioning of the rabbit NM response or adaptation of the VOR can be viewed as rather simple, uniform movements of discrete musculature that because of their relationship to external triggers must rely on temporal precision to be executed successfully. In contrast, the type of motor learning studied here can be more accurately described as a complex series of integrated movements that require moment to moment modulation of muscular force and velocity in space to achieve accuracy. In particular, complicated whole body movement patterns of this sort may require inhibition of competing movements to a much greater degree than simple modulation of reflexes involving only a few localized muscles. Indeed we have noted informally that a major aspect of the AC learning process appears to involve the suppression of interfering movements such as excessive counterbalancing attempts with the tail. We believe that parallel fiber to Purkinje cell transmission may respond bidirectionally and in the aggregate may increase, as predicted by anatomical data, to accommodate the mandates of learning this complicated motor task.

Acknowledgements

Preparation of this paper and conduct of recent work described herein was supported by MH35321, AG10154, HD 07333, an NSERC Fellowship and the Retirement Research Foundation.

References

1. Turner AM, Greenough WT. Differential rearing effects on rat visual cortex synapses. I. Synaptic and neuronal density and synapses per neuron. Brain Res 1985;329:195–203.
2. Greenough WT, Juraska JM, Volkmar FR. Maze training effects on dendritic branching in occipital cortex of adult rats. Behav Neural Biol 1979;26:287–297.
3. Greenough WT, Larson JR, Withers GS. Effects of unilateral and bilateral training in a reaching task on dendritic branching of neurons in the rat sensory-motor forelimb cortex. Behav Neurol Biol 1985;44:1–314.
4. Greenough WT, Wang X. Altered post-synaptic response in the visual cortex in vivo of rats reared in complex environments. Soc Neurosci Abstr 1993:19:164.

5. Black JE, Isaacs K, Anderson BJ, Alcantara AA, Greenough WT. Learning causes synaptogenesis whereas motor activity causes angiogenesis in cerebellar cortex of adult rats. Proc Natl Acad Sci USA 1990;87:5568–5572.
6. Kleim JA, Lussnig E, Schwarz ER, Comery TA, Greenough WT. Synaptogenesis and Fos expression in the motor cortex of the adult rat after complex motor skill acquisition. J Neurosci 1996;(In press).
7. Kleim JA, Ballard D, Vij K, Greenough WT. The persistence of experience dependent morphological plasticity in the cerebellar cortex of the rat. Soc Neurosci Abstr 1995;21:445.
8. Anderson BJ, Alacantara AA, Greenough WT. Motor learning: Changes in synaptic organization of the rat cerebellar cortex. Neurobiol Learn Mem 1996;(In press).
9. Albus JS. A theory of cerebellar function. Math Biosci 1971;10:25–61.
10. Kleim JA, Napper RMA, Swain RA, Armstrong KE, Jones TA, Greenough WT. Selective synaptic plasticity in the cerebellar cortex of the rat following complex motor learning. Soc Neurosci Abstr 1994;20:1435.
11. Greenough WT, Kleim JA, Swain RA, Czerlanis CM, Kelly J, Pipitone MA. Increased dendritic arborization of cerebellar stellate neurons following motor skill learning in the rat. Soc Neurosci Abstr 1996;(In press).
12. Federmeier K, Kleim JA, Anderson BJ, Greenough WT. Formation of double synapses in the cerebellar cortex of the rat following motor learning. Soc Neurosci Abstr 1994;20:1435.
13. Harris KM, Stevens JK. Dendritic spines of rat cerebellar Purkinje cells: serial electron microscopy with reference to their biophysical characteristics. J Neurosci 1988;8:4455–4469.
14. Fujita M. Adaptive filter model of the cerebellum. Bio Cybern 1982;45:195–206.
15. DeSchutter E, Bower JM. An active membrane model of the cerebellar Purkinje cell I. Simulation of current clamps in slice. J Neurophysiol 1994;71:375–400.
16. Sorra KE, Harris KM. Occurrence and three-dimensional structure of multiple synapses between individual radiatum axons and their target pyramidal cells in hippocampal area CA1. J Neurosci 1993;13:3736–3748.
17. Friedlander MJ, Martin KAC, Wassenhove-McCarthy D. Effects of monocular visual deprivation on geniculo-cortical innervation of area 18 in the cat. J Neurosci 1991;11:3268–3288.
18. Jones TA. Evidence for alterations in the form of synaptic connectivity in the cortex opposite unilateral sensorimotor cortex lesions: Time-dependent degenerative and synaptic structural change. Soc Neurosci Abstr 1996;(In press).
19. Perlmutter LS, Tweedle CD, Hatton GI. Neuronal/glial plasticity in the supraoptic dendritic zone: dendritic bundling and double synapse formation at parturition. Neurosci 1984;13:769–779.
20. Bendre AA, Swain RS, Wheeler BC, Greenough WT. Augmentation of cerebellar responses to parallel fiber activation following skilled motor acquisition in rats. Soc Neurosci Abstr 1995;21:444.
21. Eccles JC, Ito M, Szentagothai J. The Cerebellum as a Neuronal Machine. New York: Springer Verlag, 1967.

Integrative and Molecular Approach to Brain Function.
M. Ito and Y. Miyashita, editors.

Cerebellar long-term plasticity and gene expression

Ryoji Yano[1], Kazutoshi Nakazawa[2], Raymond T. Kado[2,4], Laddawan Karachot[2], Masao Ito[2], Sumiko Mikawa[3], Yuriko Komine[5] and Tetsuo Yamamori[5]

Laboratory for [1]Cellular Information Processing, [2]Synaptic Function, [3]Neural Systems, Frontier Research Programs, Institute for Physical and Chemical Research (RIKEN), Wako, Japan; [4]Laboratoire de Neurobiologie Cellulaire et Moleculaire, CNRS Gif sur Yvette, France; and [5]Laboratory for Speciation Mechanisms, National Institute for Basic Biology, Okazaki, Japan

Abstract. We examined immediate early gene (IEG) expression in cerebellar slices and found c-Fos and Jun-B are markedly induced by the coapplication of AMPA and 8-Br-cGMP which cause long-term desensitization of cerebellar Purkinje cells (PCs) to test AMPA stimuli. A several fold increase of the number of double immunopositive PCs was observed after the conjunctive stimulation compared to single applications of either stimulation. Replacing 8-Br-cGMP with climbing fiber stimulation in conjunction with AMPA application in vivo also induces intense Jun-B-immunoreactivity(IR) within the AMPA applied regions. Stimulation by AMPA alone or climbing fiber alone gave far less Jun-B positive PCs throughout the cerebellar sections. The induction of Jun-B-IR by conjunctive stimulation can be blocked by nitro-arginine, a nitric oxide (NO) inhibitor. The results suggest that the conjunctive stimulation of parallel fibers and a climbing fiber, which causes LTD in the dendrite region of PC, can also induce c-Fos and Jun-B in the nucleus which may form AP-1 complexes and trigger a new class of gene expression contributing to a cerebellar long-term plasticity. To further study the molecular mechanisms of cerebellar long-term plasticity, we have isolated candidate genes which are expressed in PCs following the IEG expression.

Key words: Fos, gene expression, IEG, Jun-B, LTD.

Background

Cerebellar LTD (long-term depression) is a form of plasticity in which the prolonged reduction of the efficacy between a parallel fiber and a Purkinje cell is induced by the conjunctive stimulation of the parallel fibers and the climbing fiber to the Purkinje cell [1]. Together with hippocampal LTP, cerebellar LTD has been studied as a basic neuronal mechanism underlying mammalian learning and memory [2].

Recent progress in the study of cerebellar LTD has revealed underlying second messenger systems [3]. The transmitter from the parallel fiber to the Purkinje cell is glutamate. Among the three types of glutamate receptors, AMPA/kinate glutamate receptors have proved to be essential for functional synaptic transmission from a parallel fiber to a Purkinje cell. Metabotropic glutamate receptors also have been shown to be essential for cerebellar LTD [4]. The transmitter from climbing fiber to

---

Address for correspondence: Tetsuo Yamamori, Laboratory for Speciation Mechanisms, National Institute for Basic Biology, Okazaki 444, Japan.

the Purkinje cell is possibly glutamate, but is not yet proved. Climbing fiber stimulation can cause strong depolarization which is an essential process for cerebellar LTD. The involvement of nitric oxide (NO) in cerebellar LTD is reported in slice LTD and at behavioral levels as well [5,6]. Whether the NO synthesis required for LTD formation is due to the stimulation of the parallel fiber or climbing fiber needs to be determined.

Phosphorylation of AMPA/kinate type glutamate receptor in Purkinje cells has been proposed to be a core mechanism of cerebellar LTD [7]. In this model, coincidence of 1) binding of agonist (glutamate) to the receptor; 2) increase of intracellular free calcium; and 3) NO mediated cGMP pathway is required, at least partly, to keep the AMPA receptor phosphorylated for long times. The desensitization of the AMPA receptor through phosphorylation may be the major process occurring in cerebellar LTD for up to several hours.

However, for the formation of permanent memory there should be additional mechanisms to keep the "state" for a long time (perhaps even for a lifetime). This is where gene expression may play important roles [8].

Methods

The detailed methods used in this report were described in previous reports [9,10]. In brief, the cerebellar vermis was dissected from 7—9-week-old Wistar rats after short-ether anesthetization, and cut into parasagittal 500-μm thick slices. After 2.5 h of incubation to recover from the initial damage of cutting, slices were treated with four different paradigms: A) 0.5 mM 8-Br-cGMP for 15 min followed by 10 μM AMPA for 4 min; B) AMPA (10 μM, 4 min) alone; C) 8-Br-cGMP (0.5 mM, 15 min) alone; and D) untreated. IEG expression was analyzed with in situ hybridization and immunohistochemical techniques. For in situ hybridization, antisense oligonucleotides were labeled with deoxyadenosine ^{35}S[thio]triphosphate. For immunohistochemistry, staining was performed with avidin-biotin-peroxidase method and diaminobenzidine was used as chromogen. For in vivo experiment, Wister rats of 250—350 g were used. All animals were anesthetized with intraperitoneal sodium pentobarbital injection following atropine injection. Animals were fixed in a sterotaxic frame. All pressure points and incision areas were infused with xylocaine. A small hole was opened in the skull over the cerebellum with a dental drill and the cortex of the lobule VI exposed for recording and application of AMPA. Field potentials were recorded with tungsten or glass (Ringer's filled) microelectrodes. Inferior Olive (IO) was electrically stimulated. The placement was adjusted for best recorded response, about 300 μV, with 2-Hz stimulation. For the AMPA experiments, 50 μl AMPA (10 μM in Ringer solution, 37°C) was applied three times in the skull opening over a 10-min period for AMPA alone or for the same period while conjunctively stimulating IO at 2 Hz. For the immunohistochemistry, the cerebellum was removed after perfusion and kept in a 4% paraformaldehyde solution. It was serially placed for 24 h in 10, 20 and 30% sucrose solution before freezing. Staining of every second or third section was done overnight at 4°C with anti-Jun-B antisera (Santa Cruz Biotech, Inc., 1:100).

Immediate early gene expression in cerebellar slices

IEG (immediate early gene) is the first response genes to extracellular stimulation [11]. Thus, we first examined IEG expression in cerebellar slices under the condition that desensitize Purkinje cells to a test of AMPA stimulation. We started to examine IEG expression in "wedge recording system". The "wedge recording" measures the difference in potential between the cerebellar cortex and the white matter which are isolated by a grease gap [12,13]. Since the sole output from the cortex to the white matter is that from the Purkinje cells, the recorded potential should be due to current produced by the Purkinje cells when a certain type of drug is applied to the cortical part of the wedge. The simplicity of this method has been used to advantage for pharmacological studies of second messenger systems that control cerebellar long-term desensitization [13–15].

In 10 such experiments, IEGs examined with in situ hybridization for c-fos and jun-B expression showed significant changes under the condition that causes the desensitization of Purkinje cells (Table 1). We then detected the protein expression of Fos and Jun family members using immunohistochemical techniques. In order to quantify the degree of expression, we calculated the percentage of positive Purkinje cells under each condition (Table 2). c-Fos can be induced by the AMPA treatment. However, coapplication of AMPA and 8-Br-cGMP enhances the expression of c-Fos. Expression of another member of the Fos family, for example Fos-B, is not changed. Among Jun family, Jun-B is most significantly induced by the coapplication of AMPA and 8-Br-cGMP. AMPA or 8-Br-cGMP alone can also induce Jun-B to a significant level but coapplication of AMPA and 8-Br-cGMP doubles Jun-B expression (the net increases with the treatments are in parentheses). We have confirmed that c-Fos and Jun-B are coexpressed in the same Purkinje cell by immunostaining two adjacent thin sections with either of the two antibodies. Twelve double-positive Purkinje cells of c-Fos- and Jun-B-immunoreactivity (IR) were

Table 1. IEG expression in chemically stimulated Purkinje cells.

	A) 8-Br-cGMP + AMPA	B) AMPA	C) 8-Br-cGMP	D) Untreated
c-fos	++	++	–	–
fos-B	–	–	–	–
fra-1	–	–	–	–
fra-2	(+)	–	–	–
c-jun	(+)	–	–	–
jun-B	++	(+)	+	–
jun-D	–	–	–	–
NGFI-A	++	++	++	+
NGFI-B	(+)	(+)	–	–
PC4	–	–	–	–

– background level; (+) weak positive; + positive; ++ highly positive signal (cited from [9] with permission).

Table 2. Positively reacting Purkinje cells under the four different conditions.

	A) 8-Br-cGMP + AMPA	B) AMPA	C) 8-Br-cGMP	D) Untreated
c-Fos (n = 6)	29.9 ± 5.3% (23.6 ± 2.9%)	17.9 ± 4.7% (11.6 ± 2.3%)	5.4 ± 0.6% (–0.7 ± 1.8%)	6.3 ± 2.4%
Fos-B (n = 4)	3.9 ± 2.4%	8.3 ± 1.4%	3.3 ± 0.9%	6.6 ± 2.2%
c-Jun (n = 8)	18.2 ± 4.4%	11.5 ± 3.2%	10.1 ± 2.2%	10.6 ± 3.0%
Jun-B (n = 7)	40.2 ± 5.0% (30.6 ± 3.4%)	18.4 ± 2.8% (8.8 ± 1.2%)	24.0 ± 3.9% (14.4 ± 2.3%)	9.6 ± 1.6%
Jun-D (n = 3)	0%	0%	0%	0%

The means and standard errors of the percentages of positive Purkinje cells in a fixed counting area within the section. Numbers of slices (n) examined are also shown (cited from [9] with permission).

counted among 30 cells examined in two adjacent sections. Only a few Purkinje cells were double-positive in both c-Fos-IR and Jun-B-IR following either AMPA (two out of 22 cells) or 8-Br- cGMP application alone (zero out of 17) [9].

The above results suggest that coincident inputs from the parallel fibers and the climbing fiber to a Purkinje cell, causing LTD in the dendrite, may also trigger a new set of gene expression in the nucleus. This new gene expression may underlie a cerebellar long-term plasticity. With the view supporting this model, we studied whether conjunctive stimulation to Purkinje cells can induce Jun-B in vivo.

Jun-B expression in vivo

In order to examine Jun-B expression through more physiological conjunctive stimulation to Purkinje cells, we replaced one of the conjunctive stimulation, 8-Br-cGMP, used in pharmacological treatment with climbing fiber stimulation through electrical stimulation of IO. AMPA was applied on the surface of the cerebellar cortex to substitute for parallel fiber stimulation.

The expression of Jun-B-IR was detected localized in regions conjunctively stimulated as shown in Fig. 1. The intensity of Jun-B marking was strongest in the area where AMPA was applied (Fig. 1D) and gradually decreased with distance from the area (in Fig. 1 from C to A, in one direction; from E to F, in the other direction). Since the firing of climbing fibers via IO stimulation is expected to occur in broader areas than the area of intense expression, the localization can be interpreted to reflect the area where the stimulation of AMPA and climbing fiber were conjunctive. This interpretation has been further supported by the fact that stimulation of climbing fiber or AMPA alone showed far less marking throughout the sections examined (Fig. 2C,D). Through lobules I to X, intense signals were found localized in lobule VI where AMPA was applied. These results also suggest that the conjunction of AMPA and climbing fiber stimulation is necessary for higher level of Jun-B induction.

Jun-B induction was blocked by nitro-arginine, a NO synthase inhibitor, applied under the conjunctive condition (Fig. 2B), suggesting an involvement of NO in the

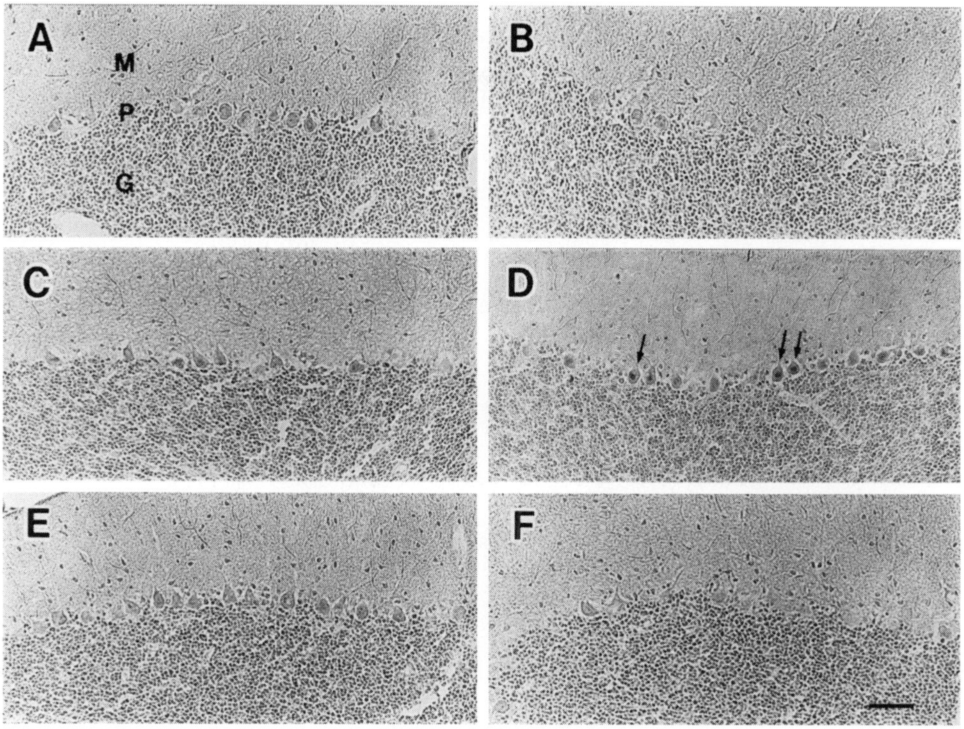

Fig. 1. Jun-B expression under conjunctive stimulation of AMPA and climbing fibers. **A** to **F** are from every 20th section spanning 1 mm on each side of C and D, approximately centered on the region of stimulation and recording. M: molecular; P: Purkinje; and G: granule layers (cited from [10] with permission).

gene expression of Jun-B. Our results suggest that activation of the NO and cGMP pathways may be important in inducing Jun-B in the nucleus. We started the experiments assuming the source of NO production to be climbing fiber stimulation [5]. It has recently been proposed that parallel fiber stimulation may be the cause of NO production [16]. If this is the case, AMPA application to the cortical surface may stimulate Purkinje cells both post and presynaptically, as well as other cells and processes in the cortex which may also produce NO. Therefore, we further examined Jun-B-IR under the conjunctive stimulation of parallel fibers and climbing fibers (climbing fibers stimulated first and parallel fiber 20 ms later at 2 Hz. Significant Jun-B expression was observed in the costimulated area (Fig. 3A, the lobule of folium VI), whereas less expression was observed in nearby area (Fig. 3B, the lobule VII).

The level of Jun-B expression in control animals sometimes varied. To know the cause of these variations, we anesthetized rats with sodium pentobarbital for several hours and compared the level of expression to nonanesthetized animals. In the nonanesthetized animals weak Jun-B expression varying among the Purkinje cells are observed. This kind of weak expression was also observed in animals receiving only

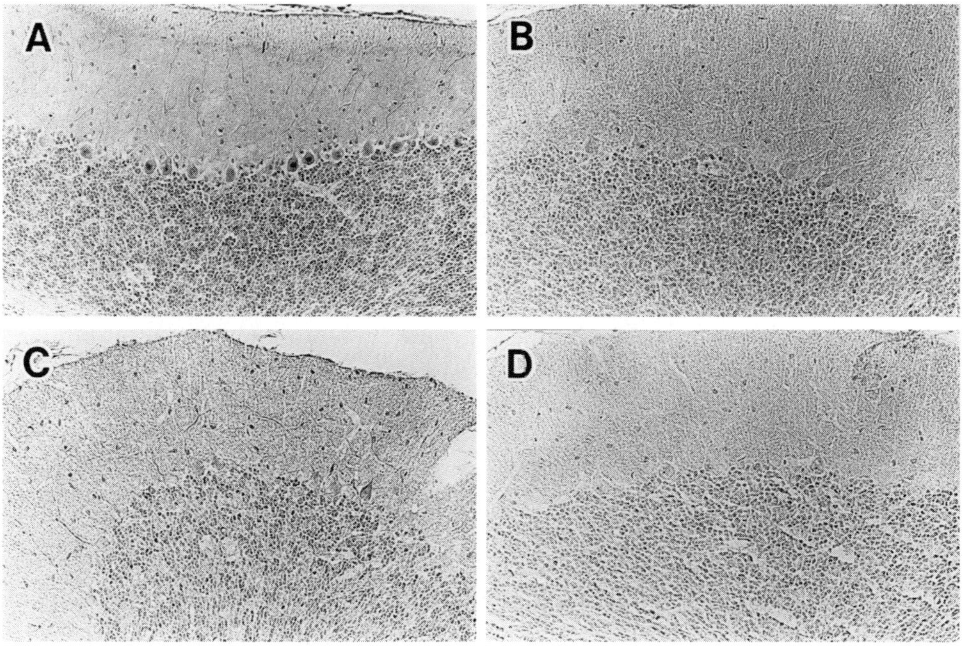

Fig. 2. Jun-B immunoreactivity under conjunctive stimulation of AMPA and climbing fiber. **A**: Conjunctive stimulation of AMPA (10 μM, ×3, 50 μl) and climbing fibre (2 Hz for 10 min). **B**: 30 μM nitro-arginine in Ringer's solution, 50 μl applied three times at 10-min intervals, before and together with AMPA during the conjunctive stimulation. **C**: Climbing fiber stimulation only. **D**: AMPA only (cited from [10] with permission). The strongest area is shown in each condition.

saline application instead of AMPA application (Fig. 4A). To the contrary, in anesthetized animals Jun-B expression was minimal (Fig. 4B). Pentobarbital is known to stimulate inhibitory neurons. Therefore, the Purkinje cells, which receive the input of inhibitory interneurons, may be inhibited during the anesthesia. This reduced activity in the Purkinje cells likely caused the shut-off of Jun-B expression in the anesthetized animals. The results obtained with in vivo stimulation paradigms thus far supports the model that the costimulation of the climbing and parallel fiber can induce a specific type of IEG in the nucleus of the Purkinje cells.

Roles of genes expressed following IEG expression in cerebellar long-term plasticity

The pathway that induces IEGs following extracellular stimulation has been revealed in a few cell types (e.g., [17]). The pathway that follows synaptic stimulation which causes hippocampal LTP also shares common pathways with those of other cell types. Some cascades of phosphorylation, including proteins of CaM kinase II, PKC and CREB following LTP has been suggested, although there may be some specific

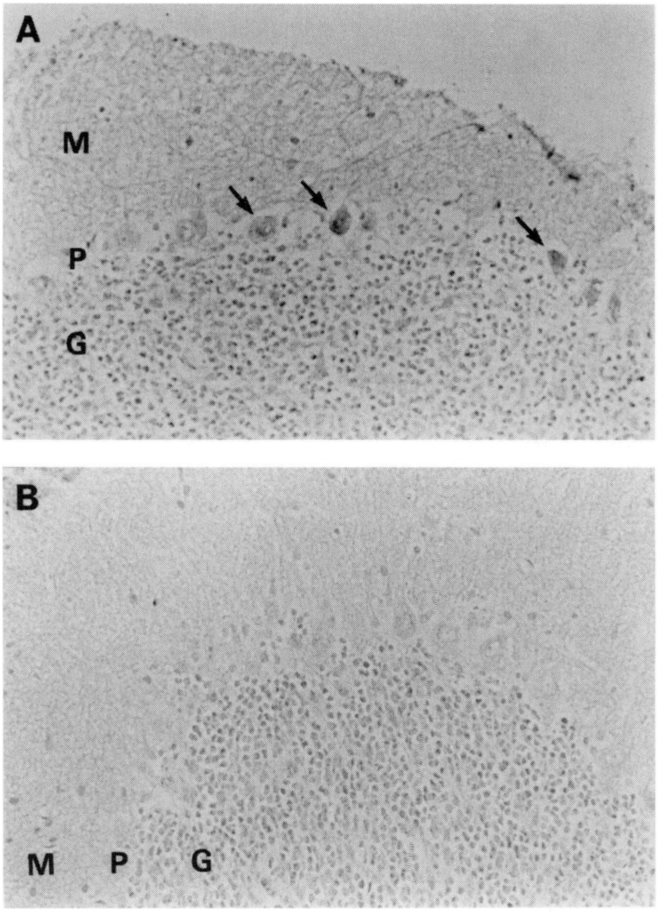

Fig. 3. Jun B-IR under conjunctive stimulation of parallel fiber and climbing fiber. Climbing fibers were stimulated through IO and parallel fibers by directly stimulating in molecular layer 20 ms later at 2 Hz. Parallel fiber and climbing fiber responses were monitored by recording the field potentials. **A**: Jun-B-IR in the area of parallel fiber stimulation (lobule VI). Arrows indicate Jun-B-IR positive Purkinje cells. **B**: Jun-B-IR in lobule VII where the Purkinje cells in the same section near but separate from the PF stimulation. M: molecular; P: Purkinje; and G: granule layers.

mechanism for sensing calcium influx, which eventually phosphorylates Ser-133 of CREB, following the synaptic current [18]. Manipulations of CREB genes suggest their involvement in long-term memory in Aplysia, *Drosophila* and mice [19]. In the cerebellum the intracellular pathway for c-Fos and Jun-B induction is not known. Our data suggest that cGMP may play an important role in Jun-B induction. The formation of AP-1 complexes including Fos/Jun-B may be a rate-limiting step to control the gene expression following conjunctive stimulation into the Purkinje cell. What genes are, then, expressed following the AP-1 forming step (Fig. 5)? Genes

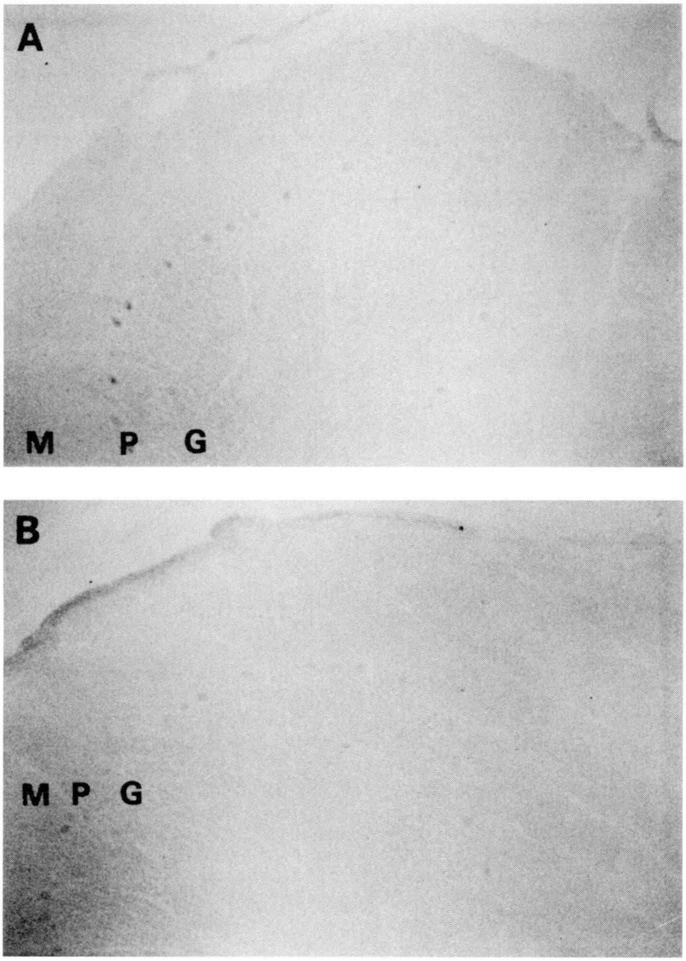

Fig. 4. Lack of Jun-B-IR in the cerebellum. Photomicrographs of cerebellar sections from animals with; **A**: Saline applied in the same way as AMPA in Fig. 3C. **B**: After 6 h deep sodium pentobarbital anesthesia. M: molecular; P: Purkinje; and G: granule layers.

whose expression is altered during the period between 0.5–3 h after hippocampal LTP have been reported [20–22]. A group of genes increases after hippocampal LTP include the β1 subunit of $GABA_A$ receptor, zif-268, CaM kinase II, tPA and *arc* etc. On the other hand, the genes for PKC β subunit, muscarinic type III receptor, trkA and BDNF decrease during this period following the LTP. The significance of the alteration of these genes in long-term synaptic plasticity is still very speculative [19].

In order to know the roles of the genes expressed following IEG expression in cerebellar long-term plasticity, we have started to isolate these kinds of genes systematically in Purkinje cells. By comparing the RNAs isolated from cultured Purkinje cells of nonstimulated and those costimulated with AMPA and 8-Br-cGMP,

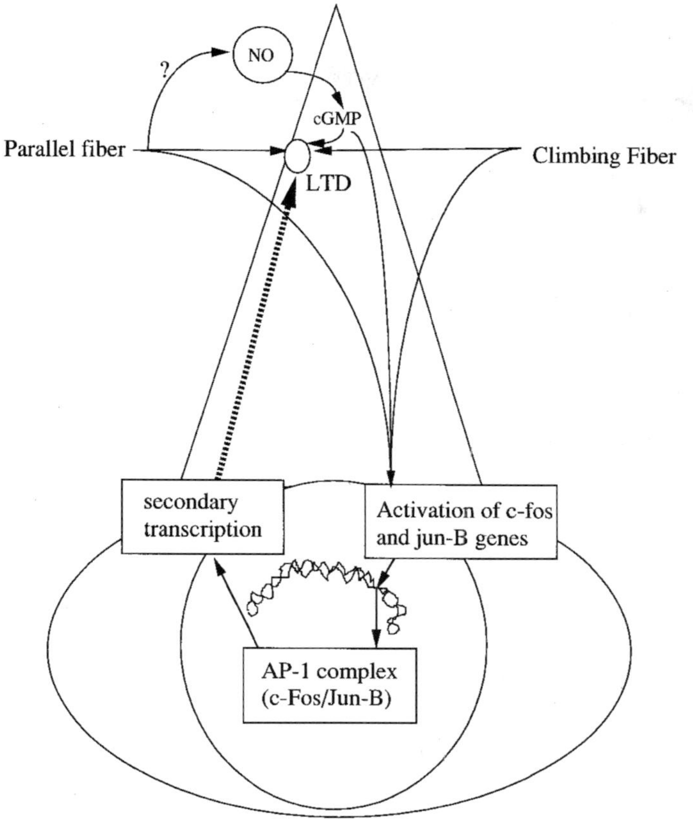

Fig. 5. Schematic model of intercellular pathways which lead to long-term cerebellar plasticity.

we have been able to isolate about 20 gene candidates which may be specifically expressed in Purkinje cells following the costimulation. Thus far, two of the genes are known to be a kinase or a kinase-related protein. The rest of them, however, are unreported genes. We are currently characterizing these genes. The analysis of the structures and the expression patterns of the isolated genes may reveal possible cellular pathways directly linking IEG expression with synaptic plasticity in long-term memory of the Purkinje cell.

Acknowledgements

We thank Drs Masaya Toyama and Hiroshi Kiyama, Osaka University, for the in situ hybridization technique, and Dr Yusaku Nakabeppu, Kyushu University, for providing the antibody against Jun-B used for slice experiments. This research has been supported by Uehara Memorial Foundation, Frontier Research Program (RIKEN), research grants from the Japanese Ministry of Education, Science, Sports and Culture, Science and Technology Agency and Tokai Science and Technology Foundation.

References

1. Ito M, Sakurai M, Tongroach P. Climbing fibre induced depression of both mossy fibre responsiveness and glutamate sensitivity of cerebellar Purkinje cells. J Physiol 1982;324:113–134.
2. Ito M. Long-term depression. Ann Rev Neurosci 1989;12:85–102.
3. Ito M. The cellular basis of cerebellar plasticity. Curr Opin Neurobiol 1991;1;616–620.
4. Aiba A, Kano M, Chen C, Stanton ME, Fox GD, Herrup K, Zwingmann TA, Tonegawa S. Deficient cerebellar long-term depression and impaired motor learning in mGluR1 mutant mice. Cell 1994;79: 377–388.
5. Shibuki K, Okada D. Endogenous nitric oxide release required for long-term synaptic depression in the cerebellum. Nature 1991;349:326–328.
6. Nagao S, Ito M. Subdural application of hemoglobin to the cerebellum blocks vestibuloocular reflex adaptation. Neuro Report 1991;2:193–196.
7. Nakazawa K, Mikawa S, Hashikawa T, Ito M. Transient and persistent phosphorylation of AMPA-type glutamate receptor subunits in cerebellar Purkinje cells. Neuron 1995;15:697–709.
8. Goele P, Castellucci VF, Schacher S, Kandel ER. The long and the short of long-term memory – a molecular framework. Nature 1986;322:419–422.
9. Nakazawa K, Karachot L, Nakabeppu Y, Yamamori T. The conjunctive stimuli that cause long-term desensitization also predominantly induce c-Fos and Jun-B in cerebellar Purkinje cells. Neuro Report 1993;4:1275–1278.
10. Yamamori T, Mikawa S, Kado R. Jun-B expression in Purkinje cells by conjunctive stimulation of climbing fibre and AMPA. Neuro Report 1995;6:793–796.
11. Morgan JI, Curran T. Stimulus-transcription coupling in the nervous system: involvement of the inducible proto-oncogenes fos and jun. Ann Rev Neurosci 1991;14:421–451.
12. Garthwaite J, Garthwaite G, Hajós F. Amino acid neurotoxicity: relationship to neuronal depolarization in rat cerebellar slices. Neuroscience 1986;18:449–460.
13. Ito M, Karachot L. Long-term desensitization of quisqualate-specific glutamate receptors in Purkinje cell investigated with wedge recording from rat cerebellar slices. Neurosci Res 1989;7:168–171.
14. Ito M, Karachot L. Messengers mediating long-term desensitization in cerebellar Purkinje cells. Neuro Report 1991;1:129–132.
15. Ito M, Karachot L. Protein kinases and phosphatase inhibitors mediating long-term desensitization of glutamate receptors in cerebellar Purkinje cells. Neurosci Res 1992;14:27–38.
16. Lev-Ram V, Makings LR, Keitz PF, Kao JP, Tsien RY. Long-term depression in cerebellar Purkinje neurons results from coincidence of nitric oxide and depolarization-induced Ca^{2+} transients. Neuron 1995;15:407–415.
17. Sheng M, Greenberg ME. The regulation and induction of c-fos and other immediate early genes in the nervous system. Neuron 1990;4:477–485.
18. Deisseroth K, Bito H, Tsien RW. Signaling from Synapse to nucleus: postsynaptic CREB phosphorylation during multiple forms of hippocampal synaptic plasticity. Neuron 1996;16:89–101.
19. Carew TJ. Molecular enhancement of memory formation. Neuron 1996;16:5–8.
20. Mackler SA, Brooks BP, Eberwine JH. Stimulus-induced coordinate changes in mRNA abundance in single postsynaptic hippocampal CA1 neurons. Neuron 1992;9:539–548.
21. Qian Z, Gilbert ME, Colicos MA, Kandel E, Kuhl D. Tissue-plasminogen activator is induced as an immediate-early gene during seizure, kindling and long-term potentiation. Nature 1993;361:453–457.
22. Lyford GL, Yamagata K, Kaufmann WW, Barnes CA, Sanders LK, Copeland NG, Gilbert DJ, Jenkins NA, Lanahan AA, Worley PF. Arc, a growth factor and activity-regulated gene, encodes a novel cytoskeleton-associated protein that is enriched in neuronal dendrites. Neuron 1995;14: 433–445.

Neural signaling and plasticity

Integrative and Molecular Approach to Brain Function.
M. Ito and Y. Miyashita, editors.

Ca^{2+} permeability and subunit composition of AMPA receptors in hippocampal neurons

Seiji Ozawa[1], Tadashi Isa[2], Keisuke Tsuzuki[1] and Masae Iino[1]
[1]*Department of Physiology, School of Medicine, Gunma University, Maebashi, Gunma; and* [2]*Department of Integrative Physiology, The National Institute for Physiological Sciences, Myodaiji, Okazaki, Japan*

Abstract. Molecular biological studies have shown that the AMPA-type glutamate receptors are composed of various combinations of four subunit proteins, GluR1, GluR2, GluR3 and GluR4. In cultured rat hippocampal neurons, two functionally different types of AMPA receptors were expressed. In the majority of neurons, referred to as type I neurons, AMPA receptors displayed an outward rectification and little Ca^{2+} permeability (type I AMPA receptors). In contrast, a small number of neurons, referred to as type II neurons, had AMPA receptors with a strong inward rectification and high Ca^{2+} permeability (type II AMPA receptors). To trace the molecular basis for the functional differences in these two types of native AMPA receptors, we adopted a method that combined the whole-cell patch-clamp recording with reverse transcription-polymerase chain reaction (RT-PCR). With this method, the GluR2 subunit was not detected in type II neurons, whereas it was expressed abundantly in type I neurons. In slice-patch experiments, type II neurons were observed in various areas of the hippocampus. These neurons stained with biocytin had round- or ellipsoidal-shaped somata and issued divergent axonal projections to the surrounding structures. Furthermore, type II AMPA receptors were involved in fast excitatory synaptic transmission in these neurons.

Key words: GluR2 subunit, glutamate receptor, inward rectification, RT-PCR, synaptic transmission.

Introduction

On the basis of agonist specificity, ionotropic glutamate receptors have been classified into three subtypes: N-methyl-D-aspartate (NMDA) receptors, α-amino-3-hydroxy-5-methyl-4-isoxazolepropionate (AMPA) receptors, and kainate receptors [1–3]. Among them, AMPA receptors are the principal mediators of fast excitatory neurotransmission in synapses in the central nervous system (CNS) [4–6].

AMPA receptors are composed of various combinations of four subunit proteins, GluR1 (GluR-A), GluR2 (GluR-B), GluR3 (GluR-C) and GluR4 (GluR-D), which are approximately 70% identical in amino acid sequences [7–9]. Expression studies have shown that the presence of the GluR2 subunit determines both the rectification properties and the Ca^{2+} permeability of the AMPA receptor channels. The AMPA receptor with the GluR2 subunit displays a linear or outwardly rectifying current-voltage (I-V) relation and little Ca^{2+} permeability. In contrast, the receptor lacking this

---

Address for correspondence: Dr Seiji Ozawa, Department of Physiology, School of Medicine, Gunma University, Showa-machi 3-39-22, Maebashi, Gunma-ken 371, Japan. Tel.: +81-272-20-7930. Fax: +81-272-20-7936.

subunit displays an inwardly rectifying I-V relation and a high Ca^{2+} permeability [7,9].

We found that two functionally different types of AMPA receptors were expressed in cultured rat hippocampal neurons. The neurons designated as type I neurons had AMPA receptors that displayed outward rectification and little Ca^{2+} permeability, whereas those designated as type II neurons had AMPA receptors with a strong inward rectification and a high Ca^{2+} permeability. To determine whether the different functional properties of AMPA receptors in type I and type II neurons are due to different subunit composition, we used a method that combined the whole-cell patch-clamp recording and reverse transcription-polymerase chain reaction (RT-PCR). Furthermore, we searched for neurons carrying inwardly rectifying and Ca^{2+}-permeable AMPA receptors in rat hippocampal slices, and examined whether they would be involved in excitatory neurotransmission.

Two distinct types of AMPA receptors expressed in cultured rat hippocampal neurons

Two functionally different types of AMPA receptors were expressed in cultured rat hippocampal neurons [10,11]. In most of the cultured neurons, AMPA receptors displayed an outwardly rectifying current-voltage (I-V) relation and little Ca^{2+} permeability. We designated these neurons as type I neurons. In contrast, in a small population of neurons AMPA receptors displayed a strong inward rectification and a high Ca^{2+} permeability. We designated these neurons as type II neurons. Type II neurons were small cells with round or elliptical somata and fine neurites, presumably interneurons. Type I neurons were morphologically heterogeneous, and prematured pyramidal cell-like neurons invariably had the properties of type I neurons [11].

Figure 1A–D shows the AMPA receptor response in a type I neuron. In the control solution, the I-V relation, obtained by plotting the peak amplitude of kainate-induced current against the membrane potential, displayed an outward rectification. In these experiments, kainate was used as a nondesensitizing agonist of AMPA receptors [12]. Current responses were recorded in the whole-cell patch clamp mode

---

Fig. 1. AMPA receptor-mediated current responses in type I (A–D) and type II (E–H) cultured hippocampal neurons. **A–C**: Current responses to kainate (a nondesensitizing agonist of AMPA receptors) applied by iontophoresis in a type I neuron. The current traces were recorded at various membrane potentials in the control (A) Na^+-free, Ca^{2+}-free (B) and Na^+-free, 10 mM-Ca^{2+} (C) solutions. The membrane potentials (in mV) are +30, +10, 0, –5, –20, –40, –60 in A, –40, –60, –80 in B and +10, 0, –20, –40, –60, –80 in C. **D**: Relation between the peak amplitude of the current response and the membrane potential in the control (●), Na^+-free, Ca^{2+}-free (■) and Na^+-free, 10 mM-Ca^{2+} (○) solutions. All the traces in A–C were taken from the same neuron. **E–G**: Current responses in a type II neuron in the control (E), Na^+-free, Ca^{2+}-free (F) and Na^+-free, 10 mM-Ca^{2+} (G) solutions. The membrane potentials (in mV) are +40, +20, 0, –20, –40, –60 in E, –60, –80, –90 in F and +20, 0, –20, –40, –60, –80 in G. H: relation between the peak amplitude of the current response and the membrane potential in the control (●), Na^+-free, Ca^{2+}-free (■) and Na^+-free, 10 mM-Ca^{2+} (○) solutions. Note that a strong inward rectification in the current response in the control solution and prominent inward Ca^{2+} currents in Na^+-free, 10 mM-Ca^{2+} solution. All the traces in E–G were taken from the same neuron. (From [10].)

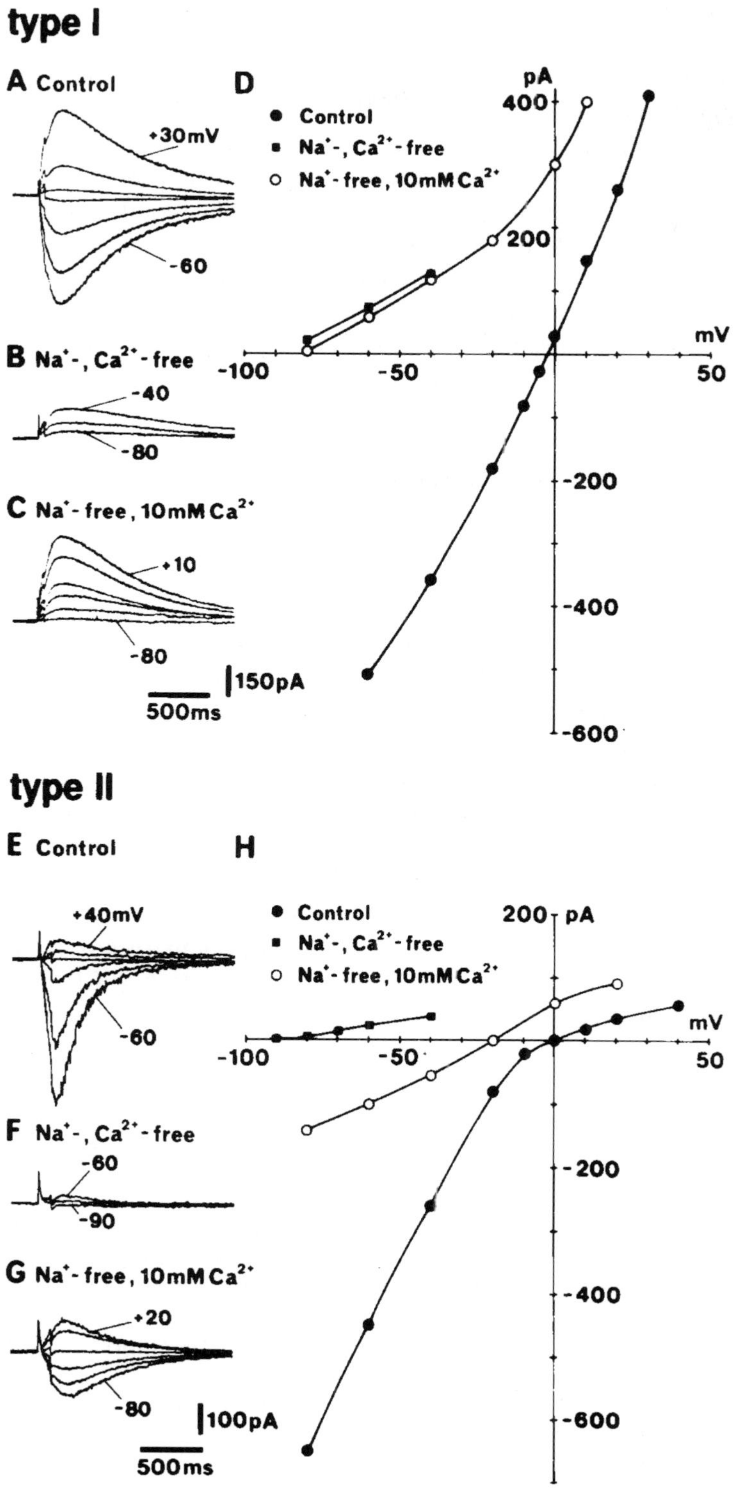
type I
A Control
+30mV
-60
B Na⁺-, Ca²⁺-free
-40
-80
C Na⁺-free, 10mM Ca²⁺
+10
-80
150pA
500ms
D
Control
Na⁺-, Ca²⁺-free
Na⁺-free, 10mM Ca²⁺
pA
400
200
-200
-400
-600
mV
-100
-50
50
type II
E Control
+40mV
-60
F Na⁺-, Ca²⁺-free
-60
-90
G Na⁺-free, 10mM Ca²⁺
+20
-80
100pA
500ms
H
Control
Na⁺-, Ca²⁺-free
Na⁺-free, 10mM Ca²⁺
200
pA
mV
-100
-50
50
-200
-400
-600

using patch pipettes containing 165 mM Cs^+. In Na^+-free, Ca^{2+}-free solution in which these cations were replaced with the impermeant cation N-methylglucamine (NMG), kainate caused no inward current. No inward current could be induced by adding 10 mM Ca^{2+} to the external solution up to –80 mV. This indicates that the permeability to Ca^{2+} of AMPA receptors in type I neurons is very low [10].

Figure 1E–H shows the AMPA receptor response in a type II neuron. The I-V relation of the kainate response showed a remarkable inward rectification in the control solution. In Na^+-free, Ca^{2+}-free solution, kainate caused no inward current. However, in Na^+-free, 10 mM Ca^{2+} solution, kainate induced prominent inward currents at potentials more negative than –20 mV. This indicates that AMPA receptors in type II neurons have substantial permeability to Ca^{2+}. In type II neurons the ratio of the permeability coefficients of Ca^{2+} and Cs^+ (P_{Ca}/P_{Cs}) of AMPA receptors was 2.3, whereas the corresponding value in type I neurons was <0.18 [10]. There was a close correlation between the degree of the inward rectification and the Ca^{2+} permeability in AMPA receptors expressed in cultured neurons [13].

Subunit composition of AMPA receptors expressed in cultured rat hippocampal neurons

The presence of the GluR2 subunit determines both rectification properties and Ca^{2+} permeability in recombinant AMPA receptors. The unique properties of GluR2 is due to difference in a single amino acid residue in the region initially assigned to the second hydrophobic region (M2). In GluR2 this residue is a positively charged arginine (R), while in the other three subunits it is a neutral amino acid, glutamine (Q). The site in M2 where these amino acids are positioned has been designated as the Q/R site [7,9]. These results suggest that the GluR2 subunit is not or, if any, very little expressed in type II hippocampal neurons in culture. To correlate the expression of each subunit with the functional properties of native AMPA receptors at the single-cell level, we have coupled the whole-cell patch-clamp recording and reverse transcription (RT) followed by polymerase chain reaction (PCR) amplification [14,15]. This method has been established by Lambolez et al. [16], and will hereafter be abbreviated as the patch-clamp RT-PCR.

Figure 2 illustrates the expression of different AMPA receptor subunits in functionally identified neurons. In gel analysis shown in Fig. 2, the PCR products from four type I and four type II neurons were digested by restriction enzymes that selectively cut either GluR1, GluR2, GluR3 or GluR4. The restriction analysis indicated that the GluR2 subunit was expressed abundantly together with GluR1 in four type I neurons shown in Fig. 2A. GluR3 and GluR4 were occasionally detected in other type I neurons together with GluR1 and GluR2. On the other hand, only the GluR1 and GluR4 subunits were expressed and the GluR2 and GluR3 subunits were not detected in type II neurons (Fig. 2B). The absence of GluR2 in type II neurons was further confirmed by Southern analysis (Fig. 2C). In gel analysis in Fig. 2C1, the PCR products from a type I neuron and five type II neurons were digested by a restriction enzyme specific for GluR2. The digestion products of GluR2 were seen

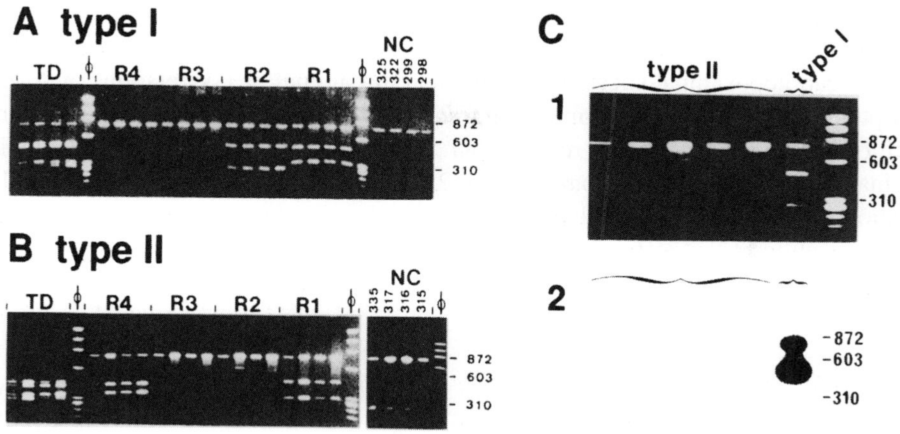

Fig. 2. Subunit composition of AMPA receptors in cultured hippocampal neurons. **A and B**: Analysis of cDNA encoding AMPA receptor subunits in four type I (A) and four type II (B) neurons that had been identified functionally by the whole-cell patch clamp recording. Individual cell numbers are indicated, and in all panels (NC, R1-R4 and TD), four lanes correspond to the four different cells. NC, agarose gel electrophoresis of uncut cDNA fragments obtained after first PCR. Other panels show results of second amplification cut by the following restriction enzymes (size of predicted fragments is indicated in bp in parentheses): R1, Bgl I specific for GluR1 (300 and 449); R2, BSP 1286I specific for GluR2 (478 and 271); R3, Eco 47III specific for GluR3 (359 and 396); R4, EcoR I specific for GluR4 (411 and 338); TD, the four enzymes in A and both Bgl I and EcoR I in B. In each lane, 749-bp (755 bp for GluR3) bands correspond to the GluR1-4 fragments resistant to enzymatic digestion and the two smaller bands to fragments of subunit cut by its specific enzyme. Note absence of digestion product with the enzyme specific for GluR2 in the four type II neurons (B) and presence of these fragments in the four type I neurons (A). Position of molecular weight markers Hae III digest of ϕX 174 (ϕ) is indicated on right. **C**: Gel analysis (1) and Southern blot of gel (2) of a type I and five type II neurons. C1: electrophoresis of second PCR products cut by BSP 1286I specific for GluR2. Bands corresponding to digestion products of GluR2 are visible only in the type I neuron. C2: 478-bp band corresponding to digestion product of GluR2 shown in C1 is labeled with a GluR2-specific oligoprobe in type I neuron together with some uncut 749-bp fragment. No labeling is seen at corresponding positions in type II neurons, indicating absence of GluR2. (From [14].)

only in the type I neuron. Southern blot of the gel indicated that the GluR2-specific oligoprobe labeled only the DNA fragments of the type I neuron but not those of type II neurons (Fig. 2C2). These results obtained by the patch-clamp RT-PCR method indicate that a strong inward rectification and a high Ca^{2+} permeability in AMPA receptors in type II neurons is due to either lack of or very little expression of GluR2 [14,15].

Neurons carrying inwardly rectifying and Ca^{2+}-permeable AMPA receptors in hippocampal slices

To clarify the functional significance of inwardly rectifying and Ca^{2+}-permeable AMPA receptors (type II AMPA receptors) in the CNS, it is necessary to know their distributions in the CNS and whether they are involved in synaptic transmission. Therefore, we searched for neurons carrying this type of AMPA receptors in 5–13-day-old rat hippocampal slices [17]. We found three types of I-V relations of AMPA receptor responses in nonpyramidal neurons in hippocampal slices: type I response with outwardly rectifying I-V relation (RI > 1.0), type II response with I-V relation of marked inward rectification (RI < 0.25), and intermediate response with I-V relation of weaker inward rectification (0.25 ≤ RI ≤1.0). Here, RI is the abbreviation of rectification index defined as follows:

$$RI = \frac{\dfrac{I_{+40}}{(40-E_{rev})}}{\dfrac{I_{-60}}{(-60-E_{rev})}} \qquad (1)$$

where I_{+40} and I_{-60} are the amplitudes of the kainate/AMPA-induced currents at +40 and –60 mV, respectively, and E_{rev} is the reversal potential. Neurons with type I, type II and intermediate I-V relations of AMPA receptors were referred to as type I, type II and intermediate neurons, respectively [17].

Figures 3A1 and 3B1 show whole-cell I-V relations of AMPA receptors in type I and type II neurons, respectively. To estimate the Ca^{2+} permeability of these neurons, a large nucleated outside-out patch was made and the I-V relation of the AMPA receptors was obtained in the Na^+-free solution containing 100 mM Ca^{2+} (Figs. 3A2 and 3B2). The reversal potentials of AMPA receptors in type II neurons were +11.8 ± 3.6 mV (n = 5, mean ± SD), being much more positive than the corresponding value in type I neurons (–44.4 ± 14.0 mV, n = 7). This value in the intermediate neuron was –8.7 ± 7.4 mV (n = 7), being between those of the type I and type II neurons [17]. Thus, the degree of the inward rectification was correlated closely with the Ca^{2+} permeability in AMPA receptors in the nonpyramidal neurons in hippocampal slices, as was the case in cultured rat hippocampal neurons. Patch-clamp RT-PCR analyses for subunit composition of AMPA receptors in the neurons in hippocampal slices remain to be conducted. We speculate that the GluR2 subunit would not be or, if any, very little expressed in the type II neurons in slices. AMPA receptors assembled with and without GluR2 would coexist in the intermediate neurons.

Morphological features of type II neurons were investigated by staining with biocytin. The somata of type II neurons were round or ellipsoidal in shape. These neurons were observed in various areas of the hippocampus, most frequently in the stratum radiatum and stratum lacunosum-moleculare of both the CA1 and CA3 regions, and in the stratum moleculare of the dentate gyrus. They issued axonal projections and terminated in the surrounding regions. These results strongly suggest that the type II neurons are interneurons [17].

To examine whether the type II AMPA receptors are involved in excitatory

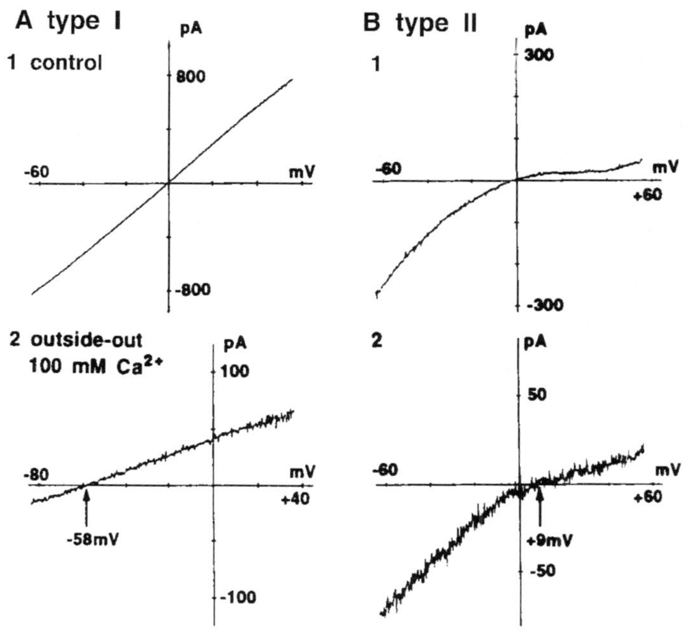

Fig. 3. Two distinct types of AMPA receptors expressed in nonpyramidal neurons in rat hippocampal slices. **A and B**: I-V relations of AMPA receptor-mediated responses in type I (A) and type II (B) neurons. (1): I-V relations of whole-cell kainate (200 (M) responses between –64 and +56 mV. (2): I-V relations of kainate responses in large nucleated outside-out patches bathed in isotonic Ca^{2+} solution containing 100 mM Ca^{2+}. The outside-out patches were excised from the somatic regions of which I-V relations of the whole-cell kainate responses are shown in (1). In B, the patch pipette was filled with the internal solution containing 300 μM spermine to maintain the inward rectification in the cell-free patch. (See [17] for more details.)

synaptic transmission, I-V relations of the excitatory postsynaptic currents (EPSCs) were investigated in neurons identified as type II by their I-V relations of the whole-cell response to kainate (Fig. 4D). The EPSCs evoked by local stimulation in the presence of 10 μM bicuculline (a GABA antagonist) in type II neurons had two kinetically distinct components (Fig. 4A). The slow component was abolished by 50 μM D-2-amino-5-phosphonovalerate (APV) (Fig. 4B). The fast component was unaffected by 50 μM APV, but was abolished completely by 5 μM 6-cyano-7-nitroquinoxaline-2,3-dione (CNQX). Thus, the EPSCs were generated by activation of both AMPA and NMDA receptors in type II neurons. When the peak amplitudes of the fast components of EPSCs recorded in the presence of 50 μM APV were plotted against the membrane potential, the I-V relation displayed a strong inward rectification (RI < 0.25) (● in Fig. 4C). This indicates that the inwardly rectifying AMPA receptors are involved in excitatory synaptic transmission [17].

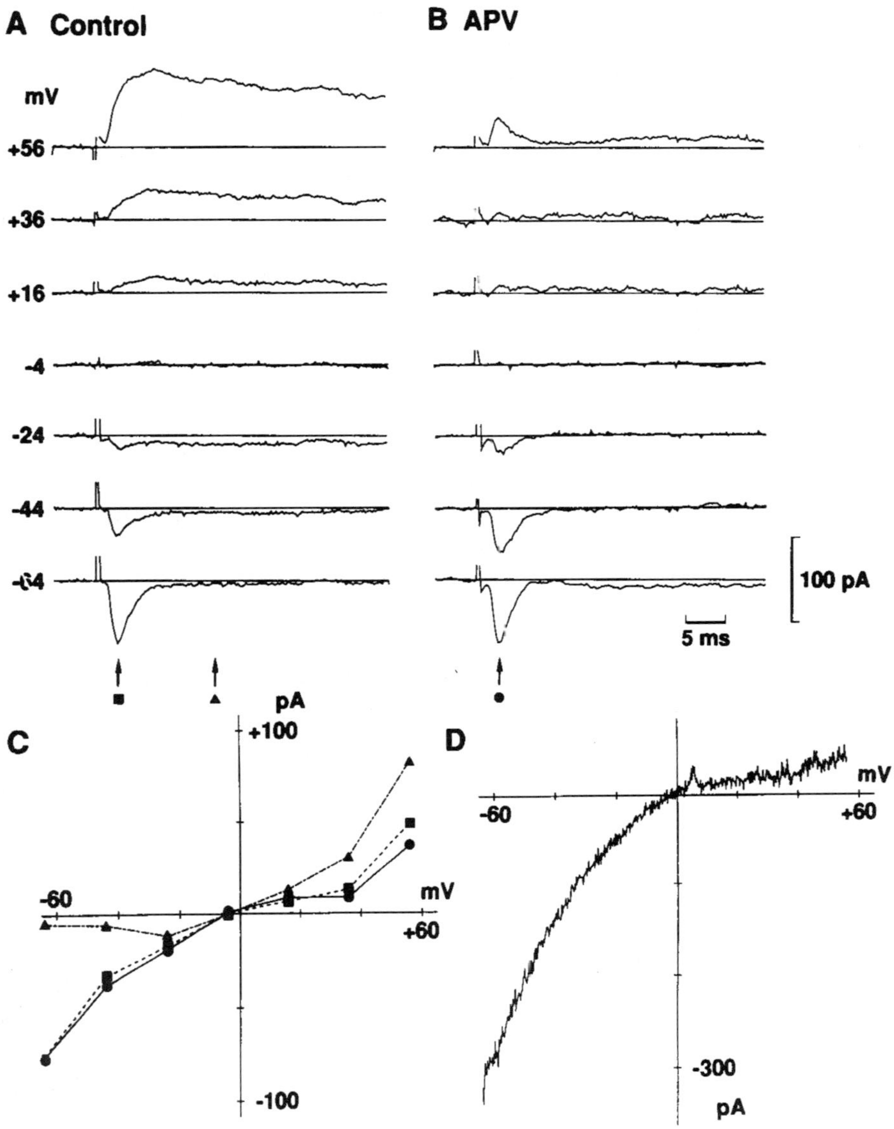

Fig. 4. EPSCs recorded in a type II neuron in the stratum radiatum of the CA1 region. **A**: EPSCs recorded at various membrane potentials between –64 and +56 mV in 20 mV steps in control solution containing 10 μM bicuculline methobromide (Bic). Bic was added in order to block synaptic activities mediated by GABA. Stimulation of the presynaptic fibers was applied at a point 80 μm distant from the soma of the recorded neuron. **B**: EPSCs recorded during addition of 50 μM APV. **C**: I-V relations of EPSCs. Squares and triangles indicate amplitudes of EPSCs at 2.9 and 14.7 ms following the stimuli, respectively, in A. (▲) represent mainly the EPSCs generated by activation of NMDA receptors. (●) indicate the amplitude of EPSCs at 2.9 ms in B, which represents the EPSCs generated by activation of AMPA receptors. The rectification index (RI) of this I-V relation was 0.16. **D**: I-V relation of the whole-cell response to kainate (200 μM) obtained in the same neuron before recording EPSCs. (From [17.)

Discussion

We have shown that two functionally different types of AMPA receptors are expressed in cultured rat hippocampal neurons. The patch-clamp RT-PCR method has allowed us to analyze the molecular basis of different functional properties of these native AMPA receptors. This method has revealed that functionally identified type II neurons in culture express only GluR1 and GluR4 subunits. This subunit composition explains the unique electrical properties of AMPA receptors expressed in type II neurons, such as their strong inward rectification and high Ca^{2+} permeability. The patch-clamp RT-PCR method will be useful for a wide range of studies to elucidate the molecular mechanisms underlying physiological functions at the single cell level.

Type II neurons carrying inwardly rectifying and Ca^{2+}-permeable AMPA receptors are distributed in various areas of the hippocampus. Morphological analysis has suggested that the type II neurons are interneurons. Since cultured type II hippocampal neurons express mRNA for GAD65, an enzyme synthesizing GABA, it is most likely that the type II neurons in the hippocampus are inhibitory interneurons [14]. We have also demonstrated that the type II AMPA receptors are involved in the excitatory synaptic transmission in type II neurons. Type II neurons carry NMDA receptors which are also involved in the synaptic transmission in these neurons. It is widely accepted that NMDA receptors participate in causing synaptic plasticity by regulating intracellular Ca^{2+} concentrations [18]. NMDA receptors are permeable to Ca^{2+} at depolarized membrane potentials [19,20], whereas type II AMPA receptors allow Ca^{2+} to permeate at more hyperpolarized membrane potentials. These two types of Ca^{2+}-permeable ionotropic glutamate receptors would play different functional roles by regulating Ca^{2+} entry in different contexts.

Acknowledgements

The authors wish to thank Drs J. Rossier, B. Lambolez, E. Audinat, P. Bochet, F. Crépel and S. Itazawa, for their collaboration. This study was supported by grants from the Uehara Memorial Foundation, the Human Frontier Science Program and the Ministry of Education, Science and Culture of Japan.

References

1. Watkins JC, Evans RH. Excitatory amino acid transmitters. Ann Rev Pharmacol Toxicol 1981; 165–204.
2. Mayer ML, Westbrook GL. The physiology of excitatory amino acids in the vertebrate central nervous system. Prog Neurobiol 1987;28:197–276.
3. Monaghan DT, Bridges RJ, Cotman CW. The excitatory amino acid receptors: their classes, pharmacology and distinct properties in the function of the central nervous system. Ann Rev Pharmacol Toxicol 1989;29:365–402.
4. Hestrin S, Nicoll RA, Perkel DJ, Sah P. Analysis of excitatory synaptic action in pyramidal cells using whole-cell recording from rat hippocampal slices. J Physiol (Lond) 1990;422:203–225.
5. Keller BU, Konnerth A, Yaari Y. Patch clamp analysis of excitatory synaptic currents in granule cells

of rat hippocampus. J Physiol (Lond) 1991;435:275–293.

6. Ozawa S, Iino M, Abe M. Excitatory synapse in the rat hippocampus in tissue culture and effects of aniracetam. Neurosci Res 1991;12:72–82.
7. Hollmann M, Heinemann S. Cloned glutamate receptors. Annu Rev Neurosci 1994;17:31–108.
8. Nakanishi S. Molecular diversity of glutamate receptors and implications for brain function. Science 1992;258:597–603.
9. Seeburg PH. The molecular biology of mammalian glutamate receptor channels. Trends Neurosci 1993;16:359–365.
10. Iino M, Ozawa S, Tsuzuki K. Permeation of calcium through excitatory amino acid receptor channels in cultured rat hippocampal neurones. J Physiol (Lond) 1990;424:151–165.
11. Ozawa S, Iino M, Tsuzuki K. Two types of kainate response in cultured rat hippocampal neurons. J Neurophysiol 1991;66:2–11.
12. Ozawa S, Iino M. Two distinct types of AMPA responses in cultured rat hippocampal neurons. Neurosci Lett 1993;155:187–190.
13. Iino M, Mochizuki S, Ozawa S. Relationship between calcium permeability and rectification properties of AMPA receptors in cultured rat hippocampal neurons. Neurosci Lett 1994;173:14–16.
14. Bochet P, Audinat E, Lambolez B, Crépel F, Rossier J, Iino M, Tsuzuki K, Ozawa S. Subunit composition at the single-cell level explains functional properties of a glutamate-gated channel. Neuron 1994;12:383–388.
15. Ozawa S, Rossier J. Molecular basis for functional differences of AMPA-subtype glutamate receptors. News Physiol Sci 1996;11:77–82.
16. Lambolez B, Audinat E, Bochet P, Crépel F, Rossier J. AMPA receptor subunits expressed by single Purkinje cells. Neuron 1992;9:247–258.
17. Isa T, Itazawa S, Iino M, Tsuzuki K, Ozawa S. Distribution of neurones expressing inwardly rectifying and Ca^{2+}-permeable AMPA receptors in rat hippocampal slices. J Physiol (Lond) 1996; 491:719–733.
18. Bliss TVP, Collingridge GL. A synaptic model of memory: long-term potentiation in the hippocampus. Nature 1993;361:31–39.
19. MacDermott AB, Mayer ML, Westbrook GL, Smith SJ, Barker JL. NMDA-receptor activation increases cytoplasmic calcium concentration in cultured spinal cord neurones. Nature 1986;321: 519–522.
20. Ascher P, Nowak L. The role of divalent cations in the N-methyl-D-aspartate responses of mouse central neurones in culture. J Physiol (Lond) 1988;399:247–266.

Integrative and Molecular Approach to Brain Function.
M. Ito and Y. Miyashita, editors.

Isoform-specific involvement of calcineurin in neuronal death of a gerbil model of brain ischemia and in Alzheimer disease pathology

C. Tanaka[1], T. Kawamata[1], T. Hashimoto[1], M. Sasaki[1] and N. Saito[2]

[1]*Hyogo Institute for Aging Brain and Cognitive Disorders, Himeji Hyogo; and* [2]*Biosignal Research Center, Kobe University, Kobe, Japan*

Abstract. Protein phosphorylation and dephosphorylation by Ca^{2+}-dependent enzymes have been shown to play key roles in modulation of the synaptic efficacy and neuronal cell death. In the present study, the involvement of calcineurin in neuronal death in a gerbil model of brain ischemia and in the autopsied brains from patients with or without Alzheimer disease (AD) was analyzed using isoform-specific antibodies. Transient ischemia induced neuronal death with DNA fragmentation in the CA1 pyramidal cells of the gerbil hippocampus. Calcineurin Aβ was markedly decreased as compared with Aα in the CA1, 7 days after the ischemia. When FK506 1 mg/kg, a calcineurin inhibitor, was injected immediately after the ischemic insult, FK506 prevented neuronal death, DNA fragmentation and alterations in immunoreactivities of both isoforms in the CA1 pyramidal cells. These findings suggest that calcineurin is likely to be involved in ischemia-induced neuronal apoptosis. In the AD brains, calcineurin Aβ positive pyramidal cells and granular cells were markedly decreased. Calcineurin Aβ was colocalized with τ protein and ubiquitin along some neurofibrillary tangles within degenerative neurons and dystrophic neurites within the senile plaques, suggesting that calcineurin Aβ is involved in dephosphorylation of τ at early stages of tangle formation in the AD brain.

Taken together, these observations suggest the specific involvement of the isoform of calcineurin in neuronal death and AD pathology.

Key words: apoptosis, calcineurin Aα, calcineurin Aβ, FK506, immunohistochemistry.

Introduction

Currently, much attention is being directed to the role of protein kinase and phosphatase in synaptic plasticity and neuronal death, events considered to be triggered by increases in intracellular Ca^{2+} concentrations. Ca^{2+}-dependent kinases and phosphatase such as protein kinase C (PKC), Ca^{2+}/calmodulin-dependent kinase II (CaMKII) and Ca^{2+}/calmodulin-dependent serine/threonine specific phosphatase (calcineurin) are present in high in the mammalian brain [1–3].

Kinases have been implicated in the mechanism of activity-dependent long-term changes in synaptic efficacy. Based on a wide range of studies, PKC and CaMKII are not part of the molecular machinery involved in production of long-term potentiation, but they are a key regulatory component [4–6]. Phosphatases participate in inverse mechanisms for regulation of synaptic strength. Calcineurin specific inhibitors block

Address for correspondence: Chikako Tanaka MD, PhD, Hyogo Institute for Aging Brain and Cognitive Disorders, 520 Saisho-ko, Himeji Hyogo 670, Japan. Tel.: +81-792-95-5511. Fax: +81-792-95-8199.

the induction of long-term depression and reduce the threshold of long-term potentiation [7].

Despite the apparent link between calcium and cell death, the immediate targets of this calcium flux remain largely unknown. One potential mediator of calcium signaling during neuronal death is calcineurin. As shown in Fig. 1, the holoenzyme of calcineurin is a heterodimer, consisting of the catalytic (calcineurin A) and regulatory (calcineurin B) subunits. Calcineurin A contains a calmodulin binding domain. Calcineurin B contains four Ca^{2+} binding domains that form a distinct helical structure called an EF hand. Calcineurin comprises a family of enzymes. We isolated cDNA clones from a rat brain library and determined the primary structure of calcineurin A isoforms, Aα and Aβ [8,9].

In the present study, the involvement of calcineurin in neuronal death in a gerbil model of brain ischemia and in the autopsied brains from patients with or without Alzheimer disease (AD) was analyzed using isoform-specific antibodies.

Materials and Methods

A gerbil model of brain ischemia

Male Mongolian gerbils weighing 60–90 g were anesthetized with halothane. Transient ischemia were made by occluding both common carotid arteries with clips for 5 min. These animals were maintained at 37°C throughout the procedure using

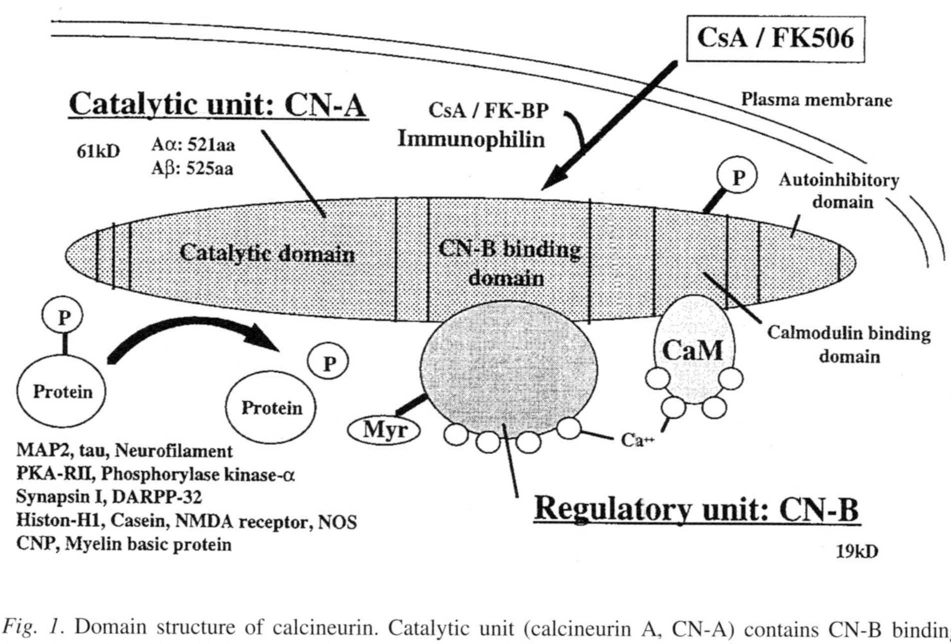

Fig. 1. Domain structure of calcineurin. Catalytic unit (calcineurin A, CN-A) contains CN-B binding domain, calmodulin binding domain, autoinhibitory domain and CsA/ FK506-immunophillin complex binding sites. Regulatory unit (calcineurin B, CN-B) contains four Ca^{2+} binding sites.

a heat pad until the gerbils had fully recovered from the anesthesia. Sham-operated controls were given the same treatment except the occlusion of the carotid artery. FK506 (tacrolimus, Fujisawa) or rapamycin was injected intravenously, after induction of transient ischemia. The gerbils were then decapitated at different times, brains were rapidly removed and frozen or immersed in the fixative containing 4% paraformaldehyde in 0.1 M phosphate buffer, pH 7.4. Coronal sections 20 μm thick were stained with cresyl violet. The number of intact neurons in the CA1 pyramidal cell layer was counted and the density of neurons per millimeter in the pyramidal cell layer was calculated. For in situ DNA nick-end labeling (TUNEL), the sections were stained using Apo-Tag Plus kits (ONCOR, MD).

Tissue collection from postmortem human brains

Thirteen brains with sporadic AD were compared with 11 control brains without dementia. All patients fulfilled clinical criteria for AD. The diagnosis of AD was confirmed in each case by postmortem neuropathological examinations. Small pieces of brain tissue were fixed in 4% paraformaldehyde in 0.1 M phosphate buffer, pH 7.4, for 2 days at 4°C and transferred to a cold solution of 15% sucrose in 0.01 M phosphate buffered saline (PBS), pH 7.4. Sections were cut on a freezing microtome at 30 μm thickness, placed in 15% sucrose solution and stored at 4°C until used.

Preparation of antibodies against calcineurin isoforms

Rabbit antisera specific to the calcineurin isoforms were prepared against synthetic oligopeptides of 15 amino acids which corresponded to the carboxyl-terminal portion of each isoform. These antisera were purified on affinity columns of the synthetic peptides used as antigens. The specificity of antibodies against calcineurin Aα and Aβ was confirmed by immunoblot analysis with recombinant calcineurin Aα and Aβ using gerbil and human brain homogenates. Immunoabsorption tests were made using immunohistochemical techniques.

Immunohistochemistry

Sections were pretreated with H_2O_2 to eliminate endogenous peroxidase. Free floating sections were incubated with primary antibody for 72 h in the cold. The primary antibody was diluted at 1:200 in PBS containing 0.3% Triton-X100 and 10% normal goat serum. After primary incubation, sections were treated with biotinylated goat anti-rabbit IgG (Vector) for 2 h at room temperature, followed by incubation in the avidin-biotinylated HRP complex (Vector) for 1 h. Peroxidase labeling was detected by incubation with solution containing 0.02% 3,3′-diaminobenzidine (DAB), 0.6% nickel ammonium sulfate, 0.05 M imidazole and 0.0003% H_2O_2 to give a purple precipitate. For double immunostaining, sections were put through a second, slightly modified cycle. After completion of the first cycle, sections were treated for 30 min with 0.5% H_2O_2 solution in PBS. The second immunohistochemical cycle was carried out in a similar manner except that nickel ammonium sulfate was omitted from the DAB solution and a brown precipitate formed.

Results and Discussion

Neuroprotective effects of FK506 on delayed neuronal death in the gerbil hippocampus

Histological observation showed no definite neuronal damage to the CA1 to CA3 subfields and the dentate gyrus until 2 days after the transient ischemic insult. Four days after the ischemia, damages were seen in the pyramidal cell layer of CA1 subfield but not in the CA2 and CA3 subfields or in the dentate gyrus. Seven days after the transient ischemia, the density of neurons in the pyramidal cell layer of CA1 subfield was reduced to 17% of the control. In lymphocytes, calcineurin serves as the target for immunosuppressive drug action. Immunosuppressants, FK506 and cyclosporine A bind to cytoplasmic immunophilin and the resultant complex inhibits the phosphatase activity of calcineurin, thereby NFAT activation and cytokine transcription are inhibited [10]. We reported [11] that the FK506-FKBP12 complex inhibited the phosphatase activity of calcineurin Aα and Aβ purified from the rat brain, indicating that calcineurin is a relevant cellular target of FK506 when bound to FKBP-12. As FKBP is co-localized with calcineurin at a high concentration in the mammalian brain [12], FK506 can be used as a specific inhibitor of calcineurin.

When FK506 1 mg/kg was injected immediately after the ischemia insult, FK506 significantly prevented neuronal loss in the CA1 subfield as shown in Fig. 2. Following injection 3 h after the ischemic insult, FK506 at the same dose did not prevent neuronal loss induced by transient ischemia. This ischemia-induced neuronal loss was not afforded protection by rapamycin 1 mg/kg which can bind to FKBP but not inhibit the phosphatase activity of calcineurin [10]. Pretreatment of rapamycin 5 mg/kg significantly antagonized the protective effect of FK506 1 mg/kg on the ischemia-induced neuronal loss in the CA1. These findings suggest that FK506 protects against transient ischemia-induced neuronal death as a result of the inhibition of calcineurin. The protective effect of FK506 against neuronal injury was noted in an in vitro model of glutamate neurotoxicity [13] and in an in vivo model of focal cerebral ischemia [14]. These findings support our concept that inhibition of the phosphatase activity of calcineurin is necessary to prevent neuronal death at the early phase of ischemia-induced cellular events.

Next, we asked whether FK506 would prevent apoptotic changes occurring in CA1 pyramidal neurons after transient ischemia. When in situ DNA nick-end labeling of the hippocampus was done, DNA fragmentation was evident in CA1 pyramidal cells, 4 days after the ischemic insult. However, DNA fragmentation was not evident in CA1 pyramidal cells in gerbils treated with 1 mg/kg FK506. Therefore, calcineurin is likely to be involved in ischemia-induced neuronal apoptosis.

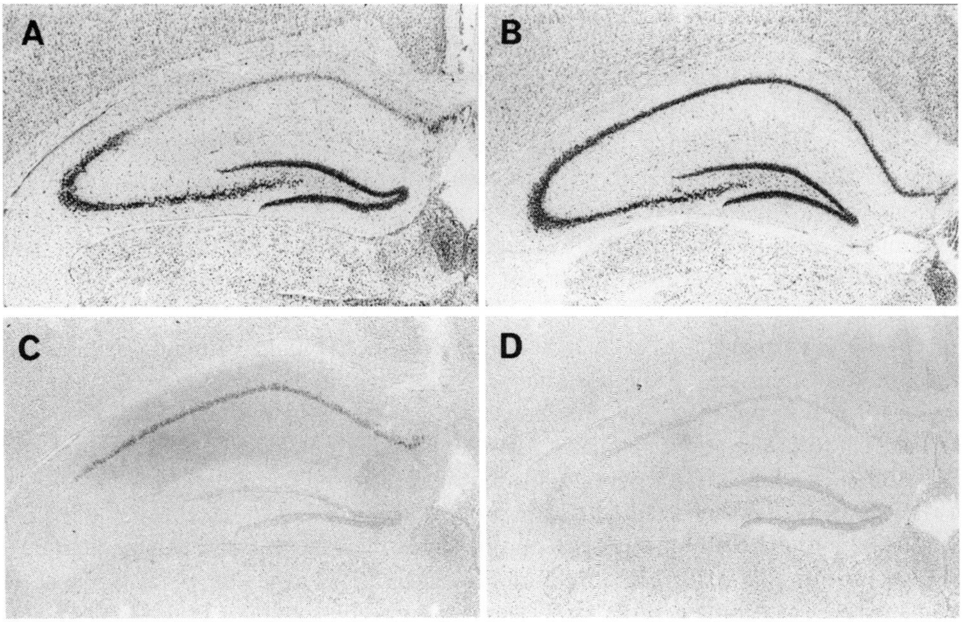

Fig. 2. Neuroprotective effect of FK506 on neuronal death and DNA fragmentation induced by transient ischemia in the gerbil hippocampus. Transient ischemia were made by occluding both common carotid arteries for 5 min. The gerbils were given intravenously with vehicle (**A** and **C**) or FK506 1 mg/kg (**B** and **D**), immediately after induction of transient ischemia. The brains were obtained 7 days after ischemic insult and coronal sections were stained with cresyl violet (**A** and **B**). For in situ DNA nick-end labeling (TUNEL), the sections were obtained 4 days after ischemia, and stained using Apo-Tag Plus Kits (**C** and **D**).

Alterations of calcineurin A isoforms in the gerbil model of brain ischemia

In situ hybridization histochemistry of calcineurin A isoforms in the rat brain showed a distinct distribution of the mRNA encoding two isoforms of the catalytic subunits Aα and Aβ, but the expression pattern of Aα and Aβ is quite similar in the pyramidal cell layer throughout the hippocampus [15]. Nevertheless, a transient ischemia induced neuronal cell death mainly in pyramidal cells of CA1 region in the gerbils hippocampus. We obtained two highly isoform-specific antibodies to determine which isoforms of calcineurin A are related to ischemia-induced apoptosis in the hippocampus. Western blots showed that the amounts of calcineurin Aα and Aβ were not significantly changed until 3 days after the ischemic episode and significantly decreased 7 days later. The amounts of calcineurin Aβ were largely decreased.

To determine the isoform-specific role of calcineurin in apoptosis of neuronal cells, the cellular localization of both isoforms was analyzed immunohistochemically, using isoform-specific antibodies. Immunoreactivities of both isoforms are enriched in the gerbil hippocampal formation, especially in the CA1 and CA2 regions. The highest density of Aα was located in neuropils in the CA1 and mossy fibers in the CA3. Aβ was also enriched in the CA1 and CA2 regions, but a low level of Aβ was noted in granular cells and in mossy fibers. Immunohistochemistry revealed that 7 days after ischemic insult, calcineurin Aα and Aβ were drastically decreased in the pyramidal cells and the lacunosum molecular layers in CA1 and a part of CA2 regions. These observations indicate that both calcineurin Aα and Aβ are present in pyramidal cells but not in the afferent fiber terminals from the thalamus and entorhinal cortex to the distal dendrites of pyramidal cells in lacunosum molecular layers.

On the other hand, it is of interest that the distinct features of both isozymes are seen in the oriens and radiatum layers. The immunoreactivity of Aα appears to increase slightly but that of Aβ to decrease significantly in the oriens and radiatum layers. These observations indicate that fibers and terminals from commissural and Schaffer collaterals of CA3 pyramidal cells are intact, and contain Aα but no or only a few Aβ. Treatment with FK506 prevented ischemia-induced alterations in immunoreactivities of both isoforms and neuronal death with DNA fragmentation in the CA1 pyramidal cells. The target of FK506-FKBP12 complex at postsynapse is probably both isoforms and the target at presynapse may be Aα, in the CA1.

Alteration of calcineurin A isoforms in the AD brain

Current studies revealed that phosphatase 1 and 2A activities were reduced by 15% in AD brains, while activities of phosphatase 2B (calcineurin) and 2C are not affected [16,17]. Alterations of calcineurin isoforms in postmortem AD brains remain the subject of ongoing study.

Histochemical analysis in the human brain revealed differential localization of two calcineurin A isoforms. In the hippocampus, the density of calcineurin Aα immunoreactivity is high in CA1/2 pyramidal cells and mossy fibers, moderate in granule cells and low in CA3/4 cells. The density of Aβ immunoreactivity is moderate in the CA3/4 pyramidal cells and granular cells but low in CA1/2 pyramidal cells.

In the CA1 subfield of the hippocampus from patients with AD, the Aα positive cells were significantly decreased but intact neurons stained more intensely than did those in controls. The Aα immunoreactivity in the granular cells markedly increased, but that in mossy fibers disappeared. Accumulation of the Aα immunoreactivity in the granular cells suggests a damage of axonal transport of calcineurin Aα within mossy fibers. Aα immunoreactivity was not associated with neurofibrillary tangles and senile plaques. In contrast, the Aβ positive pyramidal cells in CA3/4 subfields and granular cells were markedly decreased in AD brain. Aβ immunoreactivity accumulated along some neurofibrillary tangles within degenerative neurons and dystrophic neurites within the senile plaque (Fig. 3).

It has been reported that calcineurin is expressed both in the normal neurons of the AD and control brains as well as in tangles bearing neurons of the AD brain [18].

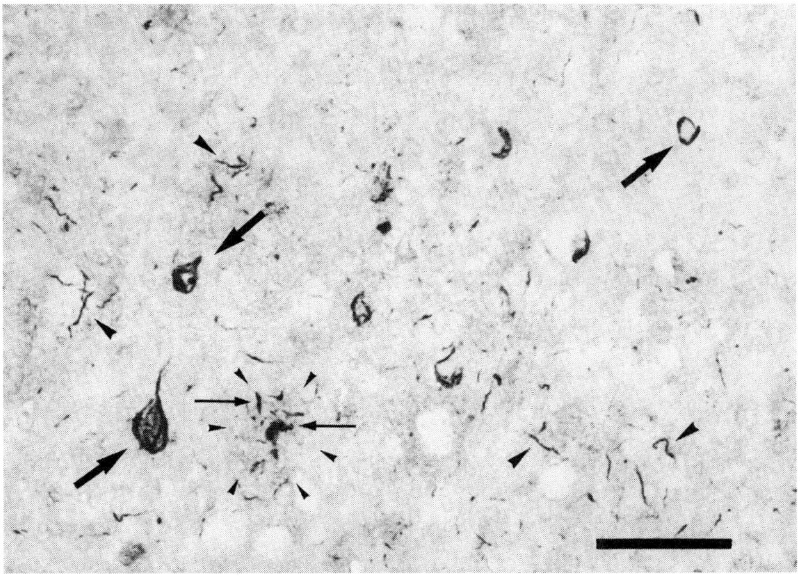

Fig. 3. Immunostaining of the angular cortex from an Alzheimer disease patient for calcineurin Aβ. The Alzheimer pathologies such as neurofibrillary tangles (thick arrow), degenerative neurites (thin arrow) within a senile plaque (surrounded by small arrow-head), and neuropil threads (large arrow-head) were intensely immunolabeled for calcineurin Aβ. Bar = 50 mm.

Recent studies clearly demonstrated that both calcineurin and phosphatase 2A can dephosphorylate recombinant human τ in vitro [19]. In addition, calcineurin and phosphatase 2A has been reported to dephosphorylate subpopulations of authentic PHFs from AD brain [20]. In the present study, double immunostaining of AD brains for calcineurin Aβ and τ-1 clearly showed that calcineurin Aβ is co-localized with τ-1 along neurofibillary tangles within degenerative neurons and even in pretangled neurons. These results indicate a possibility that calcineurin Aβ is involved in the dephosphorylation of τ protein. Furthermore, double immunostaining of CA1 hippocampus from AD cases, for calcineurin Aβ and ubiquitin showed that calcineurin Aβ accumulates on ubiquitinated abnormal fibrillary structures such as neurofibrillary tangles within the degenerated neurons and dystrophic neurites within the senile plaque, but the Aβ immunoreactivity is not associated with extracellular neurofibrillary tangles. These results indicate that calcineurin Aβ is involved in dephosphorylation of τ at early stages of tangle formation in the AD brain.

The finding that calcineurin Aβ is co-localized with neurofibrillary structures in degenerative neurons suggests that calcineurin, in particular Aβ isoform has a role in pathological events which lead to the formation of tangles and senile plaques.

In the hippocampus the PKC (α, βI, βII, γ, ε), CaM KII(α, β) and calcineurin (Aα, Aβ) are expressed in pyramidal cells and granular cells, with different intracellular localizations [1,2,15]. Ca^{2+}-dependent PKC, CaMKII and calcineurin Aβ localize predominantly at postsynaptic sites. Calcineurin Aα and ε-PKC localize in presynaptic terminals. Taken together, these observations suggest the specific involvement of the isoform of protein kinase and phosphatase in neuronal cell death and AD pathology.

Acknowledgements

We thank M. Kobayashi for pertinent advice and the gift of FK506. This work was supported by a special research grant from Uehara Memorial Foundation.

References

1. Tanaka C, Nishizuka Y. The protein kinase C family for neuronal signaling. Ann Rev Neurosci 1994; 17:551–567.
2. Quimet CC, McGuinness TL, Greengard P. Immunocytochemical localization of calcium/calmodulin-dependent protein kinase II in rat brain. Proc Natl Acad Sci USA 1984;81:5604–5608.
3. Cohen P. The structure and regulation of protein phosphatases. Ann Rev Biochem 1989;58:453–508.
4. Abeliovich A, Chen C, Goda Y, Silva AJ, Stevens CF, Tonegawa S. Modified hippocampal long-term potentiation in PKC gamma-mutant mice. Cell 1993;75:1253–1262.
5. Silva AJ, Stevens CF, Tonegawa S, Wang Y. Deficient hippocampal long-term potentiation in alpha-calcium-calmodulin kinase II mutant mice. Science 1992;257:201–206.
6. Mayford M, Wang J, Kandel ER, O'Dell TJ. CaMKII regulates the frequency-response function of hippocampal synapses for the production of both LTD and LTP. Cell 1995;81:891–904.
7. Mulkey RM, Herron CE, Malenka RC. An essential role for protein phosphatases in hippocampal long-term depression. Science 1993;261:1051–1055.
8. Ito A, Hashimoto T, Hirai M, Takeda T, Shuntoh H, Kuno T, Tanaka C. The complete primary structure of calcineurin A, a calmodulin binding protein homologous with protein phosphatases 1 and 2A. Biochem Biophys Res Commun 1989;163:1492–1497.
9. Kuno T, Takeda T, Hirai M, Ito A, Mukai H, Tanaka C. Evidence for a second isoform of the catalytic subunit of calmodulin-dependent protein phosphatase (calcineurin A). Biochem Biophys Res Commun 1989;165:1352–1358.
10. Liu J, Farmer JD, Lane WS, Friedman J, Weissman I, Schreiber SL. Calcineurin is a common target of cyclophilin-cyclosporin A and FKBP-FK506 complexes. Cell 1991;66:807–815.
11. Mukai H, Kuno T, Chang CD, Lane B, Luly JR, Tanaka C. FKBP12-FK506 complex inhibits phosphatase activity of two mammalian isoforms of calcineurin irrespective of their substrates or activation mechanisms. J Biochem Tokyo 1993;113(3):292–298.
12. Steiner JP, Dawson TM, Fotuhi M, Glatt CE, Snowman AM, Cohen N, Snyder SH. High brain densities of the immunophilin FKBP colocalized with calcineurin. Nature 1992;358:584–587.
13. Dawson TM, Steiner JP, Dawson VL, Dinerman JL, Uhl GR, Snyder SH. Immunosuppressant FK506 enhances phosphorylation of nitric oxide synthase and protects against glutamate neurotoxicity. Proc Natl Acad Sci USA 1993;90:9808–9812.
14. Sharkey J, Butcher SP. Immunophilins mediate the neuroprotective effects of FK506 in focal cerebral ischaemia. Nature 1994;371:336–339.
15. Takaishi T, Saito N, Kuno T, Tanaka C. Differential distribution of the mRNA encoding two isoforms of the catalytic subunit of calcineurin in the rat brain. Biochem Biophys Res Commun 1991; 174(1):393–398.
16. Gong CX, Singh TJ, Grundke-Iqbal I, Iqbal K. Phosphoprotein phosphatase activities in Alzheimer disease brain. J Neurochem 1993;61(3):921–927.
17. Billingsley ML, Ellis C, Kincaid RL, Martin J, Schmidt ML, Lee VM-Y, Trojanowski JQ. Calcineurin immunoreactivity in Alzheimer's disease. Exp Neurol 1994;126(2):178–184.

18. Pei JJ, Sersen E, Iqbal K, Grundke-Iqbal I. Expression of protein phosphatases (PP-1, PP-2A,PP-2B and PTP-1B) and protein kinases (MAP kinase and $P34^{cdc2}$) in the hippocampus of patients with Alzheimer disease and normal aged individuals. Brain Res 1994;655:70–76.
19. Gong CX, Singh TJ, Grundke-Iqbal I, Iqbal K. Alzheimer's disease abnormally phosphorylated τ is dephosphorylated by protein phosphatase-2B (calcineurin). J Neurochem 1994;62:803–806.
20. Drewes G, Mandelkow EM, Baumann K, Goris J, Merlevede W, Mandelkow E. Dephosphorylation of tau protein and Alzheimer paired helical filaments by calcineurin and phosphatase-2A. FEBS Lett 1993;336:425–432.

Integrative and Molecular Approach to Brain Function.
M. Ito and Y. Miyashita, editors.

Regulation of Ca^{2+}/calmodulin-dependent multifunctional protein kinases in the brain

Hitoshi Fujisawa, Sachiko Okuno, and Takako Kitani
Department of Biochemistry, Asahikawa Medical College, Hokkaido, Japan

Abstract. There are three multifunctional Ca^{2+}/calmodulin-dependent protein kinases (CaM-kinases), CaM-kinase I, II, and IV, which are distributed in different regions inside a neuronal cell; CaM-kinase I exists in the cytosol, CaM-kinase II exists in the cytosol near the cell membrane, sometimes existing as a membrane-bound form, and CaM-kinase IV exists inside the cell nucleus, each playing the respective role in response to Ca^{2+} in their working area. Among the three CaM-kinases, the activity of CaM-kinase II is regulated through self-regulatory mechanism; activation by autophosphorylation on Thr-286 and inactivation by autophosphorylation on Thr-305/306. In contrast, the other two kinases, CaM-kinase IV and I, are markedly activated through phosphorylation on Thr-196 and Thr-177, respectively, by another Ca^{2+}/calmodulin-dependent protein kinase, CaM-kinase kinase. The activated CaM-kinase IV is inactivated through phosphorylation on Ser-332, located in the calmodulin-binding domain. CaM-kinase IV kinase purified from rat cerebral cortex can activate both CaM-kinase IV and I, but it exists inside the cell nucleus, where CaM-kinase IV occurs but CaM-kinase I does not occur. Immunoblot and immunotitration analyses suggest the existence of at least two distinct isoforms of CaM-kinase kinase. Thus, Ca^{2+}-signaling in the central nervous system appears to be mediated by the regulatory system consisting of at least three multifunctional protein kinases and two or more CaM-kinase kinase cascades.

Key words: Ca^{2+} signaling, calmodulin-dependent protein kinase kinase, protein kinase kinase cascade, protein phosphorylation, regulation of enzyme activity, signal transduction.

Introduction

The calcium ion is known to play important roles in regulation of a variety of neuronal functions and most of the diverse actions of Ca^{2+} may be mediated through protein phosphorylation by three multifunctional calmodulin-dependent protein kinases, calmodulin-dependent protein kinase (CaM-kinase) I, II, and IV (see [1–4] for reviews). These three protein kinases exist relatively widely and abundantly in the central nervous system but are distributed in different regions inside a neuronal cell. Progress in studies of the molecular mechanisms for regulation of the activities of these important protein kinases is crucial to an understanding of how brain function is regulated in molecular terms. Both the Ca^{2+}/calmodulin-dependent and independent activities of CaM-kinase II are markedly activated through autophosphorylation on Thr-286 occurring in the presence of Ca^{2+}/calmodulin [5–8], and the Ca^{2+}/calmodulin-dependent activity of the activated enzyme is lost upon autophosphorylation of

---

Address for correspondence: Hitoshi Fujisawa, Department of Biochemistry, Asahikawa Medical College, Asahikawa, Hokkaido 078, Japan.

Thr-305/306 occurring in the absence of Ca^{2+}, presumably because the binding of calmodulin to the enzyme is blocked by phosphorylation of Thr-305/306 located within the calmodulin-binding site, while the Ca^{2+}/calmoduln-independent activity remains unchanged [9—11]. In contrast to CaM-kinase II which is activated through autophosphorylation on Thr-286 located in the autoinhibitory domain, we recently discovered a novel Ca^{2+}/calmodulin-dependent protein kinase which markedly activated CaM-kinase IV [12] and, shortly thereafter, Lee and Edelman [13] and Mochizuki et al. [14] independently found that CaM-kinase I was also activated by another Ca^{2+}/calmodulin-dependent protein kinase, suggesting the existence of Ca^{2+}/calmodulin/kinase/kinase cascade regulating the function of the neuronal cell. In this symposium, the regulatory mechanism for the activity of CaM-kinase IV by CaM-kinase kinase α and also the evidence for the existence of another Ca^{2+}/calmodulin-dependent protein kinase kinase (CaM-kinase kinase β) are presented.

Materials and Methods

Calmodulin was purified from *Escherichia coli* transformed with expression vector pET11d carring a cDNA encoding chicken brain calmodulin [15], essentially as described by Gopalakrishna and Anderson [16]. The cDNA encoding chicken brain calmodulin was kindly donated by A.R. Means [17]. Recombinant rat brain CaM-kinase IV expressed in *E. coli* was partially purified as described previously [12]. Recombinant rat brain CaM-kinase IV expressed in Sf9 cells was purified to apparent homogeneity as described previously [15]. CaM-kinase kinase α was purified from rat cerebral cortex as described previously [18]. Recombinant rat brain CaM-kinase kinase α expressed in *E. coli* was obtained as described [19]. Antiserum against CaM-kinase IV was obtained as described [18]. Antibody against CaM-kinase kinase α was purified from the antisera from immunized Japanese white rabbits by affinity chromatography [19].

The activity of CaM-kinase IV was determined by measuring the phosphorylation of a synthetic peptide, syntide-2 [20] or peptide-γ [21]. The standard assay mixture contained, in a final volume of 50 μl, 50 mM Mops-NaOH (pH 7.0 at 30°C), 2 mM dithiothreitol, 0.1 mM $CaCl_2$ or EGTA, 1 μM calmodulin, 5 mM $Mg(CH_3COO)_2$, 0.1 mM [γ-^{32}P]ATP (100—200 cpm/pmol), 40 μM syntide-2 (or 200 μM peptide-γ), and a suitable amount of CaM-kinase IV. After incubation for 5 min at 30°C, 30-μl aliquots were withdrawn and the incorporation of [^{32}P]phosphate into the peptide substrate was determined by the phosphocellulose paper method [22].

A number of the cDNAs coding the various mutant CaM-kinase IV were constructed with the respective mutagenic oligonucleotides using the mutan-K kit (TAKARA), essentially according to the method of Kunkel et al. [23]. The mutant cDNAs were inserted into pET11 vectors and the constructs were transformed into *E. coli* strain BL21 (DE3) [24]. The bacteria were grown to an A_{600} of 0.6 to 1.0 at 30°C in 50 ml of M9ZB medium containing 200 μg/ml ampicillin, and isopropyl-β-D-thiogalactopyranoside was added to a final concentration of 1 mM. After 3—4 h, the bacteria were harvested by centrifugation, suspended in 1 ml of 20 mM Tris-HCl (pH

7.5 at 4°C) containing 1 mM phenylmethylsulfonyl fluoride, and 10 μg/ml each of microbial protease inhibitors (leupeptin, pepstatin, antipain A, and chymostatin) (Peptide Institute, Osaka), and then disrupted by sonic oscillation. The residue was removed by centrifugation to generate crude extracts.

The activity of CaM-kinase kinase were assayed by measuring the CaM-kinase IV activity generated by incubation with recombinant CaM-kinase IV under Ca^{2+}/calmodulin-dependent protein phosphorylation conditions, as described previously [18]. The standard reaction mixture contained, in a final volume of 50 μl, 50 mM Mops-NaOH (pH 7.0 at 30°C), 5 mM $Mg(CH_3COO)_2$, 0.2 mM [γ-^{32}P]ATP (100 cpm/pmol), 0.1 mM EGTA, 0.2 mM $CaCl_2$, 2 μM calmodulin, 2 mM dithiothreitol, 40 μM peptide-γ, 2 μg of recombinant CaM-kinase IV, and a suitable amount of CaM-kinase kinase. After incubation for 5 min at 30°C, a sample of 30 μl was withdrawn and the incorporation of [^{32}P]phosphate into peptide-γ was determined by the phosphocellulose paper method [22]. Phosphorylation of CaM-kinase IV by CaM-kinase kinase α was carried out in the standard reaction mixture, except that recombinant CaM-kinase IV and peptide-γ were replaced by a suitable amount of CaM-kinase IV. After incubation for the indicated times at 30°C, the incorporation of [^{32}P]phosphate into CaM-kinase IV was determined by the 3MM paper method of Corbin and Reimann [25], except that the filter papers were washed with ice-cold 10% trichloroacetic acid containing 2 mM ATP.

Northern blot analysis was carried out essentially as described by Sambrook et al. [26]. Poly(A)$^+$ RNA was denatured with 1 M glyoxal and 50% dimethyl sulfoxide for 1 h at 50°C, electrophoresed on a 1% agarose gel, and then transferred to a nylon membrane filter (NYTRAN NY 13N, Schleicher and Schuell). Hybridization was performed with the *Pvu*II-*Pvu*II fragment (410 bp) of the CaM-kinase kinase α cDNA [19] labeled by primer extension with random primers in a solution consisting of 5× SSC, 50 mM sodium phosphate (pH 7.0), 5× Denhardt's reagent, 0.1% SDS, 250 μg/ml salmon DNA, and 50% formamide for 24 h at 45°C.

Western blot analysis was performed essentially as described by Winston et al. [27]. Crude tissue extracts were subjected to SDS-polyacrylamide gel electrophoresis on 7.5% acrylamide gel, and then the protein bands were transferred onto a polyvinylidene difluoride membrane (Fluorotrans, Pall Bio Support). The membrane was blocked with 5% nonfat milk in phosphate buffered saline for 30 min at 24°C, then incubated with 0.358 μg/ml antibody to CaM-kinase kinase α (or antiserum against CaM-kinase IV at 1:10,000 dilution) in the blocking buffer for 2 h, followed by incubation with 30 μg/ml goat anti-rabbit immunoglobulins (IgA + IgG + IgM) conjugated with peroxidase (Organon Teknika) at 4°C overnight. The positive bands were detected with diaminobenzidine tetrahydrochloride and H_2O_2 in the presence $CoCl_2$.

The tissues obtained from 10-week-old female Wistar rats were homogenized with a Potter-Elvehjem homogenizer in three volumes of 20 mM Hepes-NaOH (pH 7.5 at 4°C) containing 1 mM dithiothreitol, 0.1% Triton X-100, and 20 μg/ml each of the microbial protease inhibitors. The residues were removed by centrifugation to generate the crude extracts.

Immunotitration was carried out by incubation of crude extracts with various amounts of antiserum against CaM-kinase kinase α in a mixture (final volume of 70 μl) containing 10 mM Mops-NaOH (pH 7.0), 0.15 M NaCl, 0.05% Tween 80, 0.1 mM EDTA, 0.1 mM dithiothreitol, and 5 μg/ml each of the microbial protease inhibitors at 24°C. After incubation for 20 min, 70 μl of a 10% suspension of *Staphylococcus aureus* was added, and it was incubated for 30 min with shaking. An aliquot (4 μl) of the supernatant obtained on centrifugation for 10 min at 15,000 × *g* was assayed for CaM-kinase kinase. *S. aureus* (Cowan I strain) was cultured as described by Goding [28]. The cells were collected and fixed with formaldehyde essentially as described by Kessler [29], and stored as a 10% (wet weight/volume) suspension at –80°C. Before use, the cells were washed once and suspended in 10 mM Hepes-NaOH (pH 7.2) containing 10% glycerol, 0.05% Tween 80, and 0.1 mM EDTA, 0.1 mM dithiothreitol, 0.1 M NaCl, and 5 μg/ml each of the microbial protease inhibitors.

SDS-polyacrylamide gel electrophoresis was carried out according to the method of Laemmli [30]. Gel overlay assay by biotinylated calmodulin, which was prepared from the recombinant chicken calmodulin as described by Mangels and Gnegy [31], was carried out as described by Kincaid et al. [32]. The concentration of calmodulin was determined spectrophotometrically using an absorption coefficient, A_{280} (1 mg/ml), of 0.21 [33] and the molecular weight of 16,700 [34,35]. Other proteins were determined by the method of Lowry et al. [36], as modified by Peterson [37] with bovine serum albumin as a standard. Phosphoamino acid analysis was performed by the method of Cooper et al. [38].

Results and Discussion

When cDNA for rat brain CaM-kinase IV was expressed in *E. coli*, the enzyme was produced in a good yield, but it did not show significant activity. CaM-kinase IV kinase was initially discovered in rat brain by its activity activating the inactive recombinant CaM-kinase IV expressed in *E. coli* [12]. Since recombinant CaM-kinase IV expressed in Sf9 cells shows very similar properties to the enzyme purified from rat brain and can be much more easily purified, as compared to the enzyme expressed in *E. coli* [39], the regulatory mechanism was examined with the recombinant enzyme expressed in Sf9 cells. Table 1 shows the kinetic properties of CaM-kinase IV before and after activation by CaM-kinase kinase. The Km values for a peptide substrate and ATP decrease and the Vmax value increases markedly upon activation by CaM-kinase kinase. When CaM-kinase IV is incubated with CaM-kinase kinase under the phosphorylation conditions in the presence of Ca^{2+}/calmodulin, approximately 4 mol of phosphate are incorporated into 1 mol of CaM-kinase IV, but the activation occurs upon the initial incorporation of 1 mol of phosphate catalyzed by CaM-kinase kinase and the subsequent phosphorylation results from the action of CaM-kinase IV itself activated by CaM-kinase kinase [40].

It is important to determine the phosphorylation site involved in the marked activation of CaM-kinase IV by CaM-kinase kinase for an understanding of the

Table 1. Changes in kinetic parameters of CaM-kinase IV produced by phosphorylation by CaM-kinase kinase.

		Before activation	After activation
Km for peptide-γ	Ca^{2+} (+)	82 μM	41 μM
	Ca^{2+} (-)	167	93
Km for ATP	Ca^{2+} (+)	139	7
	Ca^{2+} (-)	200	48
Vmax	Ca^{2+} (+)	108 nmol/min/mg	11294 nmol/min/mg
	Ca^{2+} (-)	0.6	443

The activities of purified recombinant CaM-kinase IV expressed in Sf9 cells were determined with peptide-γ as substrate in the presence and absence of Ca^{2+}, before and after full activation by incubation with CaM-kinase kinase. Kinetic parameters were obtained from double reciprocal plots of the activities as functions of substrate concentrations. The V_{max} values were calculated from apparent V_{max} values on the basis of the Michaelis equation, $v = V_{max}/(1 + Km/[s])$.

regulatory mechanism of CaM-kinase IV activity. The phosphorylation of a recombinant mutant CaM-kinase IV, K71R, which has lost the enzyme activity owing to the substitution of arginine for Lys-71 involved in the binding of ATP, by CaM-kinase kinase occurred exclusively on threonine residue (data not shown), although autophosphorylation of CaM-kinase IV following the phosphorylation by CaM-kinase kinase occurs almost exclusively on serine residues [41,42]. Recently, Selbert et al. [43] reported that replacement of Thr-196 with a nonphosphorylatable alanine abolishes both the phosphorylation and activation of CaM-kinase IV by CaM-kinase Ia kinase. Table 2 shows that the mutant enzymes, T196A and T200A, in which Thr-196 and Thr-200 were replaced by alanine, respectively, showed little, if any,

Table 2. The enzyme activities of mutant CaM-kinase IV.

	Before incubation (pmol/5 min/μl)	After incubation (pmol/5 min/μl)
Wild-type	0.78	124.9
T196A	0.34	0.19
T196S	2.69	70.7
T196D	22.5	19.2
T200A	0.27	0.56
T200S	0.22	91.62
T200D	0.06	1.18

The CaM-kinase IV activity of the crude extract of bacteria transformed with expression vector pET11 carring a cDNA coding wild-type enzyme, mutant T196A (Thr-196 was replaced by Ala), T196S, T196D, T200A, T200S, or T200D was determined before and after incubation with CaM-kinase kinase α under the protein phosphorylation conditions in the presence of Ca^{2+}/calmodulin. The contents of the enzyme in the respective crude extracts were almost the same on the basis of the results of SDS-polyacrylamide gel electrophoresis.

enzyme activity both before and after incubation with CaM-kinase kinase α. The mutant enzyme, T196D, in which Thr-196 was replaced by a negatively charged aspartic acid, showed almost 30 times as high activity as the wild-type enzyme before incubation with CaM-kinase kinase, but no more activation was observed after the incubation, presumably mimicking the enzyme phosphorylated on Thr-196 by CaM-kinase kinase. In contrast to T196D, the mutant T200D showed no enzyme activity before the incubation. These results, taken together, suggest that Thr-196 of CaM-kinase IV is the phosphorylation site involved in the marked activation by CaM-kinase kinase. The results also indicate that Thr-196 and Thr-200 can be replaced by a serine residue for the activity. CaM-kinase I has been demonstrated to be activated upon phosphorylation on Thr-177, whose position in CaM-kinase I corresponds to Thr-196 in CaM-kinase IV (Fig. 1), by CaM-kinase I kinase [44]. This together with the observations that both CaM-kinase IV and I are activated by both CaM-kinase IV kinase and CaM-kinase I kinase [43,45], the question of whether CaM-kinase IV kinase and CaM-kinase I kinase are identical remaining unanswered, suggests that the activities of CaM-kinase IV and I are controlled by similar regulatory mechanism.

When CaM-kinase IV is incubated with CaM-kinase kinase under the protein phosphorylation conditions in the presence of Ca^{2+}/calmodulin, the enzyme is markedly activated upon phosphorylation on Thr-196 by CaM-kinase kinase, as described above, and after maximal phosphorylation (approximately 4 mol of phosphate/mol of enzyme) was attained, the addition of excess EGTA to remove free Ca^{2+} resulted in additional phosphorylation of the enzyme, causing a decrease in

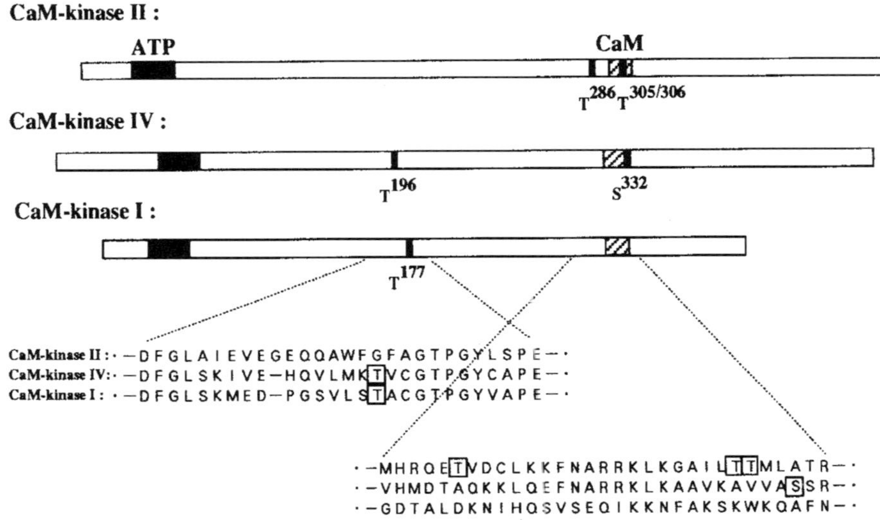

Fig. 1. Schematic representation of phosphorylation sites involved in the regulation of the activities of CaM-kinases. The ATP-binding and calmodulin-binding sites are represented by the solid and hatched bars, respectively. Amino acid sequences around the phosphorylation sites are given at the bottom. Amino acid residues the phosphorylation of which was involved in the regulation of the enzyme activity are boxed.

the Ca^{2+}/calmodulin-dependent activity without affecting the Ca^{2+}/calmodulin-independent activity, as shown in Fig. 2. Incubation for 4 h after addition of EGTA resulted in incorporation of additional 3 mol of phosphate into the enzyme and almost complete loss of the Ca^{2+}/calmodulin-dependent activity, the enzyme becoming a Ca^{2+}/calmodulin-independent form whose activity was 15 times higher than that of the original enzyme. To identify the phosphorylation site involved in the inactivation of Ca^{2+}/calmodulin-dependent activity, phosphorylated CaM-kinase IV was digested with lysyl endopeptidase, then the phosphopeptides purified by HPLC were analyzed by a protein sequenator, and four serine residues, Ser-332, Ser-333, Ser-337, Ser-341 were suggested for the candidates [41]. When these four serine residues were replaced with alanine and aspartic acid, and changes in the enzyme activity of the mutant enzymes during the incubation with CaM-kinase kinase in the presence of Ca^{2+}/calmodulin followed by the incubation in the presence of excess EGTA were examined, only the mutant enzymes of Ser-332 showed essentially different behavior from the wild-type enzyme. Figure 3 shows the time course of changes in the enzyme activity of the recombinant wild-type, S332A, and S332D enzymes during the incubation. The activity of the S332A mutant was markedly increased by incubation

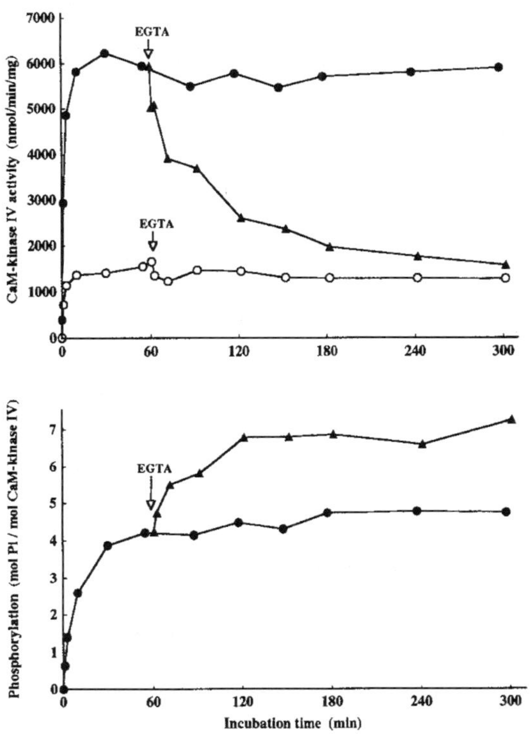

Fig. 2. Time course of activity and phosphorylation of CaM-kinase IV. CaM-kinase IV (10 μg/ml) expressed in Sf9 cells was incubated with CaM-kinase kinase (0.2 μg/ml) in the phosphorylation mixture at 30°C (●), and after incubation for 60 min, excess EGTA was added to remove Ca^{2+} (▲). At the indicated times, the CaM-kinase IV activity was determined in the presence (●, ▲) or absence (○) of Ca^{2+}, and the incorporation of phosphate into protein was determined.

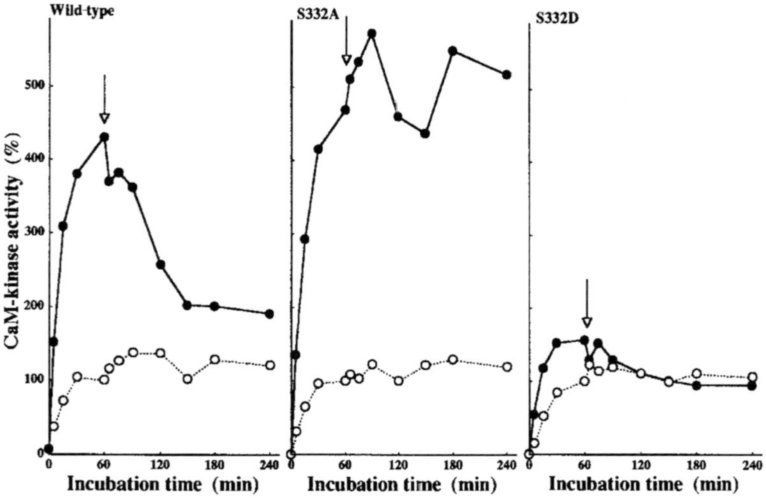

Fig. 3. Time course of changes in the enzyme activity of mutant CaM-kinase IV during the incubation under the phosphorylation conditions. The crude extract of bacteria transformed with expression vector pET11 carring a cDNA coding wild-type enzyme, or mutant S332A or S332D was incubated with CaM-kinase kinase (0.2 μg/ml) in the phosphorylation mixture at 30°C, and after incubation for 60 min, excess EGTA was added. At the indicated times, the CaM-kinase IV activity was determined in the presence (●) or absence (○) of Ca^{2+}.

with CaM-kinase kinase under the phosphorylation conditions in a similar manner to the wild-type enzyme, but the subsequent incubation in the presence of excess EGTA caused no significant decrease in the Ca^{2+}/calmodulin-dependent activity, indicating the involvement of Ser-332 in the inactivation of the enzyme after removal of Ca^{2+}. On the other hand, the incubation of the S332D mutant with CaM-kinase kinase under the phosphorylation conditions resulted in a similar activation of Ca^{2+}/calmodulin-independent activity to the wild-type enzyme, but caused no marked activation of the Ca^{2+}/calmodulin-dependent activity. These results indicate that Asp-332 makes the enzyme insensitive to stimulation by Ca^{2+}/calmodulin by mimicking phospho-Ser-332, confirming the involvement of the phosphorylation of Ser-332 in the loss of the Ca^{2+}/calmodulin-dependent activity. Since Ser-332 is located within the putative calmodulin-binding site of CaM-kinase IV [46] (Fig. 1), it is conceivable that the loss of Ca^{2+}/calmodulin-dependent activity may be due to loss of the ability of the enzyme to bind calmodulin. In fact, calmodulin overlay analysis has shown that the wild-type enzyme, unphosphorylated or phosphorylated in the presence of Ca^{2+}/calmodulin, bind calmodulin, but the wild-type enzyme phosphorylated after addition of EGTA and the S332D mutant cannot bind calmodulin [41]. Of interest, replacement of Ser-333, just adjacent to Ser-332, with alanine or aspartic acid did not affect the changes in the enzyme activity and also the ability to bind calmodulin, indicating that Ser-333 is located outside the calmodulin-binding site.

Figure 4 shows the tissue distributions of CaM-kinase IV and CaM-kinase kinase

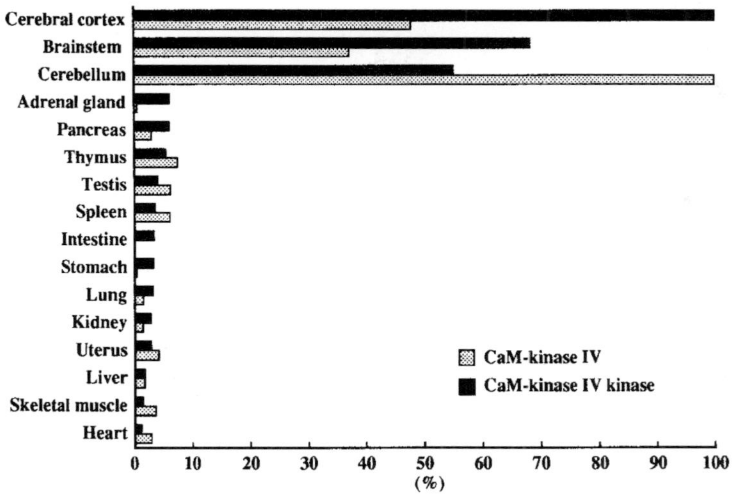

Fig. 4. Tissue distributions of CaM-kinase IV and CaM-kinase IV kinase activities. A sample of 2 μl of each tissue extract was assayed for CaM-kinase IV activity. For assay of CaM-kinase kinase, each tissue extract (0.002–0.03 μl) was preincubated in a reaction mixture (10 μl) containing 1.2 μg of recombinant CaM-kinase IV expressed in *E. coli* and then the activity of the recombinant CaM-kinase was determined. The results are expressed as percentages of the specific activities (activity/total protein) of the tissues which showed the highest activities, that is, cerebral cortex for CaM-kinase kinase activity and cerebellum for CaM-kinase IV activity.

activities in the rat. Both activities were particularly high in the brain, and the other tissues showed very little activities. Figure 5A shows the tissue distributions of immunoreactivities to antibodies raised against CaM-kinase IV and CaM-kinase kinase α. The thymus, which gave a strong immunoreactivity to CaM-kinase IV antibody in a position corresponding to CaM-kinase IV, showed very little activity of CaM-kinase IV, but the activity was remarkably increased upon incubation with CaM-kinase kinase under the phosphorylation conditions, indicating that CaM-kinase IV exists as an inactive form in thymus. In contrast, the activity of the crude extracts of cerebral cortex, brainstem, and cerebellum was not significantly activated by incubation with CaM-kinase kinase. The retina gave a significant immunoreactivity to CaM-kinase IV antibody in a position distinct from those of CaM-kinase IV and calspermin, a product of a gene within CaM-kinase IV gene [47,48], observed in testis, in agreement with the observations of the selective occurrence of a cross-reacting material in retina (O. Hatase and M. Tokuda, personal communication). With respect to CaM-kinase kinase α, cerebellum which showed a high activity of CaM-kinase IV kinase gave only a weak immunoreactivity to antibody against CaM-kinase kinase α. This together with the results of Northern blot analysis with a probe specific for CaM-kinase kinase α (Fig. 5B) that cerebellum gave a much weaker band than cerebral cortex and brainstem indicate that cerebellum, in contrast with brainstem, shows a comparable activity of CaM-kinase IV kinase but contains a much

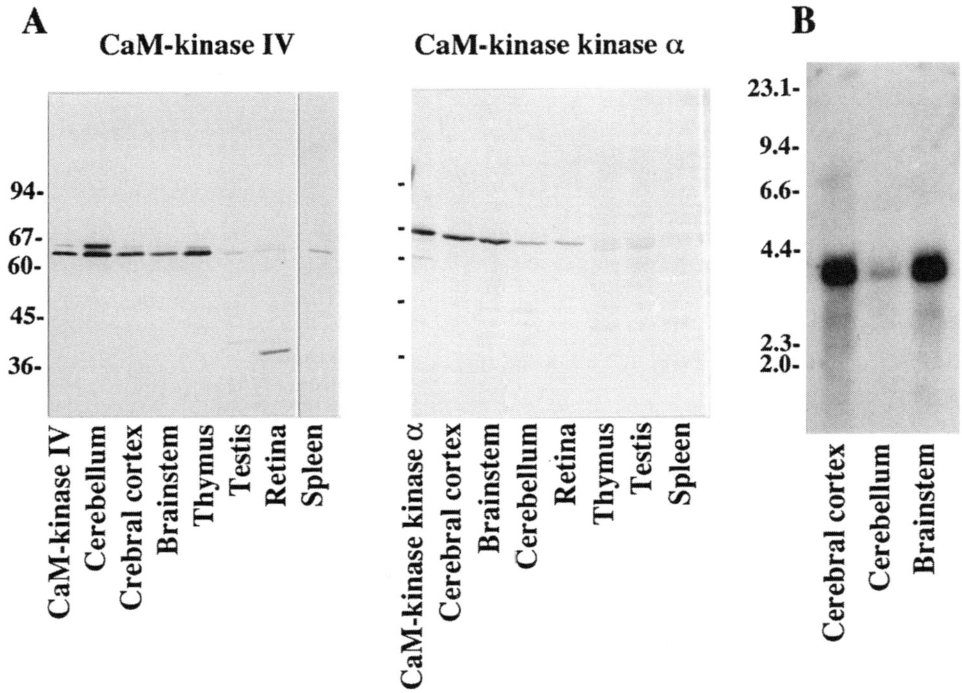

Fig. 5. Western blot (A) and Northern blot (B) analyses. **A**: For analysis of CaM-kinase IV, approximately 25 ng of CaM-kinase IV purified from rat cerebellum and 10 μg protein of the crude extracts of the indicated tissues of the rat were analyzed by Western blot analysis using antiserum against CaM-kinase IV. For analysis of CaM-kinase kinase α, approximately 20 ng of CaM-kinase kinase α and 20 μg protein of the crude extracts of the indicated tissues of the rat were analyzed by Western blot analysis using antibody against CaM-kinase kinase α. Molecular masses in kilodaltons are given on the left. **B**: Approximately 5 μg of poly(A)$^+$ RNAs from rat cerebral cortex, cerebellum, and brainstem were analyzed by Northern blot analysis with 410-bp cDNA probe [19]. RNA sizes in kilobases are given on the left.

smaller amount of CaM-kinase kinase α, suggesting the existence of another CaM-kinase kinase in cerebellum. Further strong support for this contention was provided by immunotitration of the crude brain extracts with antibody against CaM-kinase kinase α, as shown in Fig. 6. When the crude extracts of rat brainstem, cerebellum, and the bacteria transformed with CaM-kinase kinase α cDNA were immunotitrated with antibody against CaM-kinase kinase α, almost all the activity of the recombinant enzyme was immunoprecipitated, but only approximately 25 and 56% of the activity of cerebellum and brainstem, respectively, were immunoprecipitated by the addition of an excess of the antibody. Thus, most of the activity of CaM-kinase IV kinase in cerebellum appears to result from CaM-kinase kinase other than CaM-kinase kinase α.

Figure 7 summarizes signal transduction by CaM-kinases. Recent immunohistochemical studies showed that CaM-kinase I [49], CaM-kinase IV [50–52], and CaM-

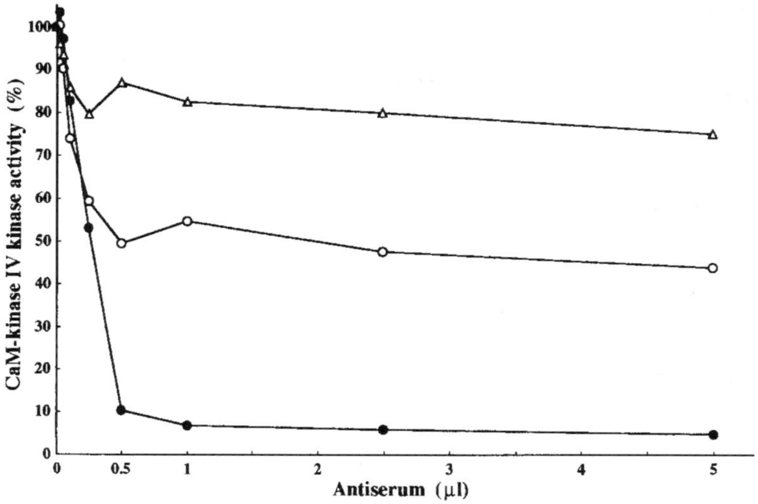

Fig. 6. Immunotitrations of crude brain extracts with antiserum against CaM-kinase kinase α. Approximately 0.45, 10.6, and 14.0 μg of the crude extracts of bacteria transformed with pET11d carring a cDNA coding CaM-kinase kinase α (●), rat brainstem (○), and cerebellum (Δ), respectively, were immunotitrated with varying amounts of the antiserum against CaM-kinase kinase α. The results are expressed as percentages of the activities obtained without antiserum.

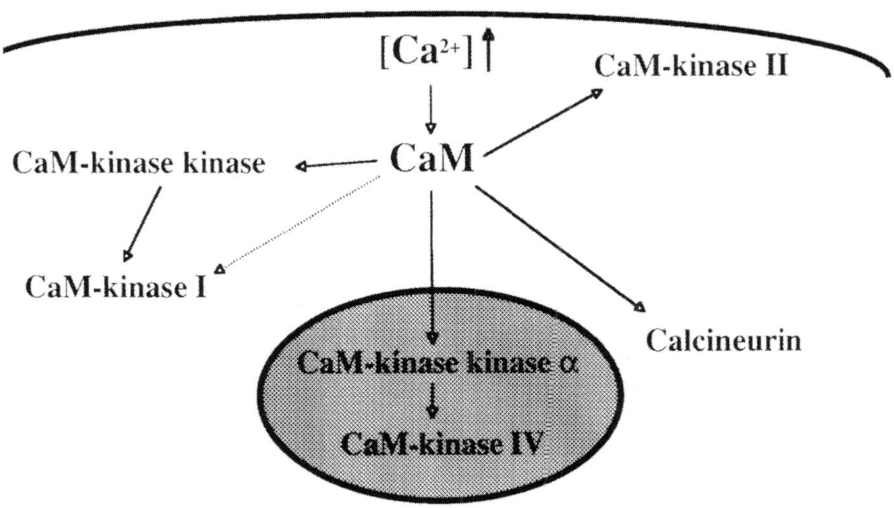

Fig. 7. Signal transduction by CaM-kinases.

kinase kinase α [53] are relatively widely distributed in the brain, and in cytosol, cell nuclei, and cell nuclei, respectively, in neuronal cells. These observations suggest that CaM-kinase kinase α may be involved in the regulation of the activity of CaM-kinase

IV, but not CaM-kinase I. A transcription factor, CREB, has recently been suggested as a physiological substrate of CaM-kinase IV in neuronal cell nuclei [54–56]. Since recent studies have suggested the involvement of CREB in the formation of long-term memory [57], the CaM-kinase kinase/CaM-kinase IV cascade may play important roles in the regulation of memory formation.

References

1. Fujisawa H. Calmodulin-dependent protein kinase II. Bioessays 1990;12:27–29.
2. Colbran RJ, Soderling TR. Calcium/calmodulin-dependent protein kinase II. Curr Top Cell Regul 1990;31:181–221.
3. Hanson PI, Schulman H. Neuronal Ca^{2+}/calmodulin-dependent protein kinases. Annu Rev Biochem 1992;61:559–601.
4. Nairn AC, Picciotto MR. Calcium/calmodulin-dependent protein kinases. Sem Cancer Biol 1994;5: 295–303.
5. Kwitatkowski AP, Shell DJ, King MM. The role of autophosphorylation in activation of the type II calmodulin-dependent protein kinase. J Biol Chem 1988;263:6484–6486.
6. Katoh T, Fujisawa H. Autoactivation of calmodulin-dependent protein kinase II by autophosphorylation. J Biol Chem 1991;266:3039–3044.
7. Ikeda A, Okuno S, Fujisawa H. Studies on the generation of Ca^{2+}/calmodulin-independent activity of calmodulin-dependent protein kinase II by autophosphorylation. Autophosphorylation of the enzyme. J Biol Chem 1991;266:11582–11588.
8. Ishida A, Kitani T, Fujisawa H. Evidence that autophosphorylation at Thr-286/Thr-287 is required for full activation of calmodulin-dependent protein kinase II. Biochim Biophys Acta 1996;I311:211–217.
9. Hashimoto Y, Schworer CM, Colbran RJ, Soderling TR. Autophosphorylation of Ca^{2+}/calmodulin-dependent protein kinase II. Effects on total and Ca^{2+}-independent activities and kinetic parameters. J Biol Chem 1987;262:8051–8055.
10. Colbran RJ, Soderling TR. Calcium/calmodulin-independent autophosphorylation sites of calcium/calmodulin-dependent protein kinase II. Studies on the effect of phosphorylation of threonine 305/306 and serine 314 on calmodulin binding using synthetic peptides. J Biol Chem 1990;265: 11213–11219.
11. Patton BL, Miller SG, Kennedy MB. Activation of type II calcium/calmodulin-dependent. Protein kinase by Ca^{2+}/calmodulin is inhibited by autophosphorylation of threonine within the calmodulin-binding domain. J Biol Chem 1990;265:11204–11212.
12. Okuno S, Fujisawa H. Requirement of brain extract for the activity of brain calmodulin-dependent protein kinase IV expressed in *Escherichia coli*. J Biochem 1993;114:167–170.
13. Lee JC, Edelman AM. A protein activator of Ca^{2+}-calmodulin-dependent protein kinase Ia. J Biol Chem 1994;269:2158–2164.
14. Mochizuki H, Sugita R, Ito T, Hidaka H. Phosphorylation of Ca^{2+}/calmodulin-dependent protein kinase V and regulation of its activity. Biochem Biophys Res Commun 1993;197:1595–1600.
15. Kitani T, Okuno S, Fujisawa H. Inactivation of Ca^{2+}/calmodulin-dependent protein kinase IV by Ca^{2+}/calmodulin and restoration of the activity by Mg^{2+}/EGTA. J Biochem 1995;117:1070–1075.
16. Gopalakrishna R, Anderson WB. Ca^{2+}-induced hydrophobic site on calmodulin: application for purification of calmodulin by phenyl-Sepharose affinity chromatography. Biochem Biophys Res Commun 1982;104:830–836.
17. Putkey JA, Ts'ui KF, Tanaka T, Lagace' L, Stein JP, Lai EC, Means AR. Chicken calmodulin genes. A species comparison of cDNA sequences and isolation of a genomic clone. J Biol Chem 1983;258: 11864–11870.
18. Okuno S, Kitani T, Fujisawa H. Purification and characterization of Ca^{2+}/calmodulin-dependent protein kinase IV kinase from rat brain. J Biochem 1994;116:923–930.

19. Okuno S, Kitani T, Fujisawa H. Evidence for the existence of Ca^{2+}/calmodulin-dependent protein kinase IV kinase isoforms in rat brain. J Biochem 1996;119:1176–1178.
20. Hashimoto Y, Soderling TR. Calcium/calmodulin-dependent protein kinase II and calcium/phospholipid-dependent protein kinase activities in rat tissues assayed with a synthetic peptide. Arch Biochem Biophys 1987;252:418–425.
21. Miyano O, Kameshita I, Fujisawa H. Purification and characterization of a brain-specific multifunctional calmodulin-dependent protein kinase from rat cerebellum. J Biol Chem 1992;267: 1198–1203.
22. Roskoski R Jr. Assay of protein kinase. Meth Enzymol 1983;99:3–6.
23. Kunkel TA, Roberts JD, Zakour RA. Rapid and efficient site-specific mutagenesis without phenotypic selection. Meth Enzymol 1987;154:367–382.
24. Studier FW, Rosenberg AH, Dunn JJ, Dubendorf JW. Use of T7 RNA polymerase to direct expression of cloned genes. Meth Enzymol 1990;185:60–89.
25. Corbin JD, Reimann EM. Assay of cyclic AMP-dependent protein kinases. Meth Enzymol 1974;38: 287–290.
26. Sambrook J, Fritsch EF, Maniatis T. In: Molecular Cloning: A Laboratory Manual 2. NY: Cold Spring Harbor Laboratory, 1989.
27. Winston SE, Fuller SA, Hurrell JGR. In: Ausubel et al. (eds) Current Protocols in Molecular Biology 2. MA: Wiley Interscience, 1987;10.8.1–10.8.6.
28. Goding JW. Use of staphylococcal protein A as an immunological reagent. J Immunol Meth 1978; 20:241–253.
29. Kessler SW. Rapid isolation of antigens from cells with a staphylococcal protein A-antibody adsorbent: parameters of the interaction of antibody-antigen complexes with protein A. 1975;115: 1617–1624.
30. Laemmli UK. Cleavage of structural proteins during the assembly of the head of bacteriophage T4. Nature 1970;227:680–685.
31. Mangels LA, Gnegy ME. Carbachol stimulates binding of a photoreactive calmodulin derivative to calmodulin-binding proteins in intact SK-N-SH human neuroblastoma cells. J Biol Chem 1992;267:5847–5854.
32. Kincaid RL, Billingsley ML, Vaughan M. Preparation of fluorescent, cross-linking, and biotinylated calmodulin derivatives and their use in studies of calmodulin-activated phosphodiesterase and protein phosphatase. Meth Enzymol 1988;159:605–626.
33. Dedman JR, Potter JD, Jackson RL, Johnson JD, Means AR. Physicochemical properties of rat testis Ca^{2+}-dependent regulator protein of cyclic nucleotide phosphodiesterase. Relationship of Ca^{2+}-binding, conformational changes, and phosphodiesterase activity. J Biol Chem 1977;252:8415–8422.
34. Dedman JR, Jackson RL, Schreiber WE, Means AR. Sequence homology of the Ca^{2+}-dependent regulator of cyclic nucleotide phosphodiesterase from rat testis with other Ca^{2+}-binding proteins. J Biol Chem 1978;253:343–346.
35. Watterson DM, Sharief F, Vanaman TC. The complete amino acid sequence of the Ca^{2+}-dependent modulator protein (calmodulin) of bovine brain. J Biol Chem 1980;255:962–975.
36. Lowry OH, Rosebrough NJ, Farr AL, Randall RJ. Protein measurement with the Folin phenol reagent. J Biol Chem 1951;193:265–275.
37. Peterson GL. A simplification of the protein assay method of Lowry et al. which is more generally applicable. Anal Biochem 1977;83:346–356.
38. Cooper JA, Sefton BM, Hunter T. Detection and quantification of phosphotyrosine in proteins. Meth Enzymol 1983;99:387–402.
39. Kitani T, Okuno S, Fujisawa H. Comparison of Ca^{2+}/calmodulin-dependent protein kinase IV from rat brain, expressed in insect cells, and expressed in *Escherichia coli*. J Biochem 1995;118:364–370.
40. Okuno S, Kitani T, Fujisawa H. Full activation of brain calmodulin-dependent protein kinase IV requires phosphorylation of the amino-terminal serine-rich region by calmodulin-dependent protein kinase IV kinase. J Biochem 1995;117:686–690.
41. Watanabe S, Okuno S, Kitani T, Fujisawa H. Inactivation of calmodulin-dependent protein kinase

IV by autophosphorylation of serine 332 within the putative calmodulin-binding domain. J Biol Chem 1996;271:6903–6910.

42. Kameshita I, Fujisawa H. Autophosphorylation of calmodulin-dependent protein kinase IV from rat cerebral cortex. J Biochem 1993;113:583–590.
43. Selbert MA, Anderson KA, Huang Q, Goldstein EG, Means AR, Edelman AM. Phosphorylation and activation of Ca^{2+}-calmodulin-dependent protein kinase IV by Ca^{2+}-calmodulin-dependent protein kinase Ia kinase. Phosphorylation of threonine 196 is essential for activation. J Biol Chem 1995;270: 17616–17621.
44. Haribabu B, Hook SS, Selbert MA, Goldstein EG, Tomhave ED, Edelman AM, Snyderman R, Means AR. Human calcium-calmodulin dependent protein kinase I: cDNA cloning, domain structure and activation by phosphorylation at threonine-177 by calcium-calmodulin dependent protein kinase I kinase. EMBO J 1995;14:3679–3686.
45. Tokumitsu H, Enslen H, Soderling TR. Characterization of a Ca^{2+}/calmodulin-dependent protein kinase cascade. Molecular cloning and expression of calcium/calmodulin-dependent protein kinase kinase. J Biol Chem 1995;270:19320–19324.
46. Sikela JM, Hahn WE. Screening an expression library with a ligand probe: Isolation and sequence of a cDNA corresponding to a brain calmodulin-binding protein. Proc Natl Acad Sci USA 1987;84: 3038–3042.
47. Ono T, Slaughter GR, Cooks RG, Means AR. Molecular cloning sequence and distribution of rat calspermin, a high affinity calmodulin-binding protein. J Biol Chem 1989;264:2081–2087.
48. Means AR, Cruzalegui F, LeMagueresse B, Needleman DS, Slaughter GR, Ono T. A novel Ca^{2+}/calmodulin-dependent protein kinase and a male germ cell-specific calmodulin-binding protein are derived from the same gene. Molec Cell Biol 1991;11:3960–3971.
49. Picciotto MR, Czernik AJ, Nairn AC. Calcium/calmodulin-dependent protein kinase I. cDNA cloning and identification of autophosphorylation site. J Biol Chem 1993;268:26512–26521.
50. Jensen KF, Ohmstede C, Fisher RS, Sahyoun N. Nuclear and axonal localization of Ca^{2+}/calmodulin-dependent protein kinase type Gr in rat cerebellar cortex. Proc Natl Acad Sci USA 1991;88: 2850–2853.
51. Sakagami S, Tsubochi H, Kondo H. Immunohistochemical localization of Ca^{2+}/calmodulin-dependent protein kinase IV in the peripheral ganglia and paraganglia of developing and mature rats. Brain Res 1994;666:173–181.
52. Nakamura Y, Okuno S, Fujisawa H. An immunohistochemical study of Ca^{2+}/calmodulin-dependent protein kinase IV in the rat central nervous system: light and electron microscopic observations. Neuroscience 1995;68:181–194.
53. Nakamura Y, Okuno S, Kitani T, Otake K, Sato F, Fujisawa H. Distribution of Ca^{2+}/calmodulin-dependent protein kinase kinase α in the rat central nervous system: an immunohistochemical study. Neurosci Lett 1996;204:61–64.
54. Cruzalegui FH, Means AR. Biochemical characterization of the multifunctional Ca^{2+}/calmodulin-dependent protein kinase type IV expressed in insect cells. J Biol Chem 1993;268:26171–26178.
55. Enslen H, Sun P, Brickey D, Soderling SH, Klamo E, Soderling TR. Characterization of Ca^{2+}/calmodulin-dependent protein kinase IV. Role in transcriptional regulation. J Biol Chem 1994; 269:15520–15527.
56. Matthews RP, Guthrie CR, Wailes LM, Zhao X, Means AR, McKnight GS. Calcium/calmodulin-dependent protein kinase types II and IV differentially regulate CREB-dependent gene expression. Molec Cell Biol 1994;14:6107–6116.
57. Carew T. Molecular enhancement of memory formation. Neuron 1996;16:5–8.

Integrative and Molecular Approach to Brain Function.
M. Ito and Y. Miyashita, editors.

Physiological roles of neocortical long-term potentiation in the auditory cortex

Katsuei Shibuki, Masaharu Kudoh, Toshiyuki Fujisaki and Masashi Sakai
Department of Neurophysiology, Brain Research Institute, Niigata University, Niigata, Japan

Abstract. *Background.* To elucidate the dynamic aspects of neural networks in the auditory cortex, we studied synaptic plasticity in the rat auditory cortex in vitro, and developed a new behavioral test to estimate sound discrimination ability.

Methods. Long-term potentiation (LTP) was studied in the slices of the rat auditory cortex. To estimate sound discrimination ability, a two-tone operant conditioning paradigm was used.

Results. White matter stimulation produced field potentials in the supragranular layers in the slice and block preparations. Tetanic stimulation evoked LTP of the transsynaptic potentials. This LTP was blocked by D-2-amino-5-phosphonovalerate (D-AP5, 50 μM). LTP of orthodromic firing probability was also observed in pyramidal units. To estimate sound discrimination ability, rats were deprived of water and were exposed to two different sounds, response to only one of which was rewarded with water. All the rats discriminated between the rewarded and unrewarded sounds in a single session of 10 h.

Conclusion. Our results obtained in the slices indicate that neural circuit is plastic in the auditory cortex. The new test for estimating sound discrimination ability may reveal the physiological role of the synaptic plasticity in the auditory cortex.

Key words: auditory perception, sound discrimination, synaptic plasticity.

Introduction

In rodents, auditory perception gives an important cue to get foods or to run away from enemies. We expected that optimization of the neural networks for auditory perception may be produced by environmental sound stimuli so that animals can use auditory perception more efficiently. Of the many circuits involved in the auditory system, those in the auditory cortex may be modulated dynamically, since it is known that synaptic plasticity in the neocortex is a cellular mechanism for development of neural response selectivity or formation of columnar organization [1–3]. In the auditory cortex, responsiveness of units is changed after training using a sound cue [4,5]. Synaptic connections between auditory cortical neurons are potentiated by correlated activities [6]. These resorts lead us to study synaptic plasticity in the slices obtained from the rat auditory cortex [7,8].

Neocortical neural circuits are organized in columnar structures [9–11]. Synaptic functions in vitro are usually analysed electrophysiologically in slice preparations. However, the oxygen supply in slices is maintained by passive diffusion, so that the

Address for correspondence: Katsuei Shibuki, Department of Neurophysiology, Brain Research Institute, Niigata University, 1 Asahi-machi, Niigata 951, Japan.

thickness of the preparation must be less than 400–500 μm [12,13]. Therefore, it is difficult to study complete neocortical circuits in slices. To overcome these difficulties, we developed cortical block preparations in which the oxygen supply is maintained by the forced injection of an oxygenated medium into the preparation [14].

To assess the function of synaptic plasticity in the auditory cortex, an estimation of sound discrimination ability in a behaving animal is required. Various behavioral tests have been developed and revealed that the auditory cortex plays essential roles for discriminating between sound stimuli of complex pattern [15,16]. It is likely that synaptic plasticity may be elicited by sound stimuli during a behavioral test so that the sound discrimination ability is changed during the test. Therefore, we developed a quick and reproducible sound discrimination test in a two-tone operant conditioning paradigm [17].

Material and Methods

Slice and block preparations

Wistar rats (4–8 weeks old) were anesthetized with ether and bathed in a mixture of ice and water for 3–5 mins to reduce brain temperature. Immediately after decapitation, the brain was removed, and slices (thickness: 400 μm) were prepared from the auditory cortex. Slices were incubated at 30°C in an artificial medium bubbled with 95% O_2 and 5% CO_2. Composition of the medium (in mM) was: NaCl 124, KCl 5, NaH_2PO_4 1.24, $MgSO_4$ 1.3, $CaCl_2$ 2.4, $NaHCO_3$ 25.9 and glucose 10. In some experiments, a block of cerebral cortex including the whole auditory area was isolated from other parts of the brain with a sharp razor blade. The block was transferred into a recording chamber, which was perfused continuously with the oxygenated medium at room temperature. The oxygen supply in the block was maintained by the injection of an oxygenated medium into the preparation through a glass micropipette. The pipette was connected with stainless steel tubing to a multibarrel syringe pump (CFV-3200, Nihon Koden). To visualize the perfused area of the brain blocks, 0.005% indigo carmine was added to the medium. The rate of injection was started with 0.8 μl/min, and was gradually increased to 5–15 μl/min in a few minutes. The injection was continued for at least 60 min before recording.

Behavioral test

To estimate sound discrimination ability of rats, a two-tone operant conditioning paradigm was adopted. A Wistar rat (5–6 weeks old) was deprived of water for 48 h, and was placed in a cage equipped with a pedal (Fig. 1A). The cage was placed in a soundproof room. The rat was exposed to a sound with a sine wave (10 kHz) whose amplitude was modulated by a saw tooth wave (100 Hz) to the depth between 0 and 100%. The stimulus sound was turned on for 200 ms with the rise/fall time of 10 ms. The modulated tone burst was repeated at 500-ms intervals throughout a period of 20 s. In a preparatory training, sound stimuli were repeated at 2-min

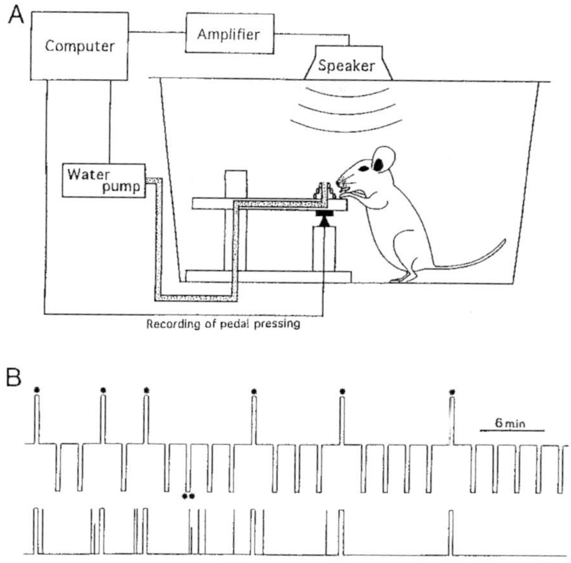

Fig. 1. **A**: Schematic drawing of the experimental setup. **B**: Example of pedal-pressing behavior of a rat in response to two different sounds. The upper trace shows timing of presentation of the rewarded (upward deflection) and the unrewarded (downward deflection) sounds. The lower trace shows pedal-pressing behavior (upward deflection). Asterisks represent trials in which the rat responded to the rewarded (*) or to the unrewarded (**) sounds. Reproduced from [17].

intervals, and pedal pressing by rats during sound presentation period of 20 s was rewarded with water through the spout at a flow rate of 3 μl/s. After rats learned to respond to sound stimuli with pedal pressing, they were deprived of water again and two different sound stimuli modulated to different depths were presented randomly to the rats. During presentation of only one of the two stimuli, pedal-pressing was rewarded with water. The difference in pedal-pressing rates between the rewarded and unrewarded sound stimuli was calculated to estimate sound discrimination ability of the rat.

Results

Long-term potentiation in the auditory cortex

White matter stimulation produced field potentials in supragranular layers of the slices. Two negative waves were found in the potentials, of which only the second component was blocked by 6-cyano-7-nitroquinoxaline-2,3-dione (CNQX, 10 μM). D-AP5 (50 μM) showed no clear effect on the field potentials. These results indicate that the first and second components represent antidromic and transsynaptic activities, respectively [7].

After constant field potentials were recorded for more than 10 min, tetanic white matter stimulation (100 Hz, 1 s) was applied twice at a 25-s interval. Although the antidromic component was unchanged, the transsynaptic component was potentiated by 45 ± 16% (mean ± SEM, n = 5; Fig. 2A and B). This LTP was completely blocked by D-AP5 (50 μM, Fig. 2C and D). Therefore it is demonstrated that tetanic white matter stimulation produces LTP in the transsynaptic potentials mediated by non-NMDA receptors, and this LTP depends on the activation of NMDA receptors during tetanic stimulation [7].

To identify the cellular counterparts of the LTP in the field potentials, intracellular potentials were recorded in supragranular neurons with a blind slice patch technique. Neurons showing antidromic spikes in response to white matter stimulation were injected with biocytin. Most of them were identified as pyramidal neurons with a staining using streptoavidine-fluoroscein or streptoavidine-Texas Red conjugates. Tetanic stimulation at the intensity lower than the threshold of antidromic spikes failed to evoke LTP in all the nine neurons tested. We tried tetanic stimulation at the intensity higher than the threshold of antidromic spikes, but only two of seven neurons showed clear LTP [8].

We suspected that intracellular dialysis during patch recording might prevent

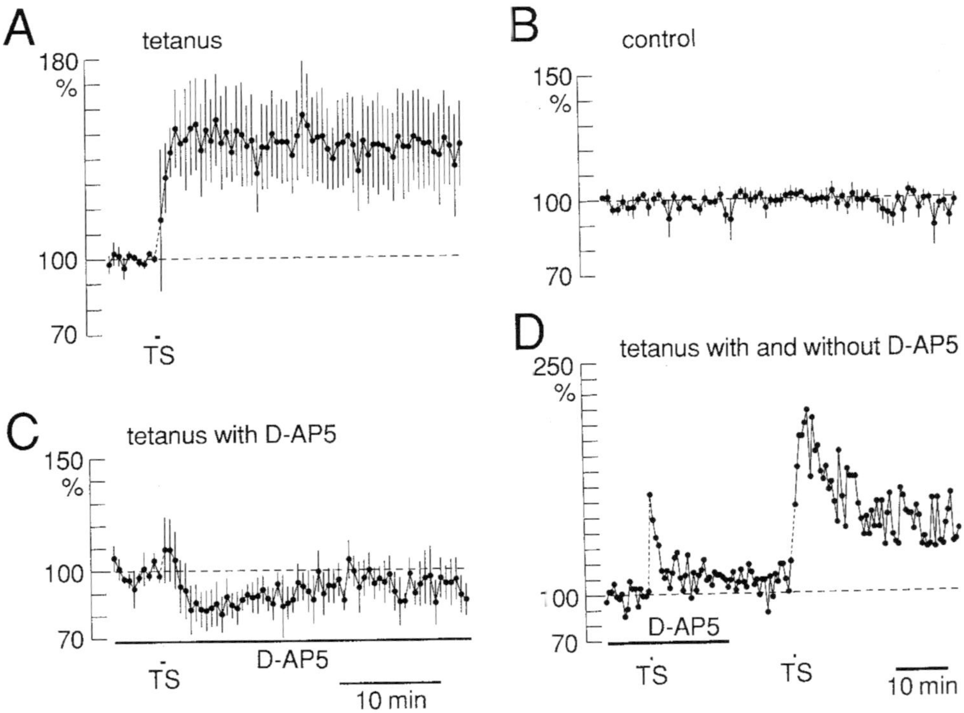

Fig. 2. LTP of trans-synaptic field potentials in supragranular layers. **A**: Tetanic white matter stimulation (TS) produced LTP. **B**: Control experiments in which no TS was given. **C**: LTP was blocked by D-AP5 (50 μM). **D**: TS did not produce LTP in the presence of D-AP5, but evoked LTP after D-AP5 was washed. Ordinate, relative amplitude of field potentials. In A–C, mean ± SEM of five experiments are plotted. Time scale in C is applicable also in A and B. Reproduced from [7].

induction of LTP [18]. Therefore, we investigated LTP by extracellular unit recording. Pyramidal units were identified by antidromic spikes following white matter stimulation. Tetanic stimulation evoking antidromic spikes produced LTP in orthodromic firing probability in all the five pyramidal units tested. This LTP was maintained for 30 min, at least. D-AP5 (50 μM) blocked the LTP. These results indicate that supragranular pyramidal outputs are potentiated by tetanic white matter stimulation [8].

Block preparation of the auditory cortex

To study properties of neocortical neural networks in the block preparation, we injected an oxygenated medium containing indigo carmine (0.005%). The dye appeared on the pial surface, and stained an area, whose diameter was 1–2 mm. Vertical cutting through the stained area revealed that the dye was spread homogeneously in a columnar region between the pial surface and the white matter. Stimulating and recording electrodes were inserted into the infra and supragranular layers, respectively. Electric stimulation produced characteristic field potentials composed of two negative waves as in the slices (Fig. 3A and B). CNQX (10 μM) selectively blocked the second negativity. The stability of field potentials in blocks was similar to that in slices. The depth profile and temperature-dependency of the potentials in blocks were similar to those in slices reported previously [19,20]. These results indicate that neural excitability and synaptic transmission are maintained stably in the perfused area of the brain blocks [14].

To confirm the role of the medium injection for the maintenance of blocks, we observed the changes in field potentials after cessation of the injection (Fig. 3A and B). Immediately after the cessation, the responses were potentiated in amplitude for several minutes. Gradual attenuation of potentials followed that. The transsynaptic potentials were decreased faster than the antidromic component. The complete

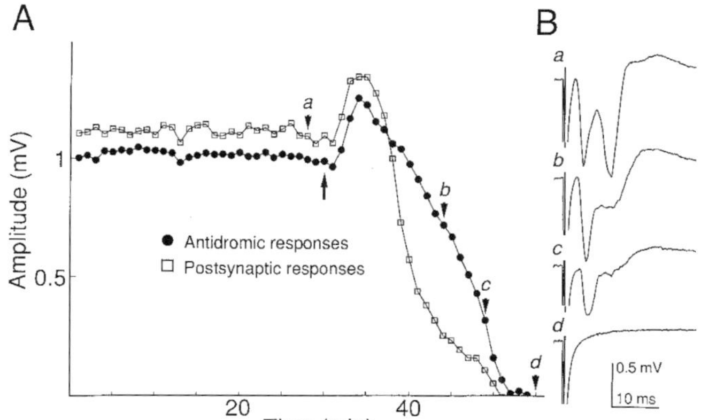

Fig. 3. Changes in field potentials after cessation of the medium injection into the brain blocks. **A**: Time course of the amplitude of antidromic and postsynaptic responses. Arrow marks the time of cessation. **B**: field potential traces recorded at each arrowhead in A. Reproduced from [14].

flattening of the responses was observed about 20 min after the injection was stopped. These results indicate that passive diffusion of oxygen from the surface is not sufficient for the functional maintenance of blocks, and that the medium injection is necessary. These data demonstrate that block preparation can be maintained by the injection and may be a useful tool for studying the neocortical neural circuits without slicing in vitro [14].

Estimation of sound discrimination ability

In the preparatory training, pedal-pressing during sound presentation was rewarded with water. The rate of pedal pressing was compared statistically in every hour between trials with or without the sound. Within 4 h of this training, significant difference in the rate (χ^2 test, $p < 0.01$) was observed in six of eight trained rats. The mean and SEM for the pedal-pressing rates of the six rats during the last 1 h were 91 ± 5% in trials with sound and 33 ± 3% in trials without sound, respectively. These six were used for the next sound discrimination test.

After free access to water, rats were deprived of water again for 48 h. This time, two different sound stimuli were presented to the rats. One was amplitude-modulated tone to the depth of 100%. The other was a tone burst of a sine wave (10 kHz) of the same intensity as the former. Pedal-pressing was rewarded with water only during presentation of the amplitude-modulated tone. In the first 2 h of the sound discrimination test, rats pressed the pedal in response to both the rewarded and unrewarded sounds. The rates were 91 ± 2 (n = 6) and 87 ± 5%, respectively. In trials conducted between 4–6 h after the initiation of the sound discrimination test, the rate of pedal-pressing in response to the rewarded sound remained as high as 93 ± 3%, while it was reduced to 45 ± 4% for the unrewarded sound. The difference in the rates was significant. After all, all the six rats tested successfully discriminated between sound stimuli within 10 h of a single session (Fig. 1B) [17].

Discussion

In this study, LTP was evoked in the transsynaptic component of field potentials. LTP of orthodromic spike firing was also observed in all the pyramidal cells tested. Sensory information is sequentially transmitted from the primary sensory cortex to the higher order cortices [21]. Cortical outputs to other cortices are discharged as supragranular pyramidal spikes, while infragranular pyramidal cells send outputs mainly to subcortical regions [22,23]. If supragranular pyramidal outputs are potentiated in the primary auditory cortex, sequential processing of auditory information is drastically modulated by the plasticity. Our results support this possibility. In preliminary experiments, we found that LTP in the auditory cortex was larger than that in the visual cortex. Therefore it is highly possible that synaptic plasticity in the auditory cortex has important roles in auditory perception.

There are at least two candidates for inputs of the potentiated synapses: afferents into the cortex and axon collaterals of pyramidal cells. LTP in synapses with afferent

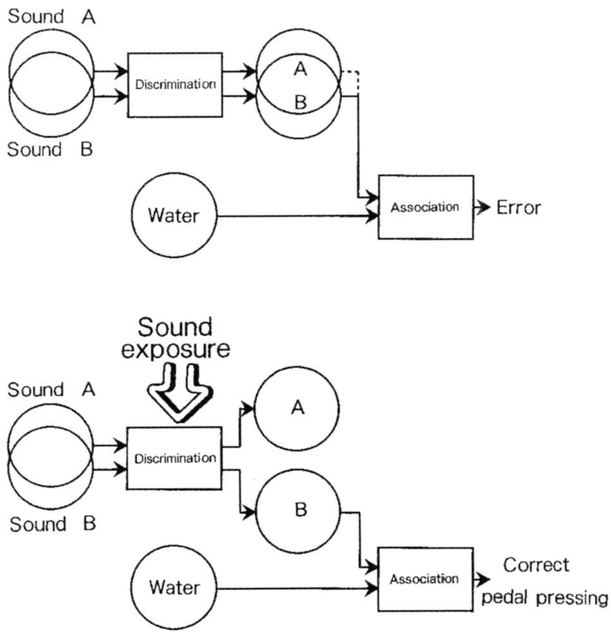

Fig. 4. Possible mechanisms for the enhancement in sound discrimination ability. Sound exposure for 48 h produces LTP in the auditory cortex so that separation between sound A and sound B is easily achieved in the auditory cortex.

inputs would change the input-output relationships of the cortex drastically. It is reported that LTP occurs also in synapses between pyramidal cells [24]. LTP of the latter type in the auditory cortex might make a group of pyramidal cells fire together to a particular sound stimulus, which might be perceived more easily as the result.

To study the neocortical neural networks in vitro, we tested viability of the brain block preparations perfused by the medium injection. In slice preparations, characteristic supragranular field potentials composed of antidromic and postsynaptic waves are recorded following white matter stimulation [7]. Similar field potentials were recorded in blocks. The columnar distribution of indigo carmine and the depth profile of the field potentials in blocks suggests that a whole columnar area between the pial surface and the white matter was functionally maintained by the injection. By using block preparations, it might be possible to study how synaptic plasticity is organized in the neural networks of the neocortex to modulate cortical functions such as sound discrimination.

To estimate sound discrimination ability of rats, we developed a new test in which all the rats successfully discriminated between sound stimuli in a single session of 10 h. Similar tests in a two-tone operant conditioning paradigm require a much longer period of daily sessions [25,26]. Our method is free from various psychological side effects of punishment [27], and the efficiency is comparable to or better than that of previous tests using active avoidance of electrical shocks for cats [15,16]. Our test allows a quick estimation of discrimination ability between particular sounds. Therefore it may be a useful tool for studying physiological roles of synaptic plasticity in sound discrimination of behaving rats. In preliminary experiments, we found that sound exposure for 48 h enhanced subsequent sound discrimination ability of the rats, and the enhancement was blocked by bilateral injection of D-AP5 into the auditory cortex. These data indicate that sound discrimination ability of behaving animals may be modulated by synaptic plasticity such as LTP induced by auditory experience (Fig. 4).

Acknowledgements

We are indebted to Mr Y. Tamura and Mr N. Taga for technical assistance. This work was supported in part by Uehara Memorial Foundation.

References

1. Hubel DH, Wiesel TN. Receptive fields of cells in striate cortex of young, visually inexperienced kittens. J Neurophysiol 1963;26:994–1002.
2. Blakemore C, Cooper GF. Development of the brain depends on visual environment. Nature 1970; 228:477–478.
3. Hirsh HVB, Spinelli DN. Visual experience modifies distribution of horizontally and vertically oriented receptive fields in cats. Science 1970;168:869–871.
4. Edeline JM, Weinberger NM. Receptive field plasticity in the auditory cortex during frequency discrimination training: selective retuning independent of task difficulty. Behav Neurosci 1993;107: 82–103.
5. Recanzone R, Schreiner CE, Merzenich MM. Plasticity in the frequency representation of primary auditory cortex following discrimination training in adult owl monkeys. J Neurosci 1993;13:87–103.
6. Ahissar E, Vaadia E, Ahissa M, Bergman H, Arieli A, Abeles M. Dependence of cortical plasticity on correlated activity of single neurons and behavioral context. Science 1992;257:1412–1415.
7. Kudoh M, Shibuki K. Long-term potentiation in the auditory cortex of adult rats. Neurosci Lett 1994; 171:21–23.
8. Kudoh M, Shibuki K. Long-term potentiation of the supragranular pyramidal outputs in the rat auditory cortex. Exp Brain Res (In press).
9. Mountcastle VB. Modality and topographic properties of single neurons of cat's somatic sensory cortex. J Neurophysiol 1957;20:408–434.
10. Hubel DH, Wiesel TN. Receptive fields, binocular interaction and functional architecture in the cat's visual cortex. J Physiol 1962;160:106–154.
11. Fujita I, Tanaka K, Ito M, Cheng K. Columns for visual features of objects in monkey inferotemporal cortex. Nature 1992;360:343–346.
12. Richards CD. The preparation of brain tissue slices for electrophysiological studies. In: Kerkut GA, Wheal HV (eds) Electrophysiology of Isolated Mammalian CNS Preparations. London: Academic Press, 1981;107–132.
13. Fujii T, Baumgärtl H, Lübbers DW. Limiting section thickness of guinea pig olfactory cortical slices studied from tissue pO_2 values and electrical activities. Pflügers Arch 1982;393:83–87.
14. Fujisaki T, Kudoh M, Shibuki K. Functional brain block preparation of the rat auditory cortex. Neurosci Res 1995;23:321–326.
15. Kelly JB, Whitfield IC. Effects of auditory cortical lesions on discriminations of rising and falling frequency-modulated tones. J Neurophysiol 1971;34:802–816
16. Cranford JL, Igarashi M, Stramler JH. Effect of auditory neocortex ablation on pitch perception in the cat. J Neurophysiol 1976;39:143–152.

17. Sakai M, Kudoh M, Shibuki K. A quick test for sound discrimination ability of rats in a single session after preparatory training. Neurosci Res 1995;21:273–276.
18. MacDonald JF, Mody I, Salter MW. Regulation of N-methyl-D-aspartate receptors revealed by intracellular dialysis of murine neurons in culture. J Physiol (Lond) 1989;414:17–34.
19. Shaw C, Teyler TJ. The neural circuitry of the neocortex examined in the in vitro brain slice preparation. Brain Res 1982;243:35–47.
20. Hiramatsu K, Kassell NF, Lee KS. Thermal sensitivity of hypoxic responses in neocortical brain slices. J Cereb Blood Flow Metab 1993;13:95–401.
21. Felleman DJ, VanEssen DC. Distributed hierarchical processing in the primate cerebral cortex. Cereb Cortex 1991;1:1–47.
22. Toyama K, Matsunami K, Ohno T. Antidromic identification of association, commissural and corticofugal efferent cells in cat visual cortex. Exp Brain Res 1969;11:187–198.
23. Wong D. Cellular organization of the cat's auditory cortex. In: Altschuler RA, Bobbin RP, Clopton BM, Hoffman DW (eds) Neurobiology of Hearing: the Central Auditory System. New York: Raven Press, 1991;367–387.
24. Hirsh JA, Gilbert CD. Long-term changes in synaptic strength along specific intrinsic pathways in the cat visual cortex. J Physiol (Lond) 1993;461:247–262.
25. Mei Z, Zheng J. Effects of exogenous gangliosides on learning and memory in rats. Jpn J Physiol 1993;43(Suppl 1):S295–S299.
26. Foster TC, Hampson RE, West MO, Deadwyler SA. Control of sensory activation of granule cells in the fascia dentata by extrinsic afferents: septal and entorhinal inputs. J Neurosci 1988;8: 3869–3878.
27. Gray JA. The Neuropsychology of Anxiety. New York: Oxford University Press, 1982.

Integrative and Molecular Approach to Brain Function.
M. Ito and Y. Miyashita, editors.

Roles of postsynaptic calcium and brain-derived neurotrophic factor in synaptic plasticity in visual cortex

Tadaharu Tsumoto, Hiroki Yasuda and Yukio Akaneya
Department of Neurophysiology, Biomedical Research Center, Osaka University Medical School, Suita City, Japan

Abstract. In this review we describe some of the results obtained from recent experiments on mechanisms underlying long-term potentiation (LTP) and long-term depression (LTD) in visual cortex of young rats. In particular, we put emphasis on the hypotheses that the induction of LTP in the visual cortex is of Hebbian type and that an input-associated Ca^{2+} rise at postsynaptic sites and subsequent phosphorylation or dephosphorylation of substrate proteins may play roles in the induction and/or expression of LTP or LTD in the developing visual cortex. Also, we deal with recent findings on a role of brain-derived neurotrophic factor in LTP and LTD in visual cortex.

Key words: calcineurin, Hebbian synapse, long-term depression, long-term potentiation, receptor tyrosine kinase, rhod-2.

In the hippocampus, visual cortex and other regions in the brain, the efficacy of synapses can be potentiated or depressed for a long time following a certain type of afferent inputs, phenomena called long-term potentiation (LTP) and long-term depression (LTD), respectively [1—3]. In the developing visual cortex, LTP and LTD are supposed to provide a synaptic basis for environmental modifiability of cortical neurons during the so-called critical period of postnatal development, although direct evidence is not available yet [3,4]. In a series of the experiments which will be mentioned briefly here, we addressed questions of what mechanisms underlie LTP and LTD in the developing visual cortex and what factors are responsible for switching the direction of long-term changes in synaptic efficiency into LTP or LTD.

Induction of LTP in visual cortex is Hebbian

In the hippocampus, depolarization of postsynaptic neurons during synaptic inputs is suggested to be a critical factor for the induction of LTP (reviewed in [5]). To see whether this is also the case in visual cortex, we prepared thin slices (140—150 μm thick) of visual cortex of young rats (postnatal days 9—18) for whole-cell voltage

Address for correspondence: Dr Tadaharu Tsumoto, Department of Neurophysiology, Biomedical Research Center, Osaka University Medical School, 2-2 Yamadaoka, Suita City, 565 Japan. Tel.: +81-6-879-3660. Fax: +81-6-879-3669. E-mail: ttsumoto@nphys.med.osaka-u.ac.jp clamp recordings from layer II/III neurons [6]. Nystatin, which makes the membrane of a cell-attached patch electrically permeable without "wash-out" of intracellular second messengers [7], was added to the pipette solution at the concentration of 200 μg/ml. Excitatory postsynaptic currents (EPSCs) evoked by test stimulation applied to an adjacent layer II/III neuron at 0.1 Hz were recorded for 5—10 min after an establishment of perforated patch-clamp recording. Then, test stimulation was paired with postsynaptic depolarization (30 or 300 ms duration) at 1 Hz for 30—60 s. In 11 of the 15 cells in which the membrane potential was depolarized to –20 mV, LTP of EPSCs was induced following pairing. LTP of EPSCs was observed in five of the 14 cells which were depolarized to –40 mV. No LTP was observed when the postsynaptic cells were clamped at –60, –70 or –90 mV. Thus, the induction of LTP may depend on postsynaptic depolarization more than about –40 mV.

We then addressed a further question of whether presynaptic activation concurrent with postsynaptic depolarization is necessary for the induction of LTP or an activation of postsynaptic receptors in conjunction with postsynaptic depolarization is just sufficient in visual cortex. For this, we carried out perforated patch-clamp recordings with nystatin from cultured neurons prepared from the visual cortex of neonatal rats [8]. Recorded neurons were activated monosynaptically by electrical stimulation of an adjacent neuron or directly by application of glutamate to dendrites of the recorded neurons through iontophoresis or pressure ejection from micropipettes. After electrophysiological recordings, cultured neurons were stained immunocytochemically with the antibody against synaptophysin. Dendritic "hot spots" which generated the largest response to glutamate applied through pipettes were found to correspond to darkly stained spots along the dendrites. This indicates that electrophysiologiclly identified "hot spots" probably represented postsynaptic sites. EPSCs evoked by test stimulation applied to the adjacent neuron at 0.1 Hz were recorded at a holding potential of –60 or –70 mV for 5—10 min after an establishment of the whole-cell recording configuration. Stimulation was then paired with postsynaptic depolarization (0 mV for 200 ms) at 1 Hz for 30 or 60 s. LTP of EPSCs was induced in seven of the 15 cells from which stable recordings were obtained for 18—30 min after pairing. When postsynaptic depolarization was paired with direct glutamate application to dendritic "hot spots" in the absence of presynaptic stimulation in 12 cells, only one showed LTP. Postsynaptic depolarization alone did not induce LTP in any of the six cells tested. Also, presynaptic stimulation alone did not induce LTP in any of the five cells tested. These results suggest that the concurrent activation of presynaptic elements with postsynaptic depolarization is necessary for the induction of LTP in visual cortex.

Measurement of Ca^{2+} entry during synaptic inputs by microscopic fluorometry with rhod-2

In the visual cortex and CA1 area of the hippocampus, it is suggested that processes for LTP induction are triggered by an entry of Ca^{2+} into postsynaptic sites mostly through N-methyl-D-aspartate (NMDA) receptor-linked channels (reviewed in [3,5]). Since 1992, we have attempted to confirm that the entry of Ca^{2+} into postsynaptic neurons though NMDA receptors is actually induced by tetanic stimulation of afferent fibers in visual cortex. To do that, we measured changes in Ca^{2+} signal during tetanic synaptic inputs in layer II/III of visual cortical slices of young rats (postnatal day 15—30) with microfluorometry using a Ca^{2+} indicator, rhod-2, and simultaneously recorded layer II/III field potentials evoked by white matter stimulation [9,10]. The reason why we used rhod-2 instead of the widely used indicator, fura-2, is that rhod-2 (K_D = 1.0 μM) has a much weaker Ca^{2+}-chelating action than fura-2 (K_D = 224 nM) and thus rhod-2 is expected to interfere with Ca^{2+}-dependent physiological processes within neurons to a lesser extent. An application of an antagonist for NMDA receptors, D,L-2-amino-5-phosphonovalerate (APV), did not significantly affect field potentials but reduced the tetanus-induced fluorescence rise to 56%, on average, of the control values. An application of a broad-spectrum antagonist for both NMDA and non-NMDA receptors, kynurenate, completely abolished the synaptically evoked component of field potentials and decreased the tetanus-induced fluorescent rise to 42%. An application of a non-NMDA receptor antagonist, 6-cyano-7-nitro-quinoxaline-2, 3-dione (CNQX), in the Mg^{2+}-free medium diminished the field potentials but did not significantly decrease the tetanus-induced fluorescence increase. Nifedipine and diltiazem, L-type Ca^{2+} channel blockers, and Ni^{2+} at 100 μM, a relatively selective blocker for T-type Ca^{2+} channels at this concentration, did not affect the tetanus-induced fluorescence rise. These results indicate that NMDA receptors play a significant role in the increase of intracellular Ca^{2+} during high-frequency synaptic inputs in the visual cortex of young rats.

In the above-mentioned experiments, the membrane-permeable type of rhod-2, rhod-2/AM, had been applied through the perfusion medium so that fluorescent signals were not exclusively derived from neurons. Subsequently, therefore, we attempted to measure the fluorescent signal which was derived only from a given neuron under microscopic observation with the method of direct intracellular injection of the dye [11]. Simultaneously we recorded excitatory postsynaptic potentials (EPSPs) evoked by test stimulation of layer IV with the whole-cell current-clamp mode through patch-pipettes used for dye injection. Thus, rhod-2 was injected directly into 12 pyramidal cell-like neurons in layer II/III through patch pipettes, and changes in Ca^{2+} concentration in apical dendrites of the neurons during tetanic stimulation of layer IV were measured simultaneously with recordings of EPSPs evoked by test stimulation of layer IV. It was found that LTP of EPSPs was induced in four cells which exhibited a strong rise of dendritic Ca^{2+} signal (197.1 ± 18.5%) while LTD was induced in other five cells which showed a weak rise of the signal (31.0 ± 4.1%). These results seem consistent with the so-called Ca^{2+}-switching hypothesis for the induction of LTP and LTD in visual cortex; an input-associated rise of Ca^{2+} at postsynaptic sites beyond a certain threshold leads to the induction of LTP while a lower rise below the threshold leads to LTD [12]. This hypothesis is schematically shown in Fig. 1.

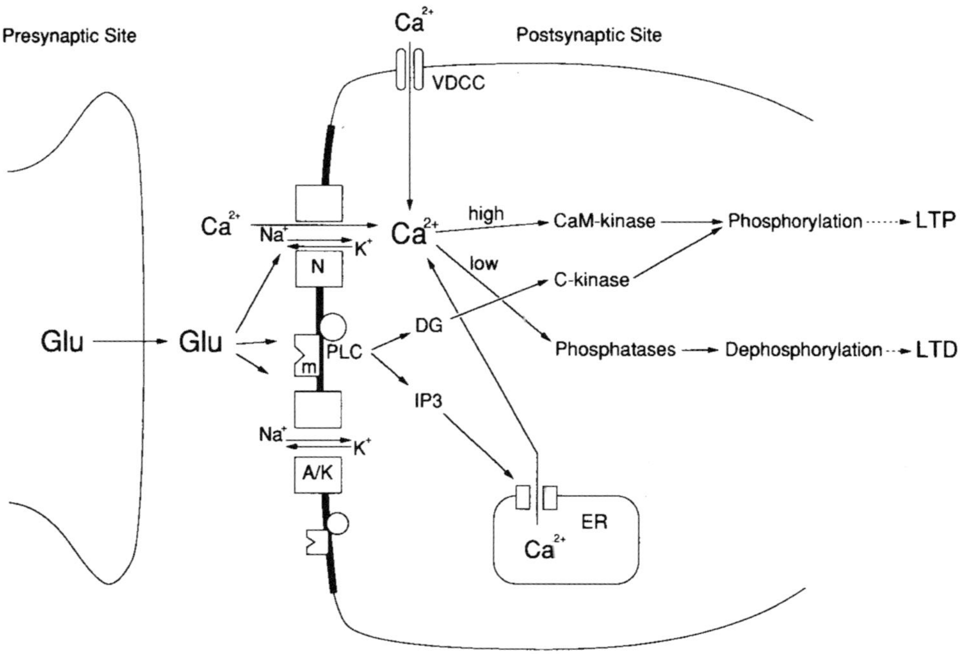

Fig. 1. Schematic depiction of possible processes inducing LTP or LTD. N, A/K and m on the postsynaptic site represent NMDA and non-NMDA types of ionotropic glutamate receptors and IP3-linked types of metabotropic glutamate receptors, respectively. The high level of Ca^{2+} rise is assumed to induce an activation of Ca^{2+}/calmodulin-dependent protein kinase II (CaM-kinase), and the lower level of Ca^{2+} rise to induce an activation of protein phosphatases. Presynaptic mechanisms which may be involved in these processes are omitted for simplicity, as are other protein kinases. Abbreviations: PLC, phospholipase C; IP3, inositol-triphosphate; ER, endoplasmic reticulum; VDCC, voltage-dependent Ca^{2+} channel. Modified from [13].

A role of phosphorylation or dephosphorylation of substrate proteins in the induction of LTP or LTD

In another series of experiments, we addressed a question of what substances or factors may play a role in the induction of LTP or LTD. Previously, we had suggested that a relatively low rise of Ca^{2+} at postsynaptic sites during tetanic synaptic inputs may activate Ca^{2+}/calmodulin-dependent protein phosphatase called calcineurin, which in turn activates processes for dephosphorylation of Ca^{2+}/calmodulin-dependent protein kinase II or other substrate proteins (reviewed in [13]). Thus, we have attempted to answer the question of whether calcineurin is involved in the induction of LTD or LTP, using a newly developed inhibitor for calcineurin called FK506. This inhibitor was applied to layer II/III neurons in visual cortical slices of young rats (postnatal day 20–40) through the perfusion medium at the concentration of 1 μM or by direct intracellular injection through intracellular sharp electrodes. We found that the inhibition of calcineurin by FK506 facilitated the induction of LTP of layer II/III field responses and of EPSPs recorded intracellularly from layer II/III neurons following tetanic stimulation of the underlying white matter [14]. We further tested whether FK506 could block the induction of LTD by a low-frequency, repetitive stimulation (1 Hz for 15 min) of layer IV in similar preparations (rats aged 20–30 postnatal days), and found that it did block the induction of LTD in most of the visual cortical slices [15]. These results suggest that calcineurin is involved in producing LTD and also it may play a role in processes antagonizing the induction of LTP in visual cortex.

An involvement of brain-derived neurotrophic factor in cortical LTP/LTD

Brain-derived neurotrophic factor (BDNF) is suggested to play a role in formation of ocular dominance columns and input-dependent modification of neuronal function in developing visual cortex [16]. Since LTP and LTD are supposed to be a synaptic basis for such experience-dependent alterations in structure and function of visual cortex as mentioned above, we have studied effects of BDNF on the induction of LTP and LTD in visual cortical slices prepared from young rats (postnatal day 15–25). Layer II/III field responses evoked by test stimulation of layer IV at 0.1 Hz were potentiated by an application of BDNF through the perfusion medium at the concentration of 200 ng/ml. This potentiation lasted for the whole of the observation period (20–30 min) after cessation of the BDNF application. At the concentration of 20 ng/ml, however, BDNF did not significantly potentiate field responses in layer II/III. On the other hand, BDNF at this concentration blocked the induction of LTD of field responses by low-frequency stimulation (1 Hz for 15 min) of layer IV [17]. These actions of BDNF were antagonized by a coapplication of K252a, an inhibitor for receptor tyrosine kinases, at the concentration of 200 nM [18]. When K252a alone was applied to slices, LTD of the stronger magnitude than in control slices was induced while the responses by test stimulation at 0.1 Hz were not significantly changed, suggesting that endogenous BDNF may prevent synapses from being depressed to some degree during low-frequency, repetitive inputs. Moreover, nerve growth factor (NGF) at the concentration of 20–200 ng/ml did not have the above-mentioned effects of BDNF. These results suggest that BDNF but not NGF may play a role in use-dependent changes in synaptic connectivity in the developing visual cortex.

Acknowledgements

This work was supported by a Grant-in-Aid (07279102) for Scientific Research on Priority Areas on "Functional Development of Neural Circuits" from the Japanese Ministry of Education, Science, Sports and Culture and in part by the Human Frontier Science Program Organization to T. Tsumoto.

References

1. Bliss TVP, Lømo T. Long-lasting potentiation of synaptic transmission in the dentate area of the

anaesthetized rabbit following stimulation of the perforant path. J Physiol (Lond) 1973;232:331–356.
2. Kuba K, Kumamoto E. Long-term potentiation in vertebrate synapses: A variety of cascades with common subprocesses. Prog Neurobiol 1990;34:197–269.
3. Tsumoto T. Long-term potentiation and long-term depression in the neocortex. Prog Neurobiol 1992; 39:209–228.
4. Singer W. Development and plasticity of cortical processing and architectures. Science 1995;270: 758–764.
5. Bliss TVP, Collingridge GL. A synaptic model of memory: long-term potentiation in the hippocampus. Nature (Lond) 1993;361:31–39.
6. Yoshimura Y, Tsumoto T. Dependence of LTP induction on postsynaptic depolarization: A perforated patch clamp study in visual cortical slices of young rats. J Neurophysiol 1994;71:1638–1645.
7. Horn R, Marty A. Muscarinic activation of ionic currents measured by a new whole-cell recording method. J Gen Physiol 1988;92:145–159.
8. Otsu Y, Kimura F, Tsumoto T. Hebbian induction of LTP in visual cortex: A perforated patch clamp study in cultured neurons. J Neurophysiol 1995;74:2437–2444.
9. Takahashi MP, Sugiyama M, Tsumoto T. Laminar difference in tetanus-induced increase of intracellular Ca^{2+} in visual cortex of young rats. Neurosci Res 1993;17:217–228.
10. Takahashi MP, Sugiyama M, Tsumoto T. Contribution of NMDA receptors to tetanus-induced increase in postsynaptic Ca^{2+} in visual cortex of young rats. Neurosci Res 1993;17:229–239.
11. Yasuda H, Tsumoto T. Long-term depression in visual cortex is associated with a lower rise of postsynaptic Ca^{2+} than LTP. Neurosci Res 1996;24:265–274.
12. Lisman J. A mechanism for the Hebb and anti-Hebb processes underlying learning and memory. Proc Natl Acad Sci USA 1989;86:9574–9578.
13. Tsumoto T. Long-term depression in cerebral cortex: A possible substrate of "forgetting" that should not be forgotten. Neurosci Res 1993;16:263–270.
14. Funauchi M, Haruta H, Tsumoto T. Effects of an inhibitor for calcium/calmodulin-dependent protein phosphatase, calcineurin, on induction of long-term potentiation in rat visual cortex. Neurosci Res 1994;19:269–278.
15. Torii N, Kamishita Y, Otsu Y, Tsumoto T. An inhibitor for calcineurin, FK506 blocks induction of long-term depression in rat visual cortex. Neurosci Lett 1995;185:1–4.
16. Thoenen H. Neurotrophins and neuronal plasticity. Science 1995;270:593–598.
17. Akaneya Y, Tsumoto T, Hatanaka H. Long-term depression blocked by brain-derived neutrophic factor in rat visual cortex. J Neurophysiol 1996;(In press).
18. Knüsel B, Hefti F. K-252 compounds: modulator of neurotrophin signal transduction. J Neurochem 1992;59:1987–1996.

Integrative and Molecular Approach to Brain Function.
M. Ito and Y. Miyashita, editors.

Multiple functional roles of brain segment-specific neuronal recognition molecule, telencephalin

Kensaku Mori[1,2], Atsushi Tamada[1,2], Hidehiko Sugino[1,2], Takeo Mizuno[2,3] and Yoshihiro Yoshihara[2,3]

[1]*Laboratory for Neuronal Recognition Molecules, Frontier Research Program, RIKEN, Hirosawa, Wako, Saitama;* [2]*Department of Neuroscience, Osaka Bioscience Institute, Furuedai, Suita, Osaka; and* [3]*Department of Biochemistry, Osaka Medical College, Takatsuki, Osaka, Japan*

Abstract. Telencephalon is the most rostral brain segment which includes the cerebral neocortex, olfactory cortex, hippocampus, striatum, amygdala, septum, and olfactory bulb. These regions take charge of higher brain functions such as memory, learning, sensory perception, and voluntary movement. We have identified a membrane glycoprotein, telencephalin (TLN), whose expression is restricted within the telencephalon. TLN is an integral membrane protein with nine tandem immunoglobulin-like domains in an extracellular region, a transmembrane domain and a short cytoplasmic tail. When coated on dish as a substrate, TLN promoted the neurite outgrowth of cultured embryonic neurons. Neuro2A cells transfected with TLN cDNA showed prominent extension of branched processes with a number of microspike-like appendages. In the adult mammalian brain, TLN is localized to the soma-dendritic membrane of subsets of telencephalic neurons, but not to the axonal membrane. Physiological experiments using hippocampal slices indicate that TLN is involved in the development of long-term potentiation at Schaffer collateral – CA1 pyramidal cell synapses.

Structure of TLN is most closely related to intercellular adhesion molecules (ICAMs)-1 and -3. ICAM-1 and -3 share a major counter-receptor, lymphocyte function-associated antigen-1 (LFA-1). Here, we demonstrate that TLN also binds to LFA-1 integrin, and will discuss the possibility that TLN on the telencephalic neurons may be a target molecule in the brain for LFA-1 expressing leukocytes and/or microglia in physiological and/or pathological conditions.

Key words: dendritic cell adhesion molecule, ICAM family. immunoglobulin superfamily, long-term potentiation, telencephalon.

Introduction

Cellular recognition among neurons and between neurons and glial cells via dendritic and axonal processes plays critical roles in the plastic changes of synaptic connections in the mature brain as well as in the formation of neuronal connections in the developing brain. During development neuronal recognition is important in the axonal elongation along specific routes and in the synapse formation with specific target neurons. In the mature brain, neuronal recognition between presynaptic axons and postsynaptic dendrites seems to be critical in the plastic modification of physiological and anatomical connections at specific synapses, which might subserve processes

Address for correspondence: Kensaku Mori, Laboratory for Neuronal Recognition Molecules, Frontier Research Program, RIKEN, 2-1 Hirosawa, Wako, Saitama 351-01, Japan.

critical in learning and memory.

Recent studies indicate that multiple cell surface adhesion molecules (CAMs) expressed on neurons mediate neuron-neuron, neuron-glia and neuron-substrate interactions and participate in the processes of axonal elongation, synapse formation, as well as plastic changes of synaptic connections [1–4]. Most of the neuronal CAMs are expressed relatively selectively either on axonal membrane or on soma-dendritic membrane. Thus, neuronal CAMs can be classified into two categories: axon-associated cell adhesion molecules (AxCAMs) [1] and dendrite-associated cell adhesion molecules (DenCAMs) [5,6], although some CAMs are known to be sorted out to both axonal and soma-dendritic membranes [7].

Here we summarize the structure, expression pattern, and function of a brain-segment-specific DenCAM, telencephalin (TLN). The following are structural characteristics of TLN: S-1 – TLN is a type I integral membrane protein and belongs to the immunoglobulin superfamily (IgSF). S-2 – Extracellular Ig-like domains of TLN are most closely related to those of intercellular adhesion molecules (ICAMs)-1 and -3.

Characteristic expression pattern of TLN includes the following: E-1 – TLN expression is restricted within the telencephalon, the most rostral brain segment. E-2 – TLN is expressed by subsets of neurons in the telencephalon, but not by glial cells. E-3 – TLN protein is localized to soma-dendritic membrane, but not to axonal membrane.

For the functional aspect of TLN, the following properties are known: F-1 – TLN has neurite-outgrowth promoting activity when used as a substrate for cultured embryonic neurons. F-2 – When neuroblastoma N2A cells are transfected with TLN cDNA, they extend long neurites with microspike-like appendages. F-3 – Application of anti-TLN antibody or recombinant TLN protein into hippocampal slices causes prominent suppression of long-term potentiation of Schaffer-collateral – CA1 pyramidal cell synapses. F-4 – TLN binds to the leukocyte integrin LFA-1 (CD11a/CD18).

Telencephalin and dendrodendritic synapses in the olfactory bulb

Because of its relatively simple cortical structure, olfactory bulb has been regarded as a good model system with which to study the structure and function of cortical neuronal circuits [8]. One of the characteristic features of the synaptic organization in the olfactory bulb is the dendrodendritic reciprocal synapses between mitral/tufted cells and granule cells. Dendrites of mitral/tufted cells make glutamatergic excitatory synapses onto the spines of granule cell dendrites. Granule cells in turn make GABAergic inhibitory synapses onto the dendrites of mitral/tufted cells.

Recent physiological study indicates a role for the dendrodendritic reciprocal synapses in enhancing odor tuning specificity of mitral/tufted cells in the main olfactory bulb [9]. The dendrodendritic synapses in both main and accessory olfactory bulbs have plastic properties. For example, an accessory olfactory system mediates pheromone recognition memory formed to male pheromones by a female mouse at mating [10]. The memory trace has been localized to the reciprocal dendrodendritic synapses between mitral/tufted cells and granule cells in the accessory olfactory bulb.

For the purpose of obtaining useful molecular markers for the dendrodendritic synapses, we produced a library of monoclonal antibodies (MAbs) against fractions rich in the dendrodendritic synaptosomes of the rabbit olfactory bulb. In the course of an immunohistochemical screening of a number of MAbs using brain sections, we noticed that one of them, MAb 271A6, labeled not only the external plexiform layer of the olfactory bulb (in which the dendrodendritic synapses distribute) but also the gray matter of all regions within the telencephalon (Fig. 1A) [11]. However, MAb 271A6 labeled no structures in the diencephalon, mecencephalon, metencephalon, and

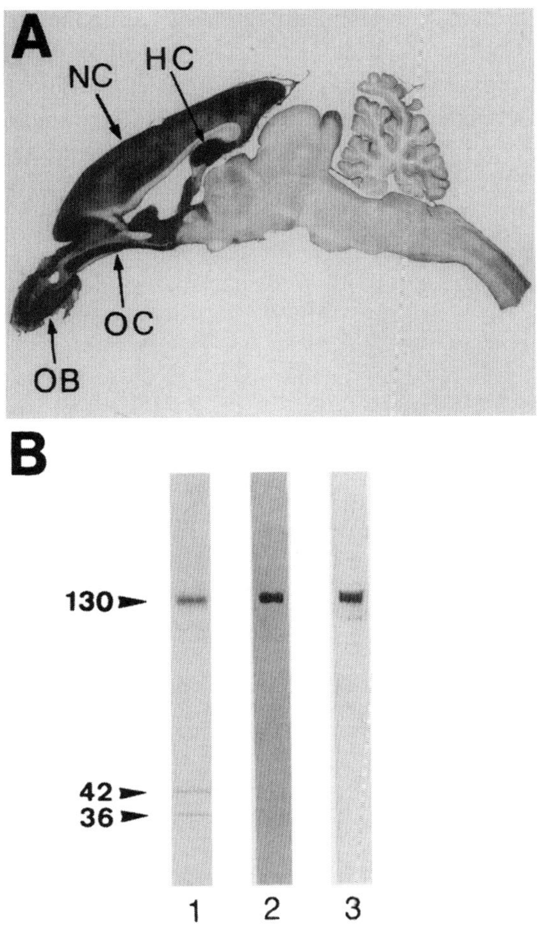

Fig. 1. Distribution (A) and purification (B) of TLN protein. **A**: Immunohistochemical staining with MAb 271A6 of a parasagittal section of a rabbit brain. Immunoreactive region (shown black) are confined within the telencephalon, which includes the olfactory bulb (OB), olfactory cortex (OC), cerebral neocortex (NC), and hippocampus (HC). **B**: Purified TLN protein (130 kD) was assessed by Coomassie blue (lane 1) and Western blotting with MAb 271A6 (lane 2) and MAb HNK-1 (lane 3). Molecular masses are shown in kilodaltons on the left. Two minor protein bands of 42 and 36 kD co-purified with TLC were identified as actin and glyceraldehyde 3-phosphate dehydrogenase, respectively. Modified from [5].

myelencephalon, the brain segments more caudal to the telencephalon. Because of the telencephalon-specific expression, the 271A6 antigen molecule was named telencephalin (TLN) [12].

Structure of telencephalin

Using immunoaffinity chromatography with a MAb 271A6, TLN protein was isolated from detergent-extracted rabbit telencephalon. Biochemical characterization of TLN protein revealed that it is a membrane glycoprotein with a molecular weight of about 130 kD containing an HNK-1 carbohydrate epitope (Fig. 1B) [12].

cDNA cloning of rabbit, mouse and human TLN revealed that TLN is a type I integral membrane protein with a signal peptide (29 amino acids in rabbits), a large extracellular region (792 amino acids), a transmembrane region (30 amino acids), and a short cytoplasmic region (61 amino acids) (Fig. 2A) [5]. The extracellular portion of TLN contains nine tandem repeats of C2-type (Ig)-like domains with intradomain disulfide bonds (Fig. 2A). The above structural features are conserved among mouse, rabbit and human TLN. Overall amino acid identity is 85% between human and mouse TLNs as well as between human and rabbit TLNs (Fig. 2B) (Mizuno et al.,

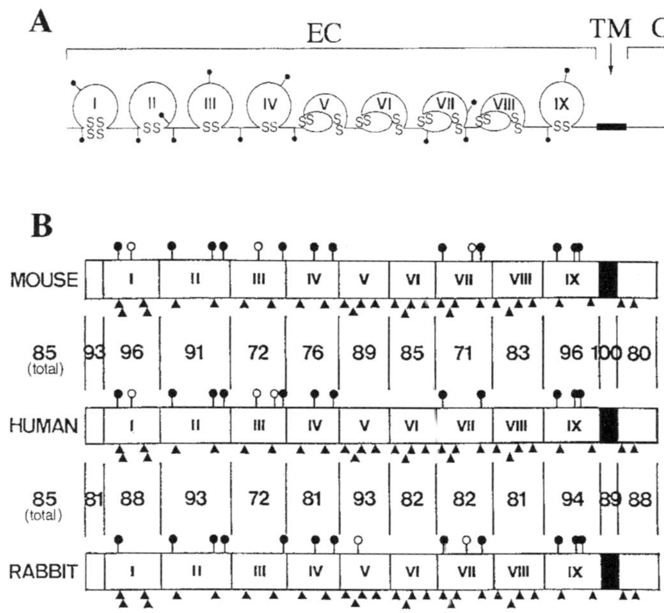

Fig. 2. Schematic representation of TLN. **A**: Proposed model of the domain structure of TLN. Individual immunoglobulin-like domains are numbered I–IX. EC: extracellular region; TM: transmembrane domain; CP: cytoplasmic region. **B**: Amino acid sequence alignment between mouse and human TLN, and between human and rabbit TLN. Roman numerals indicate immunoglobulin-like domains. The sequence conservation (percent amino acid identity) between corresponding region is shown with Arabic numerals. Putative Asn-linked glycosylation sites are indicated by lollipop symbols. Filled triangles denote conserved cysteine residues.

submitted). The tandem repeats of Ig-like domains in the extracellular region suggest that TLN may be a CAM expressed in the telencephalon.

Spatio-temporal pattern of expression of telencephalin

Telencephalon-specific expression of TLN

During prenatal development, the vertebrate brain forms five segmental enlargements (vesicles) — telencephalon, diencephalon, mesencephalon (midbrain), metencephalon (pons and cerebellum), and myelencephalon (medulla oblongata). Telencephalon is the most rostral brain segment which includes the cerebral neocortex, olfactory cortex, hippocampus, striatum, amygdala, septum, and olfactory bulb. These regions take charge of higher brain functions such as memory, learning, sensory perception, and voluntary movements.

Immunohistochemical studies using anti-TLN antibodies as well as in situ hybridization studies revealed that TLN is expressed in all regions within the telencephalon (Fig. 1A) [11,12]. Telencephalon-specific expression of TLN was observed in all the mammalian species so far examined (mouse, rat, rabbit, cat, monkey, and human).

Neuronal subset-specific expression of TLN

Within the telencephalon, the expression of TLN is restricted to neurons. Glial cells do not express TLN. In each region of the telencephalon, TLN is expressed by specific subtypes of neurons. In the rabbit olfactory bulb, for example, granule cells express TLN, whereas mitral/tufted cells do not [11]. In the visual cortex of cat and monkey brain, TLN is absent in neurons within the layer IV, while it is present in neurons of other cortical layers [14]. In the hippocampal formation, TLN is expressed by granule cells in the dentate gyrus and pyramidal cells in the CA1 and CA3 regions.

Soma-dendrite-specific expression of TLN

TLN protein seems to be sorted out selectively to the soma-dendritic membrane. Light microscopic observation of immunohistochemically stained sections showed that surface membranes of somata and dendritic processes of the telencephalic neurons are strongly positive for TLN, while TLN immunoreactivity is absent or very faint on the axonal processes [11,12]. Immuno-electron-microscopic observations in the main olfactory bulb also revealed that all the TLN-positive processes are dendrites of granule cells, and no axonal process in the bulb was observed to be TLN-immunoreactive [13].

Temporal pattern of TLN-expression

The expression of TLN is developmentally regulated. In rabbits and mice, TLN is almost absent in embryonic brain, appears just before the birth in phylogenetically old telencephalic regions such as olfactory bulb and olfactory cortex. TLN increases dramatically in a few postnatal weeks in all the telencephalic regions including cerebral neocortex, and continues to be expressed at a high level in the adult brain. Thus, the developmental appearance of TLN parallels the time of dendritic elongation and branching, and synapse formation in each region of the telencephalon [11].

Neurite outgrowth-promoting activity of TLN

Because a number of CAM of IgSF (e.g., NCAM, L1, TAG-1, BIG-1, BIG-2, and F11) show neurite outgrowth-promoting activity when used as a substrate for cultured neurons, we examined whether TLN also has neurite outgrowth-promoting activity. When a recombinant chimeric protein (TLN/Fc) composed of extracellular region of mouse TLN and Fc region of human IgG was used as a substrate for cultured embryonic neurons, it strongly enhanced the neurite outgrowth (Tamada et al., in preparation). The neurite outgrowth-promoting activity of TLN/Fc was observed not only on telencephalic neurons (e.g., cultured embryonic neurons obtained from the hippocampus and cerebral neocortex), but also on nontelencephalic neurons (e.g., cerebellar neurons). An antibody to TLN specifically blocked the neurite outgrowth-promoting activity of TLN/Fc indicating that extracellular region of TLN is responsible for the neurite outgrowth-promoting activity. Above results suggest that TLN as a substrate interacts with as yet-unknown counterreceptor(s) expressed on the cultured neurons and enhances outgrowth of their neurites.

We then checked whether expression of TLN in neurons that do not constitutively express TLN causes morphological changes of their neurites. When mouse neuroblastoma N2A cells were transfected with TLN cDNA, they extended long and branched processes that resemble dendrites (Yoshihara et al., in preparation). The long processes of TLN-transfected N2A cells bore a large number of microspike-like appendages. In contrast, transfection of N2A cells with ICAM-1 cDNA caused no morphological change. These results together with the observation that the developmental stage of TLN appearance corresponds with the time of dendritic development suggest that TLN may be involved in the formation or maturation of dendritic processes.

Telencephalin and long-term potentiation of synaptic transmission in hippocampus

Long-term potentiation (LTP) of synaptic transmission in the hippocampus is a putative model for studying the mechanism of plastic changes in synaptic connections which might underlie learning and memory [15]. Recent studies indicated a role for cell adhesion molecules in the development and/or maintenance of the LTP. For example, antibodies against IgSF CAMs, NCAM and L1, were found to block LTP in Schaffer collateral-CA1 synapses in the hippocampal slices [16]. In addition, antagonists against integrin binding sites were reported to block LTP.

Because TLN is abundantly expressed by soma-dendritic membrane of the principal neurons in the hippocampus (dentate granule cells and CA1-CA3 pyramidal cells), we examined the possibility that TLN might be involved in the development or maintenance of LTP in hippocampus. Application of anti-TLN antibody to the hippocampal slices strongly suppressed the Schaffer-collateral induced LTP in the CA1 region. Application of a recombinant chimeric protein TLN/Fc also is effective in suppressing the LTP (Sakurai et al., in preparation). These results indicate that interaction between TLN and its counterreceptor at the extracellular space of the Schaffer-collateral CA1 synapses is important for the development of the LTP. Because TLN is expressed by the dendritic membrane of CA1 pyramidal cells, interaction of TLN on postsynaptic dendrites with its yet-unknown counterreceptor on the axonal terminals might be involved in the development of the LTP (Fig. 3). If this is the case, the connection between TLN and its counterreceptor might form one of the signaling pathways for retrograde messages from postsynaptic dendrites to presynaptic axonal terminals. Alternatively, TLN on dendritic membrane might interact with its counterreceptor on glial membrane that surrounds the Schaffer collateral-CA1 synapses and contribute indirectly to the morphological change of the synaptic connection.

Telencephalin binds to a leukocyte integrin LFA-1

Amino acid sequence of rabbit, mouse and human TLN revealed that the distal eight

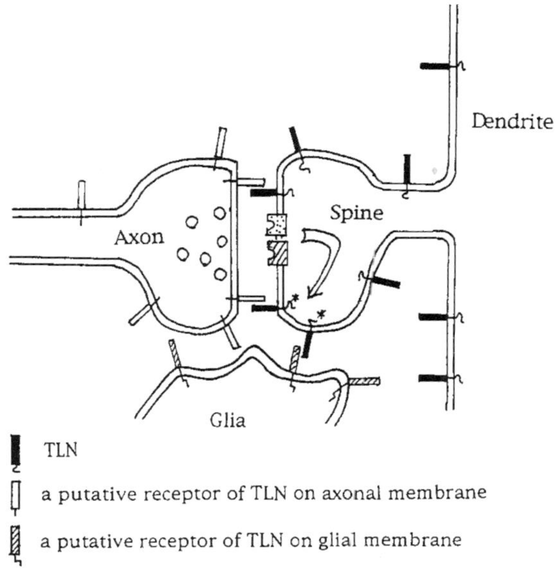

Fig. 3. Schematic diagram illustrating possible interactions during the development of LTP between TLN on postsynaptic dendrites and its counterreceptor(s) on axonal terminal and/or glial processes. An arrow in the spine indicates a possible signal to the cytoplasmic portion of TLN for retrograde messages from postsynaptic dendrite to presynaptic axonal terminal or glia.

Ig-like domains of TLN is most closely related to the Ig-like domains of intercellular adhesion molecules (ICAMs)-1 and -3, and relatively closely to ICAMs-2 and -4 (LW antigen). In particular, the domains II, III, and IV of human TLN share 64, 66, and 67% amino acid identity, respectively, with the corresponding domains of human ICAM-3. Furthermore, the N-terminal Ig-like domain of TLN contains four cysteine residues that are capable of forming two intradomain disulfide bridges, a feature conserved characteristically among the four ICAM members. Thus, TLN is a novel member of the ICAM subgroup (Fig. 4).

In the mouse chromosome, TLN gene was mapped to the proximal part of chromosome 9, close to the locus of ICAM-1 gene (Sugino et al., in preparation). In addition, human TLN gene locus was mapped to 19p13.2, which is close to the loci of ICAMs-1, -3 and -4 (Mizuno et al., submitted). These results suggest that TLN and ICAMs-1, -3, and -4 may be generated by gene duplication from a common ancestral gene. Thus, the structural similarity of TLN to ICAMs provides a new and strong link between IgSF superfamily molecules in the nervous and immune systems.

ICAMs-1, -2, and -3 are the cell surface adhesion molecules that are important in promoting a wide variety of cell-cell interactions in the immune system, including antigen-specific T lymphocyte responses, leukocytes binding to endothelium and emigration into inflammatory sites [17]. The three ICAMs differ from each other in structure, distribution, and inducibility. However, all the three share a major counter-receptor, lymphocyte function-associated antigen-1 (LFA-1, CD11a/CD18, $\alpha_L\beta_2$),

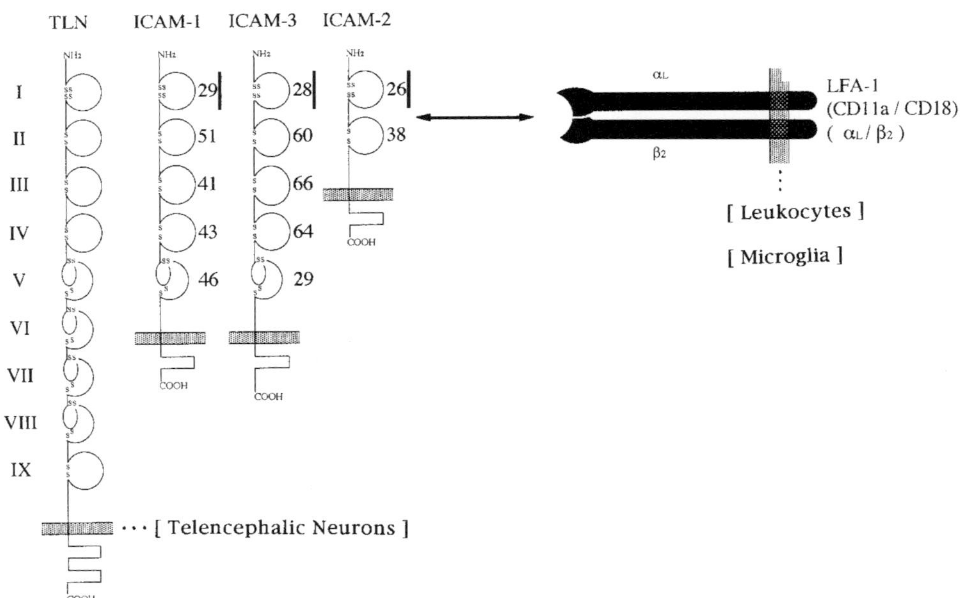

Fig. 4. Schematic diagram illustrating the binding of LFA-1 integrin to ICAM family members including TLN. The number on the right side of individual Ig domains of ICAMs indicates the amino acid identity (%) with corresponding domains of mouse TLN. TLN on the telencephalic neurons binds to LFA-1 expressed by leukocytes and microglia.

which is expressed on leukocytes and belongs to the integrin family (Fig. 4).

This raised a possibility that TLN also binds to LFA-1. To check the possibility we examined whether TLN binds HL-60 cells (human leukemia cell line) that express LFA-1. HL-60 cells strongly bound to the plastic plate coated with TLN/Fc chimeric protein. The binding was inhibited by the application of anti-LFA-1 monoclonal antibodies, indicating that TLN binds to LFA-1 (Mizuno et al., submitted).

LFA-1 is an integrin expressed selectively by leukocytes. In the brain, microglia (also called brain macrophage) weakly but constitutively express LFA-1. When microglia are activated (by, for example, brain trauma or neurological diseases), they change their shape, migrate to the damaged neurons, show phagocytic activity and up-regulate LFA-1 [18]. Thus, binding between TLN and LFA-1 may mediate interaction between telencephalic neurons and microglia in physiological and/or pathological conditions. Since LFA-1 is abundantly expressed by leukocytes such as neutrophils and T lymphocytes, TLN/LFA-1 interactions might mediate neuron/leukocyte interactions in pathological conditions in which the blood brain barrier is destroyed. The finding of binding between TLN and LFA-1 may provides a basis for detailed understanding of molecular mechanisms underlying cell-cell interactions between central nervous system and immune system.

Acknowledgements

This work was supported by grants from the Uehara Memorial Foundation, by the Special Coordination Funds for Promoting Science and Technology from the Science and Technology Agency of Japan, and by Grants-in-Aid from the Ministry of Education, Science, and Culture of Japan.

References

1. Rutishauser U, Jessell TM. Cell adhesion molecules in vertebrate neural development. Physiol Rev 1988;68:819–857.
2. Brümmendorf T, Rathjen F. Protein Profile: Cell Adhesion Molecules 1: Immunoglobulin Superfamily. London: Academic Press, 1994.
3. Doherty P, Fazeli MS, Walsh FS. The neural cell adhesion molecule and synaptic plasticity. J Neurobiol 1995;26:437–446.
4. Yoshihara Y, Oka S, Ikeda J, Mori K. Immunoglobulin superfamily molecules in the nervous system. Neurosci Res 1991;10:83–105.
5. Yoshihara Y, Oka S, Nemoto Y, Watanabe Y, Nagata S, Kagamiyama H, Mori K. An ICAM-related neuronal glycoprotein, telencephalin, with brain segment-specific expression. Neuron 1994;12: 541–553.
6. Yoshihara Y, Mori K. Telencephalin: a neuronal area code molecule? Neurosci Res 1994;21: 119–124.
7. Persohn E, Pollerberg GE, Schachner M. Immunoelectron-microscopic localization of the 180 kD component of the neural cell adhision molecule N-CAM in postsynaptic membranes. J Comp Neurol 1989;288:92–100.
8. Shepherd GM, Greer CA. Olfactory bulb. In: Shepherd GM (ed) The Synaptic Organization of the Brain. New York: Oxford University Press, 1990;139–169.
9. Yokoi M, Mori K, Nakanishi S. Refinement of odor molecule tuning by dendrodendritic synaptic

inhibition in the olfactory bulb. Proc Natl Acad Sci USA 1995;92:3371–3375.

10. Kaba H, Nakanishi S. Synaptic mechanisms of olfactory recognition memory. Rev Neurosci 1995; 6:125–141.
11. Mori K, Fujita SC, Watanabe Y, Obata K, Hayaishi O. Telencephalon-specific antigen identified by monoclonal antibody. Proc Natl Acad Sci USA 1987;84:3921–3925.
12. Oka S, Mori K, Watanabe Y. Mammalian telencephalic neurons express a segment-specific membrane glycoprotein, telencephalin. J Neurosci 1990;35:93–103.
13. Murakami F, Tada Y, Mori K, Oka S, Katsumaru H. Ultrastructural localization of telencephalin, a telencephalon-specific membrane glycoprotein, in rabbit olfactory bulb. Neurosci Res 1991;11: 141–145.
14. Imamura K, Mori K, Oka S, Watanabe Y. Variations by layers and developmental changes in expression of telencephalin in the visual cortex of cat. Neurosci Lett 1990;119:118–121.
15. Bliss TVP, Collingerridge GL. A synaptic model of memory: long-term potentiation in the hippocampus. Nature 1993;361:31–39.
16. Luthi A, Laurent J-P, Figurov A, Muller D, Schachner M. Hippocampal long-term potentiation and neural cell adhesion molecules L1 and NCAM. Nature 1994;372:777–779.
17. Springer TA. Traffic signal for lymphocyte recirculation and leukocyte emigration: the multistep paradigm. Cell 1994;76:301–314.
18. Streit WJ, Kincaid-Colton CA. The brain's immune system. Sci Am 1995;273:54–61.

Integrative and Molecular Approach to Brain Function.
M. Ito and Y. Miyashita, editors.

Identification of reelin: The gene responsible for the mouse neurodevelopmental mutation reeler

Gabriella D'Arcangelo[1], Shu-Cheng Chen[2], James I. Morgan[1] and Tom Curran[1]
[1]St. Jude Children's Research Hospital, Memphis, Tennessee; and [2]Hoffmann La Roche, Nutley, New Jersey, USA

Abstract. *Background.* The reeler strain of mice is ataxic as a consequence of a severe developmental mutation. Essentially all laminar structures in the brain are disrupted because of a failure in neuronal migration.

Methods. Positional cloning methods were used to isolate the gene responsible for the reeler defect. cDNA cloning and nucleotide sequence analysis was carried out to elucidate its predicted structure. In situ hybridization was used to identify sites of expression.

Results. We cloned a gene termed reelin that is responsible for the reeler mutation. Its predicted protein product has structural features in common with extracellular matrix proteins. It is expressed in Cajal Retzius cells in the developing cerebral cortex.

Conclusions. The reeler defect arises from the loss of an extracellular protein that is expressed early in neurodevelopment and is required for cell migration. These findings provide the first molecular foothold on the mechanisms responsible for laminar patterning of the mammalian brain.

Key words: neurodevelopment, reelin, cell migration.

Introduction

The development of a multicellular organism requires a complex series of cellular interactions that often involve dramatic reorganization and cell migration during specific time periods. A common theme in development is the formation and specialization of laminar structures. This is particularly evident in the mammalian brain which comprises over 100 billion cells aligned precisely in several multilayered structures. Defective or inappropriate cell migration can contribute to developmental abnormalities such as Miller-Dieker lissencephaly [1], to cognitive impairment and to several disease states. For example, cerebral cortical dysplasia are often associated with pediatric epilepsy [2] and cortical abnormalities have also been linked to schizophrenia [3]. Thus, an understanding of the molecular mechanisms that govern cell migration would have broad implications for many devastating human diseases.

The study of naturally occurring mutations in the mouse has made several significant contributions to our understanding of normal mammalian development. In addition to providing insights into complex biological processes, these mutations also

Address for correspondence: Tom Curran, Department of Developmental Neurobiology, St. Jude Children's Research Hospital, 332 North Lauderdale, Memphis, TN 38105, USA. Tel.: +1-901-495-2255. Fax: +1-901-495-2270. E-mail: fos1@aol.com provide an avenue for the isolation and characterization of genes that control development. In particular, neurological mutant mice represent a unique opportunity for elucidation of the events responsible for the formation and function of the mammalian central nervous system [4]. Among the known neurological mutations reeler is notable for the complexity of its associated phenotype and its potential for unraveling the mechanisms involved in neural cell patterning in the developing brain [5].

The reeler mutation was first described as an autosomal recessive trait resulting in impaired motor coordination, tremors and ataxia [6]. This behavioral phenotype is apparent in the mouse 12 days after birth, and it is associated with a widespread disorganization of the cerebellar and cerebral cortices, as well as disruption of other laminar structures in the brain. Anatomical and physiological studies of reeler led to the conclusion that the mutation affects the ability of postmitotic neurons to assume correct positions in the developing brain. In the cerebral cortex, for example, young neurons fail to migrate outward, past layers of pre-existing cells. However, despite malpositioning, various classes of neurons are able to differentiate and, to a large extent, establish the appropriate synapses on their targets. There are also more subtle defects in reeler that may be a consequence of the initial dysplasia, such as a multiplicity of innervation, a reduction of specific cell populations and altered synaptogenesis. Thus, proper laminar segregation may be a requirement for optimal neuronal connectivity and brain function. Given the crucial role played by the reeler locus in regulating the migration of many neuronal cell types, there has been a great deal of interest in isolating the gene responsible for this defect.

The first anatomical defects are detected in the reeler embryo at an early stage of corticogenesis [7]. At this stage, the cortical plate in normal mice thickens between two separating layers of the preplate: the marginal zone which contains Cajal Retzius cells and the subplate. In reeler mice, the first population of postmitotic cells reaches the external level of the cortical anlage in an apparently normal manner, although the Cajal Retzius cells appear to end up in a more superficial position in reeler. Subsequently, following the appearance of the cortical plate, the cortical primordium in reeler does not divide into two distinctive populations in the same way as the normal cortex. Instead, the cells are displaced externally as later generated neurons fail to migrate past their predecessors. This results in an apparent inverted orientation of cortical layers in the reeler mouse. It has also been suggested that a late postnatal cell death of displaced neurons located in the deep cortex may also contribute to this phenotype [8].

The cerebellum of reeler mice is markedly reduced in size and exhibits very poor foliation. The first signs of a defect are apparent after E14 when the Purkinje cells are not arranged in a regular plate but are dispersed throughout the periphery of the cortex and the intermediate zone. The marginal zone is not clearly defined and the fiber plexus of the sub-Purkinje plate is not visible [5]. Aberrant laminar organization becomes increasingly apparent in the cerebellum during the first postnatal week, at the time of migration of granule neurons. Furthermore, Purkinje cells fail to arrange into a tangential layer, several settle into ectopic positions and they display an aberrant dendritic morphology.

Several other regions of the reeler brain exhibit dramatic alterations in structure. The hippocampus is clearly affected, the dentate gyrus is not compact and the pyramidal layer is disrupted. The inferior olive exhibits a subtle defect in cellular arrangement at E14/E15 and the process of olivary foliation is defective [5]. Consistent structural abnormalities have also been reported in the reeler olfactory bulb [9]. It is reduced in size and instead of the usual single layer of cells separating the glomeruli and the external plexiform layer, there is a zone four to five cells wide. Thus, it has been suggested that essentially all laminated brain structures are affected by the reeler mutation.

The original reeler strain was described in 1951 by Falconer and has been designated the "Edinburgh" strain (rl^{ed}) because of its site of origin [6]. This strain has also been referred to as the Jackson strain since it is available from Jackson laboratories. A second allele was reported in 1981 [10] and designated the Orleans strain (rl^{ol}). A third allele (rl^{alb}) was generated in the course of a study of chlorambucil-induced mutations [11]. We isolated a fourth allele of reeler during the course of a transgenic mouse study [12]. We were in the process of creating mouse strains carrying mutated c-fos genes and we noticed that one of the lines, when bred to homozygosity, was ataxic. Neuroanatomical analyses revealed striking similarities with reeler mice and crossbreeding studies demonstrated that our mutation was allelic to reeler. Therefore, we designated our strain reeler-transgene (rl^{tg}). Thus, the reeler mutation has arisen on several independent occasions in laboratory mice. The anatomical defects in reeler mice have been the subject of numerous studies over the past 25 years and a number of models have been proposed to explain the mechanisms responsible for the defects observed in laminar organization. For example, it was suggested that during corticogenesis the earliest generated neurons remained tightly associated with radial glial fibers in reeler mice, thereby blocking the subsequent migration of later generated neurons. However, it was clear that progress in the field awaited the isolation and characterization of the gene responsible for the reeler defect. Chromosomal mapping studies had assigned reeler to a region of chromosome 5 between the engrailed-2 gene (En-2) and the P-glycoprotein (Pgy-1) locus. However, the estimated distance to the nearest marker was at least 7 cm in 1991 [13]. The rl^{tg} strain provided a possible molecular marker for the reeler gene based on the transgene DNA inserted into the locus [12].

Results and Discussion

We decided to take advantage of the opportunity to identify the reeler gene by the fortuitous insertion of a transgene into the locus [12]. However, given the difficulties experienced by others in using insertional mutagenesis as a cloning strategy we established several strict criteria at each phase of the project. First we ensured that the anatomy of our reeler mutation was identical to that of rl^{ed} and that the mutations were allelic. We cloned the junction sequences flanking the inserted transgene and mapped one to chromosome 5 and the other to chromosome 3 by interspecific backcross analysis. This meant that a chromosomal translocation had occurred at the transgene insertion site. Although this complicated the issue, we were encouraged that the 5′ flanking DNA sequences mapped to the same region as rl on mouse chromosome 5. Therefore, we initiated an attempt to isolate the reeler gene using the 5′ probe associated with rl^{tg}.

Cloning of the reeler gene

We isolated several P1 clones containing large inserts of mouse genomic DNA using the 5′ rl^{tg} probe and we identified several potential exons in this region by the exon trapping procedure [14]. Using these exons we demonstrated the presence of a large (>12 kb), predominantly brain-specific, mRNA in normal mice that was missing in both rl^{ed} and rl^{tg} (Fig. 1). We named this gene reelin. Nucleotide sequence analysis revealed a long open reading frame of 10,383 bases that is capable of encoding a protein of 3,461 amino acids. Kyte-Doolittle analysis of hydrophobicity suggested the presence of an N-terminal signal peptide containing a consensus cleavage site and no

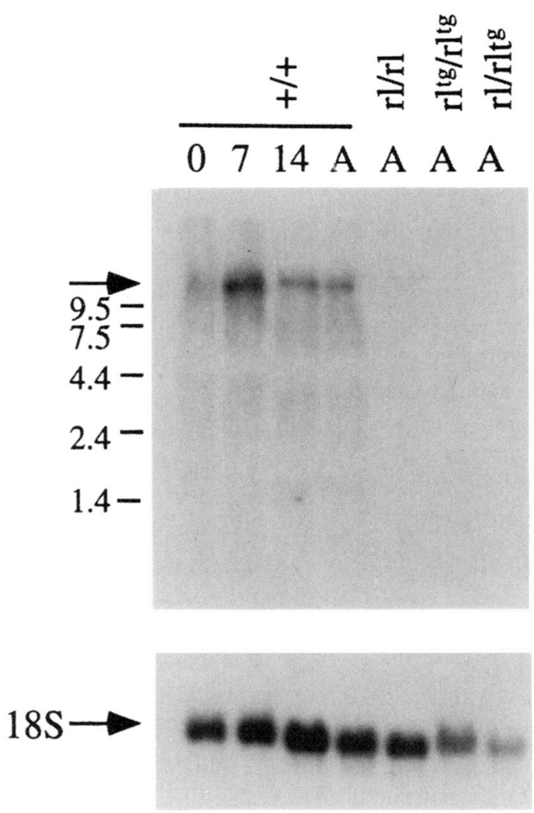

Fig. 1. Identification of a 12 kb mRNA missing in reeler. Total brain RNA was purified from normal mice at P0, P7 and P14 and from adult normal and reeler mice. The RNA was analyzed by electrophoresis and hybridized with a probe cloned from the reeler locus. Reprinted with permission from [14].

other hydrophobic stretches, implying that reelin is a secreted protein. The most striking feature of the reelin sequence is the presence of a series of eight consecutive repeats comprising 350–390 amino acids (Fig. 2). Each repeat consists of two related subdomains separated by a stretch of 30 amino acids harboring an EGF-like motif. The C-terminus of reelin is immediately preceded by a short stretch of amino acids that has a high basic charge (seven out of 15 amino acids are arginine).

Reelin has features of an extracellular protein

DNA sequence comparison with the GenBank database revealed no significant regions of homology to any known gene, with the exception of small (~50 bp) regions encoding EGF-like repeats. Amino acid sequence comparison revealed a 25% identity between the N-terminus of reelin and that of F-spondin. F-spondin is a protein secreted by floor plate neurons that is thought to be involved in cell adhesion [15]. The reelin repeats do not match any known sequence in the SwissProt, PIR or GenBank databases.

The EGF-like domains in reelin are very similar to each other and they are most related to those of the extracellular matrix proteins, Tenascin C and X, and the transmembrane receptor subunits of the integrin β chain family. This suggests that reelin is a novel gene encoding a modular protein that, in overall structural organization and amino acid composition, resembles known extracellular matrix molecules involved in neuronal cell adhesion.

Reelin is expressed during development

The primary defect in the reeler cerebral cortex occurs during embryonic life, at the time of cortical plate formation. Therefore, we analyzed expression of reelin during this critical period. In situ hybridization analysis revealed that reelin is predominantly expressed in marginal zone cells at E13.5 (Fig. 3). The location and morphology of the positive cells suggested that they were Cajal-Retzius cells. Cajal Retzius cells are the first population of neurons to appear in the developing cortical plate. Interestingly, no reelin expression was detected in radial glia or in migrating cortical neurons. This

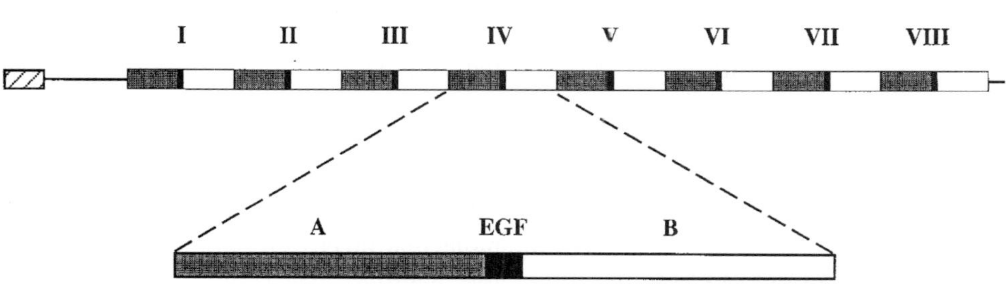

Fig. 2. Diagrammatic representation of the structure of reelin The extent of the reelin open reading frame is indicated schematically. The region of sequence similarity with F-Spondin is illustrated by a hatched box. The eight reelin repeats comprise related A and B repeated sequences flanking an EGF-like repeat. Reprinted with permission from [14].

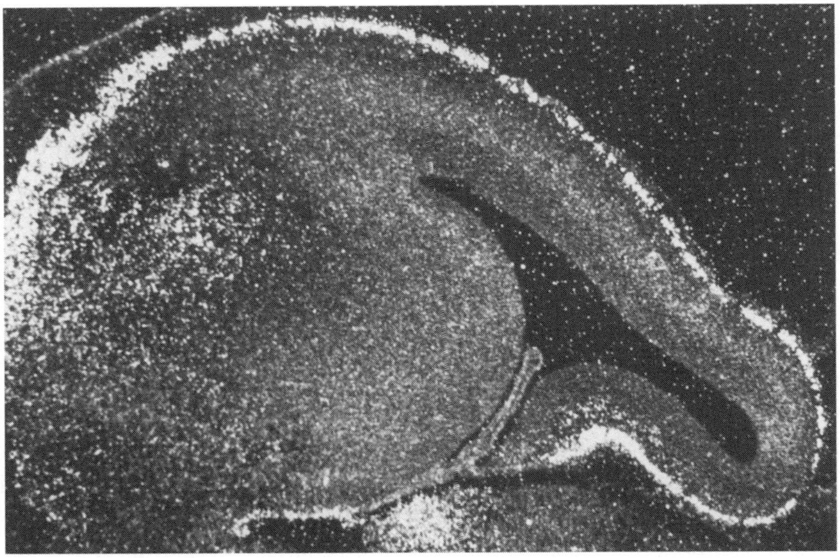

Fig. 3. Reelin expression in the developing brain. In situ hybridization analysis of reelin expression in a parasagittal section of a mouse brain at E13.5. The hybridization of a reelin antisense probe, indicated by bright silver grains, can be detected in the marginal zone of the cortex, the developing olfactory bulb, the hippocampus and the striatum. Reprinted with permission from [14].

indicates that the primary defect in reeler does not lie in the association of migrating neurons with radial glia fibers.

In addition to the cortex, reelin is also expressed in several other structures known to be affected by the reeler mutation, particularly in the hippocampus and the cerebellum. In the embryonic hippocampus, reelin expression occurs at the time of hippocampal invagination, suggesting that the postnatal anatomical defects reported in this structure may be a consequence of perturbations occurring during the earliest phase of development. A characteristic deficit in the reeler mouse, that is thought to underlie its behavioral phenotype, is the malpositioning of Purkinje cells and the marked loss of cerebellar granule cells. During prenatal development, reelin expression was observed in the rhombic lip, a region that contains precursor cells destined to form the olivary nucleus and the granule cell layer of the cerebellum. Thus, reelin expression in this instance is associated with migrating neurons and neuroblasts. Later during cerebellar development (P10), reelin is expressed in the deepest region of the external granular layer (egl) and throughout the internal granule layer (igl). In the egl reelin mRNA is associated with postmitotic, premigratory granule neurons that are about to cross the molecular layer to form the igl. However, in the igl reelin is also present in postmigratory neurons. These data demonstrate that reelin may be expressed in both migratory and postmigratory neurons. Indeed, reelin expression in other brain regions persists long after neuronal migration is complete. This implies that reelin may provide a function in the adult brain, perhaps in stabilizing the cytoarchitectonic organization. The distribution of reelin RNA is most consistent with its expression in neurons and neuronal precursors and it does not appear to be expressed in radial or Bergmann glia.

Nature of the reeler mutation

In addition to the insertion of a transgene and a region of chromosome 3, the rl^{tg} locus has a partial deletion. The deletion in the rl^{tg} allele is intragenic which provides strong circumstantial evidence supporting the identity of reelin as the reeler gene. There is also a reelin deletion in the original Edinburgh strain, rl^{ed}, which results in a loss of approximately 8 kb of 3′ exon sequences. In both cases, although the 5′ region of the gene is intact, no reelin transcripts were detected. Therefore, the genetic lesion in these strains represents a loss of function mutation. The Orleans strain of reeler (rl^{or}), has a small out-of-frame deletion that results in a predicted reelin protein product lacking 200 C-terminal amino acids [16]. The original paper describing this mutation indicated that the normal reeler gene product was a much smaller protein because of sequencing errors [17].

Future directions

The identification of the protein responsible for reeler provides an opportunity to investigate the molecular basis of neuronal migration during development. It will now be possible to express reelin protein in vitro and in vivo to assess its effect on neuronal migration and adhesion. A first step will be to determine whether reelin is a secreted protein. One of the major questions to be addressed is to identify the proteins with which reelin interacts. These proteins could be potential receptors or cofactors. Hopefully, such studies will lead to the elucidation of a signaling pathway that controls some of the critical cellular interactions required for the formation of the mammalian brain.

Acknowledgements

This work was supported in part by NIH Cancer Center Support (CORE) grant P30 CA21765, an NRSA from NINDS to G. D'Arcangelo and the American Lebanese Syrian Associated Charities (ALSAC).

References

1. Reiner O, Carrozzo R, Shen Y et al. Isolation of a Miller-Dieker lissencephaly gene containing G protein β-subunit-like repeats. Nature 1993;364:717–721.
2. Mischel PS, Nguyen LP, Vinters HV. Cerebral cortical dysplasia associated with pediatric epilepsy. Review of neuropathologic features and proposal for a grading system. J Neuropathol Exp Neurol 1995;54:137–153.
3. Gould JM, Weinberger DR. Cognitive deficits and the neurobiology of schizophrenia. Curr Opin

Neurobiol 1995;5:225–230.
4. Caviness VSJ, Rakic P. Mechanisms of cortical development: a view from mutations in mice. Ann Rev Neurosci 1978;1:297–326.
5. Goffinet AM. Events governing organization of postmigratory neurons: studies on brain development in normal and reeler mice. Brain Res 1984;319:261–296.
6. Falconer DS. Two new mutants "trembler" and "reeler", with neurological actions in the house mouse. J Genet 1951;50:192–201.
7. Goffinet AM. An early developmental defect in the cerebral cortex of the reeler mouse. Anat Embryol 1979;157:205–216.
8. Hoffarth RM, Johnston JG, Krushel LA, van der Kooy D. The mouse mutation Reeler causes increased adhesion within a subpopulation of early postmitotic cortical neurons. J Neurosci 1995;15: 4838–4850.
9. Wyss JM, Stanfield BB, Cowan WM. Structural abnormalities in the olfactory bulb of the Reeler mouse. Brain Res 1980;188:566–571.
10. Guenet J-L. A new allele of reeler. Mouse News Lett. 1981;41.
11. Flaherty L, Messer A, Russell LB, Rinchik EM. Chlorambucil-induced mutations in mice recovered in homozygotes. Proc Natl Acad Sci USA 1992;89:2859–2863.
12. Miao GG, Smeyne RJ, D'Arcangelo G et al. Isolation of an allele of reeler by insertional mutagenesis. Proc Natl Acad Sci USA 1994;91:11050–11054.
13. Dernoncourt C, Ruelle D, Goffinet AM. Estimation of genetic distances between "reeler" and nearby loci on mouse chromosome 5. Genomics 1991;11:1167–1169.
14. D'Arcangelo G, Miao GG, Chen S-C, Soares HD, Morgan JI, Curran T. A protein-related to extracellular matrix proteins deleted in the mouse mutant reeler. Nature 1995;374:719–723.
15. Klar A, Baldassare M, Jessel TM. F-spondin: a gene expressed at high levels in the floor plate encodes a secreted protein that promotes neural cell adhesion and neurite extension. Cell 1992;69: 95–110.
16. Hirotsune S, Takahara T, Sasaki N et al. The reeler gene encodes a protein with an EGF-like motif expressed by pioneer neurons. Nature Genet 1995;10:77–83.
17. Curran T, D'Arcangelo G, Goffinet A. Reeler gene discrepancies (correspondence). Nature Genet 1995;11:12.

Dynamic organization of cerebral neural network

Integrative and Molecular Approach to Brain Function.
M. Ito and Y. Miyashita, editors.

Dynamics of cortical memory mechanisms

P.S. Goldman-Rakic
Section of Neurobiology, Yale University School of Medicine, New Haven, Connecticut, USA

Key words: dopamine receptors, dopamine modulation, prefrontal cortex, rhesus monkey, single-unit recording.

The prefrontal cortex is the area of the brain most commonly associated with executive processes. Although this area has traditionally been regarded as the most complex and least understood regions of the brain, this view is changing owing to a vigorous and rapidly developing field of experimental and clinical investigation. Basic research on nonhuman primates is beginning to reveal the nature of prefrontal functions and the anatomical, physiological and pharmacological substrates of prefrontal interactions with the sensory, motor and limbic regions of the brain. Behavioral, noninvasive brain imaging and electroencephalographic analysis of human mentation are extending these findings to normal subjects and patients with a variety of neurological and psychiatric diseases. Executive functions can now be conceptualized in terms of an interplay between the basic psychological processes of long-term memory stores and on-line information processing networks engaged in the dynamic operations of working memory. The special nature of working memory was first identified in studies of human cognition [1,2] and modern neurobiological methods have identified a specific population of neurons, patterns of their intrinsic and extrinsic circuitry, and signaling molecules that are engaged in this process in animals. In this article I will first define key features of working memory and then describe its biological basis in primates.

Working memory and "on line" processing

It is now well established that the dorsolateral prefrontal cortex in the rhesus monkey (particularly Walker's area 46) qualifies as a mental sketch pad and central processor of visuospatial information. Lesions restricted to this region have been shown repeatedly to impair performance on spatial delayed-response tasks which tax an animal's working memory ability, i.e., to hold an item of information "in mind" for a short period of time and to update information from moment to moment. The

---

Address for correspondence: P.S. Goldman-Rakic, Section of Neurobiology, Yale University School of Medicine, 333 Cedar Street, New Haven, CT 06520-8001, USA. Tel.: +1-203-785-4808. Fax: +1-203-785-5263.

impairments are selective in two critical respects; performance on tasks which engage memory for the features of objects such as visual discrimination, object reversal, learning set, match-to-sample is not affected by the same lesions nor do these lesions impair performance which relies on associative memory [3–6] or sensory-guided responses [7–9]. In general, neither the consistent rules of a task nor its sensorimotor requirements cause a problem for the prefrontally lesioned animal. The monkey's difficulty lies in recalling information and using it to guide a correct response. Thus, on the basis of neuropsychological evidence, the brain obeys the distinction between working and associative memory, and that prefrontal cortex is pre-eminently involved in the former while other areas of the neocortex and hippocampus are likely to be the necessary critical substrates of memory consolidation and long-term storage [10].

The significance of working memory for higher cortical function is not necessarily self-evident. Perhaps even the quality of its transient nature misleads us into thinking it is somehow less important than the more permanent archival nature of long-term memory. However, the brain's working memory function, i.e., the ability to bring to mind events in the absence of direct stimulation, may be its inherently most flexible mechanism and its evolutionarily most significant achievement. At the most elementary level, our basic conceptual ability to appreciate that an object exists when out of view depends on the capacity to keep events in mind beyond the direct experience of those events. For some organisms, including most humans under certain conditions, "out of sight" is equivalent to "out of mind". However, working memory is generally available to provide the temporal and spatial continuity between our past experience and present actions. Working memory has been invoked in all forms of cognitive and linguistic processing and is fundamental to both the comprehension and construction of sentences. It is essential to the operations of mental arithmetic, to play chess, to play the piano (particularly without music), to deliver speech extemporaneously and finally to fantasize and to plan ahead.

Cellular correlates of working memory: neurons with "memory fields"

Single neuron recording has been used extensively to dissect the neuronal elements involved in working memory processes. This approach can also provide fresh insights into issues of functional allocation as well as deliver convergent validation of their essential nature. In the oculomotor delayed response paradigm utilized for this purpose, briefly presented visuospatial stimuli are remembered in order to provide guidance from memory for subsequent saccadic eye movements. The essential feature

---

Fig. 1. Prefrontal neurons in the region of the principal sulcus exhibit a variety of patterns of activation during the oculomotor tasks. Some respond phasically to the occurrence of a cue (upper); some respond in relation to the delay (middle); and some are activated in relation to the occurrence of a response (lower). In all cases, neuronal activity is time locked to the events of the task and is spatially tuned. Note that the delay period activity in neuron #5018 begins at the end of the cue, then exhibits tonic firing and returns to baseline as soon as the response is initiated. The class of neurons with delay period activity is the focus of the present essay. (Based on [12,13,65,66].)

1. Cue period activity

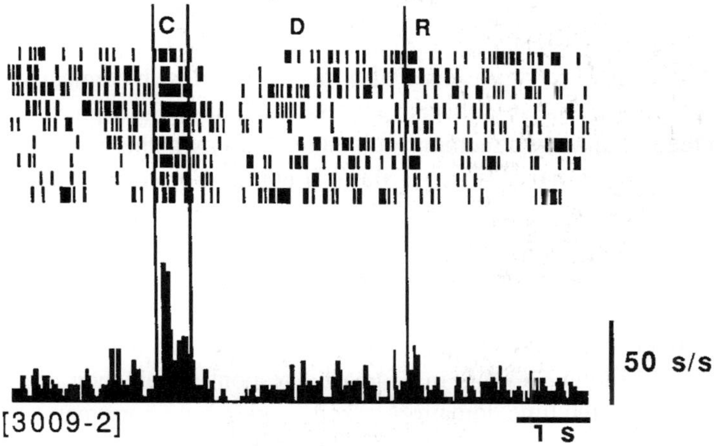

2. Delay period activity

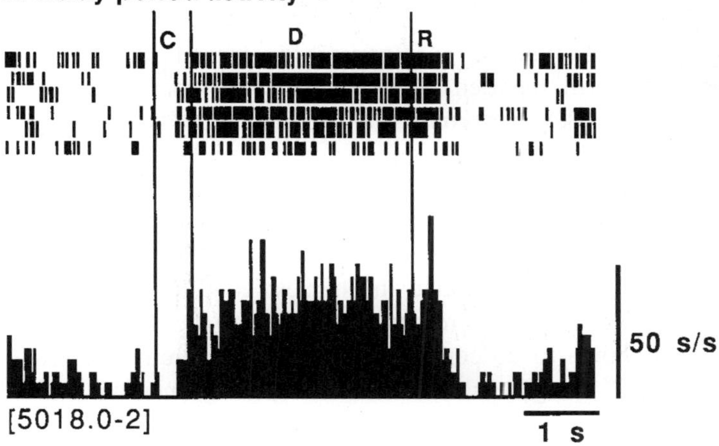

3. Response period activity

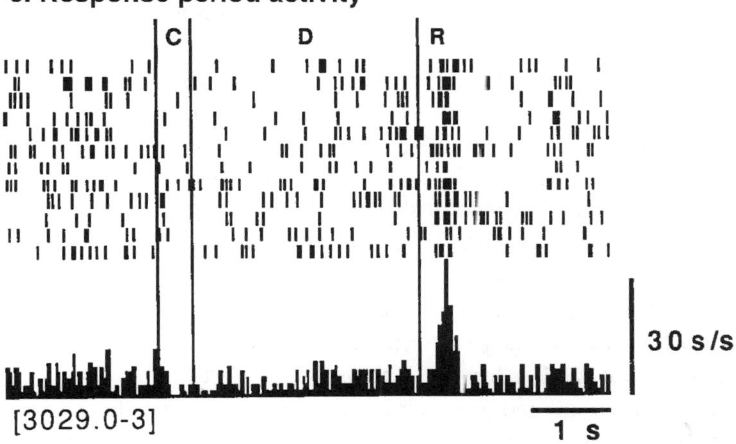

of this task is that the item to be recalled (in this case, the location of an object) has to be updated on every trial as in the moment-to-moment process of human mentation. Subsets of prefrontal neurons in the area of the principal sulcus are activated: a) phasically in the presence of a visual stimulus; b) tonically during the delay period over which the stimulus is kept "on line"; or c) show phasic reactivation in relation to the initiation of a memory-guided response, whether it be a manual [11] or oculomotor [12,13] response (Fig. 1). Many, if not most prefrontal neurons respond in more than one phase of the trial, i.e., during the cue, delay and/or response periods and their composite profile may be due to inputs from neurons whose activation is simpler and related to only one phase (for review see [13]). The firing profiles of prefrontal neurons are related to the subfunctions of registration, memory and motor control, respectively. These prefrontal neuronal activities are not only differentially time locked to the running events in a delayed response trial, they are temporally phased so as to bridge the time domain. As such, working memory may constitute the basis of temporal integration as often discussed by Fuster [14].

Very little work has been carried out on the temporal parameters of working memory. Is it possible for neurons to remain activated for longer than a few seconds in the absence of a stimulus or a response? Fuster and Jervey [15], and later Miyashita and Chang [16], reported delay-period activation lasting more than 15 s in the temporal lobe of monkey performing short-term memory tasks and similarly long delay-period activation has also been observed in prefrontal neurons [17,12]. Recent studies have provided evidence that the delay-period activity recorded in inferotemporal cells during memory tasks may reflect afferents from the prefrontal areas [18,19]. Interestingly, the area in the ventral bank of the superior temporal sulcus (STS) where Fuster and Jervey found the highest concentration of neurons with prolonged delay-period activity corresponds precisely to the portion of the STS that has recently been shown to be connected with inferior prefrontal visual working memory centers [20,21]. In general, it is doubtful that single neurons in the prefrontal or inferotemporal cortex will remain active over the minutes, hours or days that many memories can be retained and without controls for anticipatory responses, it is not always possible to rule out that this prolonged activity is related to a sustained anticipatory response in the delay period rather than to an iconic trace. Finally, neuronal activity of a given cell during the delay period is labile and can expand and contract as the delay is lengthened or foreshortened [12,17], as would be expected if a neuron is engaged in dynamic mnemonic processing. These responses take place within a narrow range of delays (less than 20 s. at present) but further studies are needed to determine the limits and constraints on the working memory system. In my view, information that is retained by monkeys over tens of seconds, enters intermediate or long-term memory stores and likely depends upon mechanisms beyond working memory, perhaps involving LTP in the hippocampal formation. The neuronal activation observed in prefrontal neurons is best viewed as a reflection of information that is "on line".

A major advance in our understanding of prefrontal cortex is the evidence that prefrontal neurons have directional biases [22] and "memory fields", defined as maximal firing of a neuron to the representation of a target in one or a few locations of the visual field – with the same neuron always coding the same location [12]. Neuronal activity in such neurons rises sharply at the end of the preferred stimulus, remains tonically active during the delay (in the absence of the stimulus or a response) and then ceases abruptly at the end of the 3–5 s delay, as the response is initiated. Importantly, the activation occurs every time the animal has to remember the preferred location, e.g., the 270° location, but not when the animal is remembering targets presented at other locations (e.g., 135, 180, and 225°). In addition, many prefrontal neurons have opponent memory fields, i.e., their rate of firing in the delay period is enhanced for one target location and inhibited during the delay on trials with target stimuli of opponent polarity. This functional distinction provides a valuable clue to how the neural circuitry subserving working memory might be organized and I will return to this question below.

The microarchitecture of prefrontal cortex

Pyramidal and nonpyramidal cells constitute the major cellular constituents of the cerebral cortex and it is widely believed that understanding the interaction of these two principal components of cortical architecture holds the key to a mechanistics understanding of cortical function. Recently, we have examined the local circuit connections of pyramidal neurons by placing small injections of cholera toxin B-subunit into circumscribed layers in area 46 [23]. The retrograde transport of cholera toxin B-subunit allowed us to examine the intrinsic or local connections of the injected cells. Following injections confined to layer IIIc, we found multiple clusters of labeled pyramidal cells at spaced intervals over 7–8 mm spans of cortex [23,24]. Figure 2 illustrates this modular architecture which may form the structure of a local spatial working memory network in which columns of pyramidal neurons with like "best directions" (e.g., 90, 180, and 270°, etc.) are interconnected in a manner analogous to the orientation column system of primary visual cortex [25].

As mentioned above, pyramidal-nonpyramidal interactions could be considered

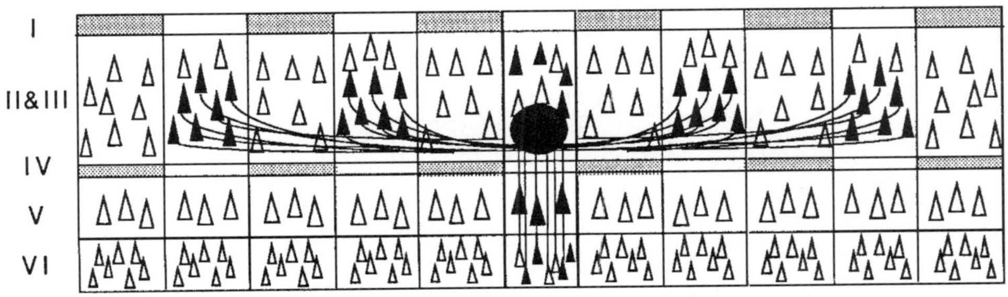

Fig. 2. Summary diagram illustrating the layer-specific patterns of intrinsic connections in prefrontal cortex (Walker's areas 46 and 9) as retrogradely labeled with Cholera Toxin-B Subunit. Clusters of labeled neurons in layer IIIc are connected by local axon collaterals of pyramidal neurons. We hypothesize that the interconnected clusters may share "best directions" or "memory fields". (From [23].)

critical to cortical function. From Cajal on, it has been appreciated that multiple types of interneurons populate the cerebral cortex and interact with pyramidal cells. While pyramidal neurons utilize the excitatory amino acid neurotransmitter, glutamate, for transmission, interneurons utilize the inhibitory neurotransmitter, γ-amino-butyric acid (GABA). Wilson et al. in this laboratory [26] have succeeded in using wave form analysis to classify functionally characterized neurons as interneurons (thin spiking neurons) or pyramidal neurons (broader and higher amplitude spikes) in monkeys as they performed the oculomotor delayed response task. This study showed that: 1) interneurons, like pyramidal neurons, express directional preferences (FS neuron in Fig. 3); and 2) the patterns of activity expressed by closely adjacent pyramidal and nonpyramidal neurons are often complementary, such that as a nonpyramidal neuron increases its rate of discharge, a nearby pyramidal neuron decreases its rate (Fig. 3) [26]. These findings provide suggestive evidence that feedforward inhibition may play a role in the construction of a memory field in prefrontal neurons. Based on these findings, Fig. 4 shows a GABAergic interneuron interconnecting two pyramidal cells with opposite "best" directions. According to this scheme, pyramidal cells with opposite best directions communicate via inhibitory interneurons such that a pyramidal neuron with a 90° memory field exhibits enhanced firing during the delay of trials in which the monkey is recalling a 90° target but is inhibited on trials when the memorandum is at the 270° location. A reciprocal pathway allows for a pyramidal neuron with a 270° memory field to inhibit one with a 90° memory field. This model suggests that nonpyramidal interneurons could provide essential feedforward inhibitory influences between pyramidal cells with opponent memory fields. It is clear from electron microscopic evidence that pyramidal cells innervate interneurons in the prefrontal cortex [27], and basket cells innervate pyramidal cells [28] in the manner illustrated.

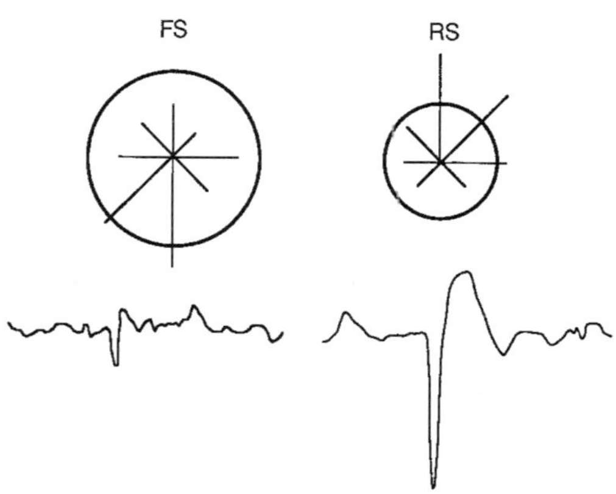

Fig. 3. Diagram illustrating complementary responses of a fast-spiking (FS) and regular spiking (RS) pair of neurons recorded simultaneously from the same electrode. Increases in RS-cell firing correspond to graded decreases in the FS-cell firing, and vice versa. In the vector plot, each vector represents response magnitude plotted relative to a stimulus location for the FS/RS pair of neurons shown. Firing rates are normalized so that the maximum vector length is 100%. The circles represent spontaneous firing rates. (From [26].)

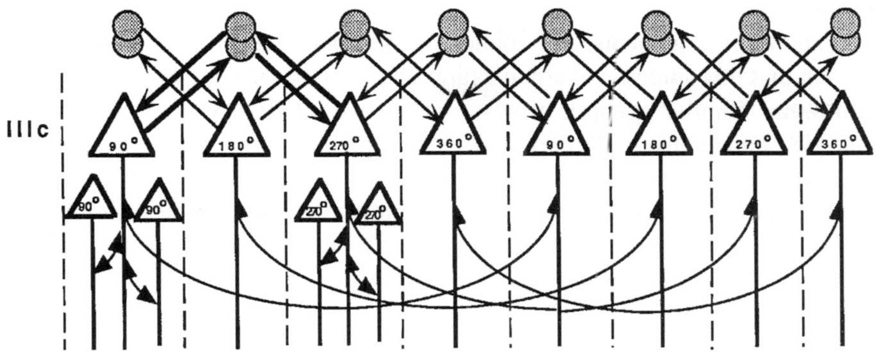

Fig. 4. Model of working memory modules consisting of clusters of tuned pyramidal neurons (red and black triangles) arrayed by target location and directly interconnected with each other by their local excitatory axon collaterals. Clusters of pyramidal neurons with like "best directions" are interconnected in a manner similar to iso-orientation columns in visual cortex. Two inhibitory interneurons (circles, presumed basket cells in the diagram) provide the reciprocal interconnections between pyramidal cells with opposite "best directions" that could explain the opponent memory fields observed by Funahashi et al. (1989). For simplicity, only the 90–270° and 270–90° ensemble is highlighted. For now, the organization of pyramidal cells with particular memory fields is hypothetical as is the reciprocity of the excitatory-inhibitory units. Further analysis of these local circuits is essential for analyzing the neural substrates of working memory.

Dopamine modulation of the canonical E-I unit

The prefrontal cortex in primates is a major target of the brainstem dopamine afferents [29–31]. Working memory deficits are present in Parkinson patients [32,33] and have been shown to result from experimental depletion of dopamine in prefrontal areas in rhesus monkeys [34]. In pursuit of the anatomical and functional characteristics of the dopamine system in the cerebral cortex generally, and the prefrontal cortex, in particular, we have discovered that dopamine axons whose cell bodies lie in the brainstem specifically target the pyramidal neurons of the cortex and moreover, form terminal specializations that are indistinguishable from traditional synaptic specializations [35,36]. In our study of human cortex obtained at biopsy, we have estimated that about 60% of dopamine synapses are on spines of target neurons [37]. Studies of the dopamine innervation at both the light and electron microscopic levels thus indicate that the dopamine innervation of prefrontal cortex, though highly ramified in dense axonal plexuses within specific layers, is directed to a specific target – the pyramidal cell – within these layers.

Dopamine-glutamate interactions in the prefrontal cortex

A major advance in our appreciation of dopamine function has, in my view, been sparked by the discovery that the spines of pyramidal neurons are often the targets of paired dopamine and excitatory terminals, the latter presumed to arise from other cortical or thalamic sources (see Fig. 5 and [35]). Synaptic triads, as we have termed these synaptic arrangements, are found in prefrontal, premotor and motor cortex, suggesting that this anatomical architecture may be widespread and common to many cortical areas. Similar synaptic triads have also been observed in the striatum of rodents [38] and monkey (Smiley and Goldman-Rakic, unpublished observations). Moreover, the distribution of dopamine afferents can now be wedded to the concentration of D1 and D2 receptor families on a layer-by-layer basis. We have used the dopamine D1-specific ligand, SCH23390, and the D2-specific antagonist, raclopride, to label D1 and D2 receptors in the prefrontal cortex of both monkeys [39] and humans [40] with the method of quantitative receptor autoradiography. Together, the location of these sites in the cortical layers corresponds remarkably to the heaviest concentration of dopamine axons, in both species. An important and unexpected difference between the D1 and D2 receptor sites, however, is that the D1-receptor sites are approximately 100-fold more abundant than the D2 sites. Most significant, in several studies, we have found that D1 receptors are present in pyramidal neurons in all layers of the prefrontal cortex, supporting our previous observation that the pyramidal cell is the preferred target of the dopamine innervation. Ultrastructural analysis showed that the spines of pyramidal cells were the most prominent cellular element labeled by our subtype-specific antibodies [41,42]. These labeled spines also received an asymmetric synaptic specialization characteristic of nerve terminals engaged in excitatory neurotransmission [42].

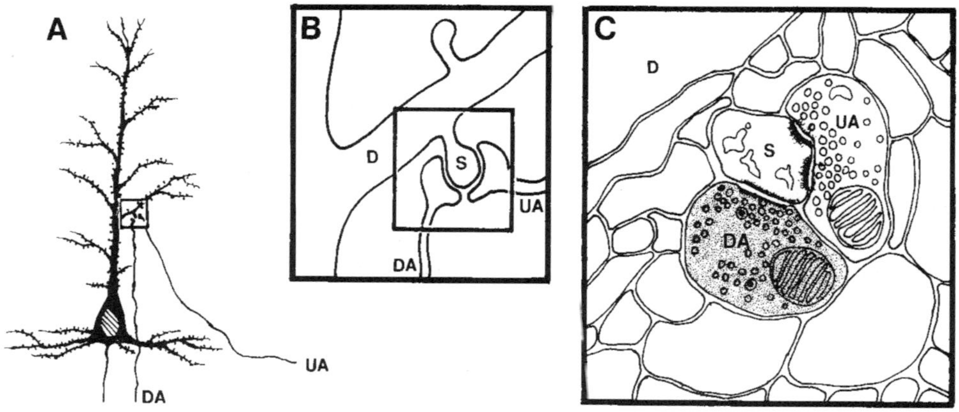

Fig. 5. Diagram of synaptic arrangements involving the dopamine input to the cortex. **A**: Afferents labeled with a dopamine (DA) specific antibody terminate on the spine of a pyramidal cell in the prefrontal cortex, together with an unidentified axon (UA). **B**: Enlargement of axospinous synapses illustrated in A showing opposition of the DA input and a presumed excitatory input (UA) that makes an asymmetrical synapse on the same dendritic (D) spine. **C**: Diagram of ultrastructural features of the axospinous synapses illustrated in B; the dopamine terminal (darkened profile representing DA immunoreactivity) forms a symmetrical synapse, the unidentified profile forms an asymmetrical synapse; the unidentified profile forms an asymmetrical synapse with the postsynaptic membrane. (Diagram modified from data presented [35].)

Dopamine modulation of mnemonic function in prefrontal neurons

The cognitive deficits associated with dopamine dysfunction in the prefrontal cortex are equally selective: they are expressed in performance which is memory-guided and not in tasks which are guided by external stimuli [10]. Evidence from studies of experimentally induced dopamine deficits in nonhuman primates as well as in patients with Parkinson's disease supports the view that short-term (working) memory is a mnemonic process especially vulnerable to dopamine loss. This dependence can be understood if dopamine afferents modulate the responses of pyramidal neurons to excitatory input, particularly via axospinous synapses on neurons that show sustained activation during mnemonic processing in the prefrontal cortex. This hypothesis has recently obtained support from single-cell recording studies in monkeys performing working memory tasks and will be discussed in further detail below.

The availability of selective dopamine antagonists has opened the possibility of testing the contribution of several dopamine receptors to cognitive function. We now have evidence that activation of D1 receptors, in particular, may play an important role in the mnemonic, predictive functions of the primate prefrontal cortex. Monkeys were trained on the same oculomotor delayed-response task used in the neurophysiological studies described above. Graham Williams has succeeded in iontophorizing dopamine agonists and antagonists onto single neurons that were isolated and functionally characterized during performance of the task. A major finding from our studies is that the D1 receptor modulates excitatory transmission in neurons which are involved directly in the mnemonic component of the task [43]. Williams found that partial block of D1 receptors (with a D1-specific antagonist, SCH 39166) enhanced the memory fields of many prefrontal neurons without altering their general excitability (Fig. 6). Thus, the neuron's activity was enhanced only in the delay period and only for the neuron's preferred direction; neuronal firing was often inhibited for the direction opposite to the preferred and unaltered for all other directions. The specificity of the drugs action on single cell activity can best be accounted for by the specific synaptic arrangement of D1 receptors in spines that receive excitatory inputs, presumably from the visual pathway carrying visuospatial coordinate information. Our finding of higher density of D1 receptors in prefrontal cortex draws attention to the potential functional significance of these receptors for cognitive processes in normal individuals and for deficits such as the thought fragmentation characteristic of diseases like schizophrenia.

Another finding in these experiments was that specific neuronal enhancement was observed only at low doses of antagonist; at higher doses, or in the presence of dopamine agonist, neuronal firing was reduced or strongly inhibited. The latter results indicated that dopamine influences mnemonic function through D1 stimulation and that either too little dopamine [34] and too much dopamine [44], as in amphetamine psychosis and experimental situations, may result in cognitive dysfunction through excessive or insufficient D1 stimulation. It is not surprising that working memory is vulnerable to the influence of dopamine dysfunction, and a common process may underlie the cognitive deficits in Parkinson's disease, in aged individuals and in

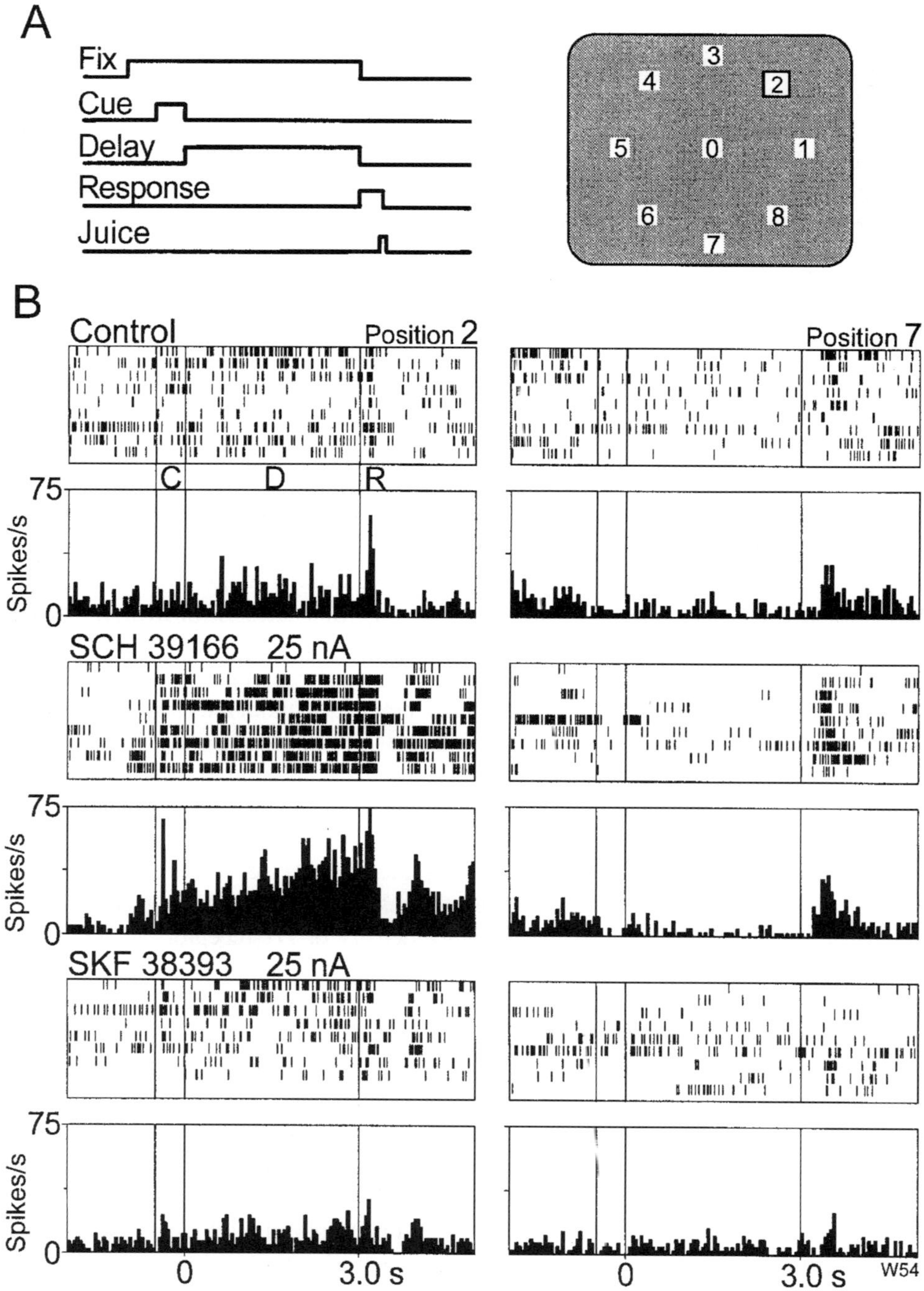
A
Fix
Cue
Delay
Response
Juice
3
4
2
5
0
1
6
8
7
B
Control
Position 2
Position 7
C
D
R
75
Spikes/s
0
SCH 39166 25 nA
75
Spikes/s
0
SKF 38393 25 nA
75
Spikes/s
0
0
3.0 s
0
3.0 s
W54

Fig. 6. **A**: Diagram of the task events in the oculomotor delayed-response paradigm. **B**: The upper panels reveal that neuron W54 has a "memory field" or preferred direction. The neuron responds strongly in the delay period when the period when the target to be called is in position #2 as shown in A (left panel). At the #7 position, the neuron's activity is depressed while at all other locations (not shown), the neuron's activity does not deviate from baseline. As shown in the next lower row of histograms, iontophoresis of SCH 39166 (Schering-Plough), a selective D_1 antagonist, clearly and consistently enhanced this memory field when applied at low ejection currents (20–25 nA) (left) but did not enhance activation for other target positions outside the memory field, including position #7 (right). Indeed, SCH 39166 slightly accentuated the reduction in activity during the delay period which often occurs on trials with targets in the location opposite to the memory field (an "opponent memory field"). Finally, the enhanced activation for target #2 during the delay could be reversed by iontophoresis of the partial D_1 agonist SKF 38393 (RBI) onto the cell and the reversal was also selective to the neuron's memory field (lower histogram). These findings provide evidence of a unique dopaminergic modulation which can be extraordinarily precise, affecting a specific component of the neuron's afferent (excitatory and inhibitory) input without affecting its general excitability.

---

schizophrenics.

The effect of cortical dopamine may thus be to modulate and/or adjust the gain on descending pathways possibly enhancing, possibly dampening, possibly disrupting information flow in the basal ganglia-thalamotectal and thalamocortical loop systems that ultimately control motor action.

Multiple working memory domains and distributed neuronal networks

The mechanisms described above can now be extended to other working memory domains such as feature working memory [20]. It has recently been shown that the prefrontal neurons which code visuospatial memoranda are located in a separate area than those that code simple, complex or categorical features of stimuli. Moreover, individual neurons that code the location of targets rarely, if ever, code object qualities or vice versa [20]. Further, physiologically informed or physiologically guided injections of pathway tracers in the spatial and object memory centers have shown them to be connected to the appropriate visual centers via relays in the parietal [45] and temporal lobes, respectively [21].

Working memory is considered a major component of the machinery of executive function [46] and it is not surprising that PET and functional MRI studies in human subjects have been focused on this function. Recent studies in healthy human subjects, for example, have shown that the middle frontal gyrus, the dorsolateral region corresponding to the areas from which recordings have been made in macaque monkeys, is activated when human subjects carry out analogous spatial working memory tasks [47,48]. Moreover, other regions of the dorsolateral prefrontal cortex are activated for verbal and other nonspatial working memory functions [49–51]. Finally, as might be expected if working memory were essential to executive function, working memory deficits and correlated prefrontal dysfunction have been demonstrated in schizophrenics [52–54], in Parkinson patients [32,33], in age-related memory decline [55] and in many other neuropathological conditions in which impairments of higher cortical processing are expressed.

It is of interest that in the monkey, the hippocampal formation is activated along with prefrontal areas during performance of working memory tasks as are the posterior parietal regions which transmit visuospatial information to the dorsolateral regions of the prefrontal cortex [56]. Recent studies employing multiple unit recording techniques in the CA1 field of the rodent hippocampus have likewise demonstrated that particular patterns of neuronal activity are associated with particular responses of the animal in a spatial delayed-response task [57]. Studies of delayed recall in humans also activate both prefrontal cortex and the hippocampal formation [58]. All of these results speak to the re-entrant network organization of the prefrontal cortex and hippocampal formation outlined at the beginning of the chapter (for further discussion, see [59]).

Clinical significance

The significance of observations on the cortical dopamine innervation for cognition is that they may suggest a variety of ways in which dopamine transmission in the cortex may alter cognitive function. Possibly we might be able to consider predictions of optimal functioning depending on the availability of dopamine in the cortical synaptic cleft and the availability, affinity and concentration of specific receptor sites in the cortex. Recent findings on the high density of D1 receptors in prefrontal cortex draw attention to the potential functional significance of these receptors for the cognitive deficits in disorders like schizophrenia. Since multiple subtypes of the D1 receptor family of receptors have now been cloned, future work will have to define the specific subtype most critical for the cognitive phenomena addressed here. Evidence that dopamine and putative glutamate profiles in prefrontal cortex are apposed to the membrane surface of the same dendritic spine [35] gives rise to hypotheses concerning the interactions of glutamate and dopamine receptors in higher cortical processes. Recent studies of human cortical "slices" have shown that dopamine enhanced NMDA-induced spiking in cortical neurons and the effect could be blocked by SCH23390, another D1 antagonist [60].

D1 receptors, along with other monoaminergic receptors, have been explored as possible targets of atypical neuroleptics. We might expect changes in these and/or related receptors to be present in the cortex of schizophrenics and/or as a function of neuroleptic treatments. These predictions have been indirectly supported by studies of receptor regulation after chronic exposure to the atypical neuroleptic, clozapine, in experimental animals [61]. The cardinal cognitive, emotional and motivational syndromes consistently associated with schizophrenia bear strong resemblance to the thought disorders, attentional problems, inappropriate or flattened affect and lack of initiative, plans and goals that characterize patients with physical prefrontal damage [62]. Most recently, schizophrenics have been tested on the spatial oculomotor working memory task that we have used to study working memory in rhesus monkeys [54]; conversely, rhesus monkeys with prefrontal lesions exhibit the same type of predictive tracking disorder that is observed in nearly 80% of schizophrenic patients

[63]. If the prefrontal cortex is the part of the cortex most responsible for working memory function and this process is dysfunctional in schizophrenia, as we think, then probing how the memory cells of the prefrontal cortex are influenced by dopamine, glutamate and other neurotransmitters is essential for understanding dysfunction in schizophrenia. The effects of these neuromodulators have received less attention in the neocortex than in the basal ganglia and now that specific receptors have been implicated in the working memory functions mediated by the prefrontal cortex, the study of their role in cognitive function would seem to be a promising line of study.

Concluding remarks

In a recent review of neural mechanisms of form and motion processing in the visual system in this journal, Van Essen and Gallant have pointed out that "the organization of the primate visual system is far more complex than most neuroscientists appreciated as recently as a decade ago" [64], the present review hopefully will convey that the prefrontal areas, though no less complex, are not necessarily more complex. Benefiting from the seminal work in the visual system that has preceded it, the function of the association cortices have taken their place with the accessible topics in neurobiology. Indeed, the work reviewed here demonstrates that the cerebral cortex is a unified structure with the mnemonic processes of its frontal lobe grafted in part upon the architecture of its sensory systems.

An important quest for the future will be to learn which specific cells and which specific receptors are the targets of most of the dopamine terminals and whether these target cells have special physiological characteristics and influence specific intrinsic or extrinsic circuits. For example, neurons that discharge tonically during mnemonic processing and thus hold "on line" information that would otherwise be lost to conscious consideration may bear one receptor subtype while those that discharge phasically after a response is made may bear another subtype [12]. If neurons with different functional topologies are indeed selectively laden with neurotransmitter receptors, then understanding the cellular basis of short-term memory would become a more realistic goal. Sophisticated animal models, combined with powerful neurobiological techniques, are now available for rigorous testing of these cellular hypotheses.

References

1. Norman DA. Models of Human Memory. New York: Academic Press, 1970.
2. Baddeley A. Working Memory. New York: Oxford University Press, 1986.
3. Jacobsen CF. Studies of cerebral function in primates. Comp Psychol Monogr 1936;13:1–68.
4. Goldman PS, Rosvold HE, Vest B, Galkin TW. Analysis of the delayed-alternation deficit produced by dorsolateral prefrontal lesions in the rhesus monkey. J Comp Physiol Psychol 1971;77:212–220.
5. Passingham RE. Delayed matching after selective prefrontal lesions in monkeys (*Macaca mulatta*). Brain Res 1975;92:89–102.
6. Mishkin M, Manning FJ. Non-spatial memory after selective prefrontal lesions in monkeys. Brain Res 1978;143:313–323.
7. Funahashi S, Bruce CJ, Goldman-Rakic PS. Dorsolateral prefrontal lesions and oculomotor delayed-

response performance: Evidence for mnemonic scotomas. J Neurosci 1993;13:1479–1497.
8. Sawaguchi T, Goldman-Rakic PS. The role of D1-dopamine receptor in working memory: Local injections of dopamine antagonists into the prefrontal cortex of rhesus monkeys performing an oculomotor delayed-response task. J Neurophysiol 1993;71:515–528.
9. Chafee MV, Goldman-Rakic PS. Prefrontal cooling dissociates memory-and sensory-guided oculomotor delayed response functions. Soc Neurosci Abstr 1994;20:335.1.
10. Goldman-Rakic PS. Circuitry of the prefrontal cortex and the regulation of behavior by representational knowledge. In: Plum F, Mountcastle V (eds) Handbook of Physiology. Bethesda, MD: American Physiological Society, 1987;373–417.
11. Fuster JM, Alexander GE. Neuron activity related to short-term memory. Science 1971;173:652–654.
12. Funahashi S, Bruce CJ, Goldman-Rakic PS. Mnemonic coding of visual space in the primate dorsolateral prefrontal cortex. J Neurophysiol 1989;61:331–349.
13. Goldman-Rakic PS, Funahashi S, Bruce CJ. Neocortical memory circuits. Q J Quant Biol 1990;55: 1025–1038.
14. Fuster JM. The prefrontal cortex, mediator of cross-temporal contingencies. Human Neurobiol 1985; 4:169–179.
15. Fuster JM, Jervey JP. Inferotemporal neurons distinguish and retain behaviorally relevant features of visual stimuli. Science 1981;212:952–955.
16. Miyashita Y, Chang HS. Neuronal correlate of pictorial short-term memory in the primate temporal cortex. Nature 1988;33:68–70.
17. Kojima S, Goldman-Rakic PS. Delay-related activity of prefrontal cortical neurons in rhesus monkeys performing delayed response. Brain Res 1982;248:43–49.
18. Fuster JM, Bauer RH, Jervey JP. Functional interactions between inferotemporal and prefrontal cortex in a cognitive task. Brain Res 1985;330:299–307.
19. Miller EK, Desimone R. Parallel neuronal mechanisms for short-term memory. Science 1994;263: 520–522.
20. Wilson FAW, O'Scalaidhe SP, Goldman-Rakic PS. Dissociation of object and spatial processing domains in primate prefrontal cortex. Science 1993;260:1955–1958.
21. Bates JF, Wilson FAW, O'Scalaidhe SP, Goldman-Rakic PS. Area TE connections with inferior prefrontal regions responsive to complex objects and faces. Soc Neurosci Abstr 1994;20:1054.
22. Niki H. Differential activity of prefrontal units during right and left delayed response trials. Brain Res 1974;70:346–349.
23. Kritzer MK, Goldman-Rakic PS. Intrinsic circuit organization of the major layers and sublayers of the dorsolateral prefrontal cortex in the rhesus monkey. J Comp Neurol 1995;359:131–143.
24. Leavitt JB, Lewis DA, Yoshioka T, Lund JS. Topography of pyramidal neuron intrinsic connections in macaque monkey prefrontal cortex (areas 9 and 46). J Comp Neurol 1993;338:360–376.
25. Gilbert CD. Circuitry, architecture and functional diversity of visual cortex. Cereb Cortex 1993;3: 373–386.
26. Wilson FAW, O'Scalaidhe SP, Goldman-Rakic PS. Functional synergism between putative g-aminobutyrate-containing neurons and pyramidal neurons in prefrontal cortex. Proc Natl Acad Sci USA 1994;91:4009–4013.
27. Williams SM, Goldman-Rakic PS, Leranth C. The synaptology of parvalbumin-immunoreactive neurons in the primate prefrontal cortex. J Comp Neurol 1992;320:353–369.
28. Somogyi P, Kisvardy ZF, Martin KAC, Whitteridge D. Synaptic connections of morphologically identified and physiologically characterized basket cells in the striate cortex of the cat. Neuroscience 1983;10:261–294.
29. Brown RM, Crane A, Goldman PS. Regional distribution of monoamines in the cerebral cortex and subcortical structures of the rhesus monkey: Concentrations and in vivo synthesis rates. Brain Res 1979;168:133–150.
30. Lewis DA, Foote SL, Goldstein M, Morrison JH. The dopamine innervation of monkey prefrontal cortex: a tyrosine hydroxylase immunohistochemical study. Brain Res 1988;449:225–243.
31. Williams SM, Goldman-Rakic PS. Characterization of the dopaminergic innervation of the primate

frontal cortex using a dopamine-specific antibody. Cereb Cortex 1993;3:199–222.
32. Gotham AM, Brown RG, Marsden CP. "Frontal" cognitive function in patients with Parkinson's disease "on" and "off" levodopa. Brain 1988;111:299–321.
33. Levin BE, Labre MM, Weiner WJ. Cognitive impairments associated with early Parkinson's disease. Neurology 1989;39:557–561.
34. Brozoski T, Brown RM, Rosvold HE, Goldman PS. Cognitive deficit caused by regional depletion of dopamine in prefrontal cortex of rhesus monkey. Science 1979;205:929–932.
35. Goldman-Rakic PS, Leranth C, Williams MS, Mons N, Geffard M. Dopamine synaptic complex with pyramidal neurons in primate cerebral cortex. Proc Natl Acad Sci USA 1989;86:9015–9019.
36. Smiley JF, Goldman-Rakic PS. Heterogeneous targets of dopamine synapses in monkey prefrontal cortex demonstrated by serial section electron microscopy: A laminar analysis using the silver-enhanced diaminobenzidine sulfide (SEDS) immunolabeling technique. Cereb Cortex 1993;3: 223–238.
37. Smiley JF, Williams SM, Szigeti K, Goldman-Rakic PS. Light and electron microscopic characterization of dopamine-immunoreactive processes in human cerebral cortex. J Comp Neurol 1992;321:325–335.
38. Smith AD, Bolam JP. The neural network of the basal ganglia as revealed by the study of synaptic connections of identified neurons. Trends Neurosci 1990;13:259–265.
39. Goldman-Rakic PS, Lidow MS, Gallager DW. Overlap of dopaminergic, adrenergic, and serotonergic receptors and complementarity of their subtypes in primate prefrontal cortex. J Neurosci 1990;10(7):2125–2138.
40. Goldman-Rakic PS, Lidow MS, Smiley JF, Williams SM. The anatomy of dopamine in monkey and human prefrontal cortex. J Neurol Transm 1992;36:163–177.
41. Smiley JF, Levey AI, Ciliax BJ, Goldman-Rakic PS. D1 dopamine receptor immunoreactivity in human and monkey cerebral cortex: Predominant and extrasynaptic localization in dendritic spines. Proc Natl Acad Sci USA 1994;91:5720–5724.
42. Bergson C, Mrzljak L, Smiley JF, Pappy M, Levenson R, Goldman-Rakic PS. Regional, cellular, and subcellular variations in the distribution of D1 and D5 dopamine receptors in primate brain. J Neurosci 1995;15:7821–7836.
43. Williams GV, Goldman-Rakic PS. Modulation of memory fields by dopamine D1 receptors in prefrontal cortex. Nature 1995;376:572–575.
44. Arnsten AFT, Cai JX, Steere JC Goldman-Rakic PS. Dopamine D2 receptor mechanisms contribute to age-related cognitive decline: The effects of quinpirole on memory and motor performance in monkeys. J Neurosci 1995;15:3429–3439.
45. Cavada C, Goldman-Rakic PS. Posterior parietal cortex in rhesus monkey: II. Evidence for segretated corticocortical networks linking sensory and limbic areas with the frontal lobe. J Comp Neurol 1989; 287:422–445.
46. Shallice T. Specific impairments in planning. Phil Trans R Soc Lond B 1982;298:199–209.
47. McCarthy G, Blamire AM, Puce A, Nobre AC, Bloch G, Hyder F, Goldman-Rakic PS, Shulman RG. Functional MR imaging of human prefrontal cortex activation during a spatial working memory task. Proc Natl Acad Sci USA 1994;91:8690–8694.
48. Jonides J, Smith EE, Koeppe RA, Awh E, Minoshima S, Mintun MA. Spatial working memory in humans as revealed by PET. Nature 1993;363:623–625.
49. Frith CD, Friston K, Liddle PF, Frackowiak RSJ. Willed action and the prefrontal cortex in man. Proc R Soc Lond B 1991;244:241–246.
50. Petrides M, Alivisatos B, Evans AC, Meyer E. Dissociation of human mid-dorsolateral from posterior dorsolateral frontal cortex in memory processing. Proc Natl Acad Sci USA 1993a;90:873–877.
51. Petrides M, Alivisatos B, Meyer E, Evans AC. Functional activation of the human frontal cortex during the performance of verbal working memory tasks. Proc Natl Acad Sci USA 1993b;90: 878–882.
52. Weinberger DR, Berman KF, Zec RF. Physiological dysfunction of dorsolateral prefrontal cortex in schizophrenia: 1. Regional cerebral blood flow (rCBF) evidence. Arch Gen Psychiat 1986;43:

114–125.
53. Fukushima J, Fukushima K, Chiba T, Tanaka S, Yamashita I, Kato M. Disturbances of voluntary control of saccadic eye movements in schizophrenic patients. Biol Psychiat 1988;23:670–677.
54. Park S, Holzman PS. Schizophrenics show spatial working memory deficits. Arch Gen Psychiat 1992; 49:975–982.
55. Salthouse TA. Mediation of adult age differences in cognition by reductions in working memory and speed of processing. Psychol Sciences 1991;2:170–183.
56. Friedman HR, Goldman-Rakic PS. Coactivation of prefrontal cortex and inferior parietal cortex in working memory tasks revealed by 2DG functional mapping in the rhesus monkey. J Neurosci 1994; 14:2775–2788.
57. Hampson RE, Heyser CJ, Deadwyler SA. Hippocampal cell firing correlates of delayed-match-to-sample performance in the rat. Behav Neurosci 1993;107:715–739.
58. Squire LR, Ojemann JG, Meizin FM, Petersen SE, Videen TO, Raichle ME. Activation of the hippocampal formation in normal humans: A functional anatomical study of memory. Proc Natl Acad Sci USA 1992;89:1837–1841.
59. Goldman-Rakic PS, Friedman HR. The circuitry of working memory revealed by anatomy and metabolic imaging. In: Levin HS, Eisenberg HM, Benton AL (eds) Frontal Lobe Function and Dysfunction. New York: Oxford University Press, 1991;72–91.
60. Cepeda C, Radisavljevic Z, Peacock W, Levine MS, Buchwald NA. Differential modulation by dopamine of responses evoked by excitatory amino acids in human cortex. Synapse 1992;11: 330–341.
61. Lidow MS, Goldman-Rakic, PS. A common action of clozapine, haloperidol and remoxipride on D1- and D2-dopaminergic receptors in the primate cerebral cortex. Proc Natl Acad Sci USA 1994;91: 4353–4356.
62. Goldman-Rakic PS. Prefrontal cortical dysfunction in schizophrenia: The relevance of working memory. In: Carroll BJ, Barrett JE (eds) Psychopathology and the Brain. American Psychopathological Association, New York: Raven Press, 1991;1–23.
63. MacAvoy MG, Gottlieb JP, Bruce CJ. Smooth pursuit eye movement representation in the primate frontal eye field. Cereb Cortex 1991;1:95–102.
64. Van Essen DC, Gallant JL. Neural mechanisms of form and motion processing in the primate visual system. Neuron 1994;13:1–10.
65. Funahashi S, Bruce CJ, Goldman-Rakic PS. Visuospatial coding in primate prefrontal neurons revealed by oculomotor paradigms. J Neurophysiol 1990;63:814–831.
66. Funahashi S, Bruce CJ, Goldman-Rakic PS. Neuronal activity related to saccadic eye movements in the monkey's dorsolateral prefrontal cortex. J Neurophysiol 1991;65:1464–1483.

Integrative and Molecular Approach to Brain Function.
M. Ito and Y. Miyashita, editors.

Origin of visual imagery: role of temporal cortex and top-down signal

Yasushi Miyashita[1,3], Machiko Ohbayashi[1], Masashi Kameyama[1], Yuji Naya[2] and Masatoshi Yoshida[3]

[1]Department of Physiology, The University of Tokyo School of Medicine, Tokyo; [2]Laboratory of Cognitive Neuroscience, National Institute of Physiological Sciences, Okazaki; and [3]JRDC Mind Articulation Project, Yushima, Tokyo, Japan

Abstract. Neuronal origin of visual imagery was examined from various viewpoints. We have tested a hypothesis ("backward-projection hypothesis") that visual perception mainly relies on the forward, bottom-up flow of information and that imagery experience, in contrast, highlights the backward projections as a possible anatomical substrate of top-down mental operations. Several lines of empirical evidence supported the hypothesis. The pair-association with color switch (PACS) task, in which the necessity for image generation and its initiation time were controlled by a color switch, provided particularly compelling evidence for the role of the temporal cortical neurons. The hypothesis predicted that backward signals from higher-order representations would reach topographically organized visual areas when an imagery task requires reconstruction of the detailed local geometry of the image. This prediction was confirmed by psychophysical experiments. It was also noted that humans use imagery in more flexible ways than the backward-projection hypothesis predicts: e.g., we can combine or arrange objects in novel configurations in images, or we can mentally draw patterns that we have never actually seen. Nevertheless, we believe imagery to be an important topic since it is one of the first higher cognitive abilities that will be firmly rooted in the brain.

Key words: feedback signal, inferotemporal cortex, medial temporal lobe, memory, neuronal activity, primate.

Introduction

"Create an image of the Pantheon in your mind. Can you count the columns that support its pediment?" [1] (Fig. 1). Our ability to "see with the mind's eye", mental imagery, has been of interest to philosophers and scientists for a long time. When the nature of the reflective consciousness was investigated by reflection, it was often asked whether "(strong) image" and "(weak) perception" are distinguishable in both psychological and epistemological dimensions [2]. The Pantheon's column is among the most known litmus tests for such distinction. Figure 1 does not indicate that mental images are actual pictures in the brain; obviously we cannot assume any homunculus in the brain who looks at the images. Nevertheless, we find imagery an important topic since it is one of the first higher cognitive abilities that will be firmly rooted in the brain and since understanding this ability would provide some

---

Address for correspondence: Prof Yasushi Miyashita, Department of Physiology, The University of Tokyo School of Medicine, 7-3-1, Hongo, Bunkyo-ku, Tokyo 113, Japan. Tel.: +81-3-3812-2111 (ext. 3459). Fax: +81-3-5800-6847. E-mail: physiol@m.u-tokyo.ac.jp

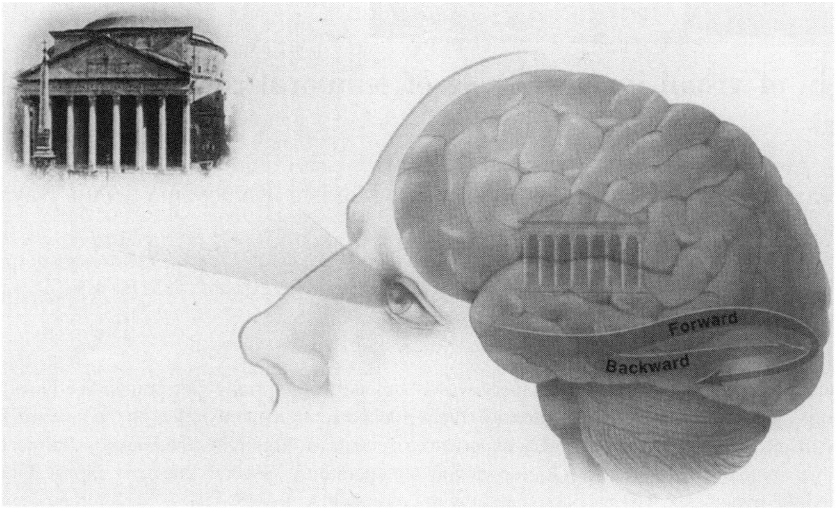

Fig. 1. Visual perception mainly relies on the forward, bottom-up flow of information. Imagery experience, in contrast, highlights the backward projections as an anatomical substrate of top-down mental operations. We may predict that, if an imagery task requires reconstruction of the detailed local geometry of the image (as counting the columns of the Pantheon), backward signals from higher-order representations would reach topographically organized visual areas. This cartoon does not indicate that mental images are actual pictures in the brain; obviously we cannot assume any homunculus in the brain who looks at the images. Adapted from Miyashita [8]; artwork by K. Sutliff.

constraints on the neural basis of consciousness.

Historically, imagery played a major role in theories of conscious phenomena, especially during the 18th century with the British Associationists or during the birth of philosophical psychology with William James and Wilhelm Wundt [3–5]. While radical behaviorism dominated experimental psychology and physiology, imagery was not regarded as an appropriate subject for scientific study. With the advent of cognitive psychology after the 1960s, it once again became acceptable. The discovery of "mental rotation" [6] convinced us that complex properties of imagery could be studied with the methodological purity of empirical science. Today there are many approaches (including neuroimaging, electrophysiological, psychophysical and neuropsychological ones) asking where and how in the brain the images of objects, scenes and living beings are generated, stored and maintained [5,7,8]. Even the question of to what degree the processes involved in visual perception and imagery share a common neural substrate was partially answered (see below). In this chapter we will sketch the current status of our understanding on imagery.

Visual system and backward-projection hypothesis of imagery

The visual system in the primate consists of a mosaic of more than 30 visual areas, starting from the primary visual cortex (V1 or striate cortex), which are functionally heterogeneous and hierarchically organized [9]. A visual stimulus on the retina is analyzed in terms of its elementary features, such as orientation, color, texture and depth, along low-level visual processing in the striate cortex and the prestriate areas, most of which are organized retinotopically. The analyzed features are integrated, along high-level visual processing in the temporal and parietal cortex, into a unique configuration that functions as the internal representation of an object and its spatial properties. A notable point here is that the connections among the visual areas are reciprocal via both forward and backward projections [9,10]. We have been testing a hypothesis ("backward-projection hypothesis") [8,10–12] that visual perception mainly relies on the forward, bottom-up flow of information and that imagery experience, in contrast, highlights the backward projections as a possible anatomical substrate of top-down mental operations (Fig. 1). The hypothesis predicts that, if an imagery task requires reconstruction of the detailed local geometry of the image (such as counting the columns of the Pantheon), backward signals from higher order representations would reach topographically organized visual areas [3,5,8].

The backward-projection hypothesis of imagery contains several components to be substantiated and tested. The first is on the nature of internal representation. There has been debate as to whether depictive (quasi pictorial) or propositional representation lies in the deepest structure for image generation [4,5,13]. The depictive theory assumes that the processes which produce and utilize visual images most likely rely on the high-level visual mechanisms. At present there are several lines of evidence for involvement of the high-level visual mechanisms. Studies performed using positron emission tomography (PET) consistently revealed activation of the dorsal (area 19) and ventral (fusiform gyrus) visual association areas, superior and inferior parietal cortex, as well as other nonvisual cortices such as the dorsolateral prefrontal cortex and angular gyrus, e.g., when subjects imagined that they started at their front door and then walked alternately to the left or the right each time they reached a corner [7,14] or when subjects imagined a letter in a grid and decided whether the letter would have covered a mark presented in the grid [5]. Perhaps one of the strongest supports for depictive representation in image generation was given by single-unit recording from visual association cortex. In monkey inferior temporal cortex, which is a candidate for the monkey homologue of the human visual area in fusiform gyrus, single neurons revealed picture-specific discharges when the monkey generated from its long-term memory store the image (paired associate) that was instructed by a previous cue stimulus [12,15]. Before considering the neuronal correlate of image generation, we briefly look at neuronal representation of visual long-term memory in the temporal cortex.

Neuronal representation of visual long-term memory in the temporal cortex

Along the occipitotemporal visual pathway from the primary visual cortex (Fig. 1), the physical properties of a visual object are analyzed in the multiple subdivisions of the prestriate-posterior temporal cortices, and the inferotemporal cortex synthesizes the analyzed attributes into a unique configuration [9,10,16]. The inferotemporal cortex has been proposed to be the memory storehouse for object vision on the basis of behavioral experiments [16,17]. But how the memories of objects are represented among neuronal networks in the inferotemporal cortex?

Most of our long-term memories of episodes are organized so that we can retrieve them by association. We propose that visual memory is organized by the same principle and encodes forms of objects as structured bundles of associations between elementary views of objects [18]. Evidence for the associative mechanism was obtained by training monkeys to memorize artificial associative relations among visual patterns, and then by examining whether picture-selective activities of inferotemporal neurons encode the stimulus-stimulus association imposed in the learning [10,19]. One example of such studies used the pair-association learning task [11,15] developed by Murray et al. [20]; computer-generated colored pictures were sorted randomly into pairs (Fig. 2A). The combination of the paired associates cannot be predicted without memorizing them beforehand, and macaque monkeys (*Macaca fuscata*) were trained to memorize these pairs. In each trial of the experiment, a cue stimulus was presented on a video monitor for 0.5 s. After a delay period, two choice stimuli, the paired associate of the cue (correct choice) and a distracter (incorrect choice), were shown. The monkey obtained a fruit juice reward when a correct choice was made. It is noted that this task is essentially the memory recall task, which demands memory retrieval and thus generation of images from long-term memory.

In the inferotemporal cortex, we recorded single neuron activity using a microelectrode and tested whether a cell preferentially responded to both of the paired associates. Figure 2B shows stimulus selectivity of a cell which responded reproducibly to only a few pictures. The strongest and the second strongest responses were ascribed to a particular pair which had no apparent geometrical similarity. We called this type of neuron a "pair-coding neuron" that manifested selective responses to both of the paired associates. Some other cells showed broader tuning and responded to more than three pictures. Nevertheless, paired pictures were found to be among the most effective stimuli for these cells (Fig. 2C).

We analyzed sampled cells by calculating response correlation with pair index or correlation coefficient [11,15]. The frequency distribution of either index demonstrated that the paired associates elicited significantly correlated responses. This result provides strong evidence that inferotemporal neurons acquire stimulus selectivity through associative learning and that the selectivity reflects a stimulus-stimulus

---

Fig. 2. Associative representation of pictures in neurons of the temporal cortex. Stimulus-selectivity of neurons encodes stimulus-stimulus associations among geometrically different complex forms, which were learned in the visual pair association task. **A**: Twelve pairs of Fourier descriptors (1 and 1′ to 12 and 12′) used as stimuli in the pair-association task. When one member of each pair is shown, trained monkeys can retrieve and choose the other member of the paired associates. **B**: Stimulus-selectivity of a pair-coding neuron. Ordinate shows mean discharge rates for each cue presentation relative to the spontaneous discharge rate (denoted by an arrowhead). Cue stimuli are labeled as "pair No." on the abscissa (light histogram bar in No. 1: 1, dark histogram bar in No. 1: 1′, etc.). This neuron selectively responded to both pictures of the paired associates 6 and 6′. **C**: Similar to B but for another neuron.

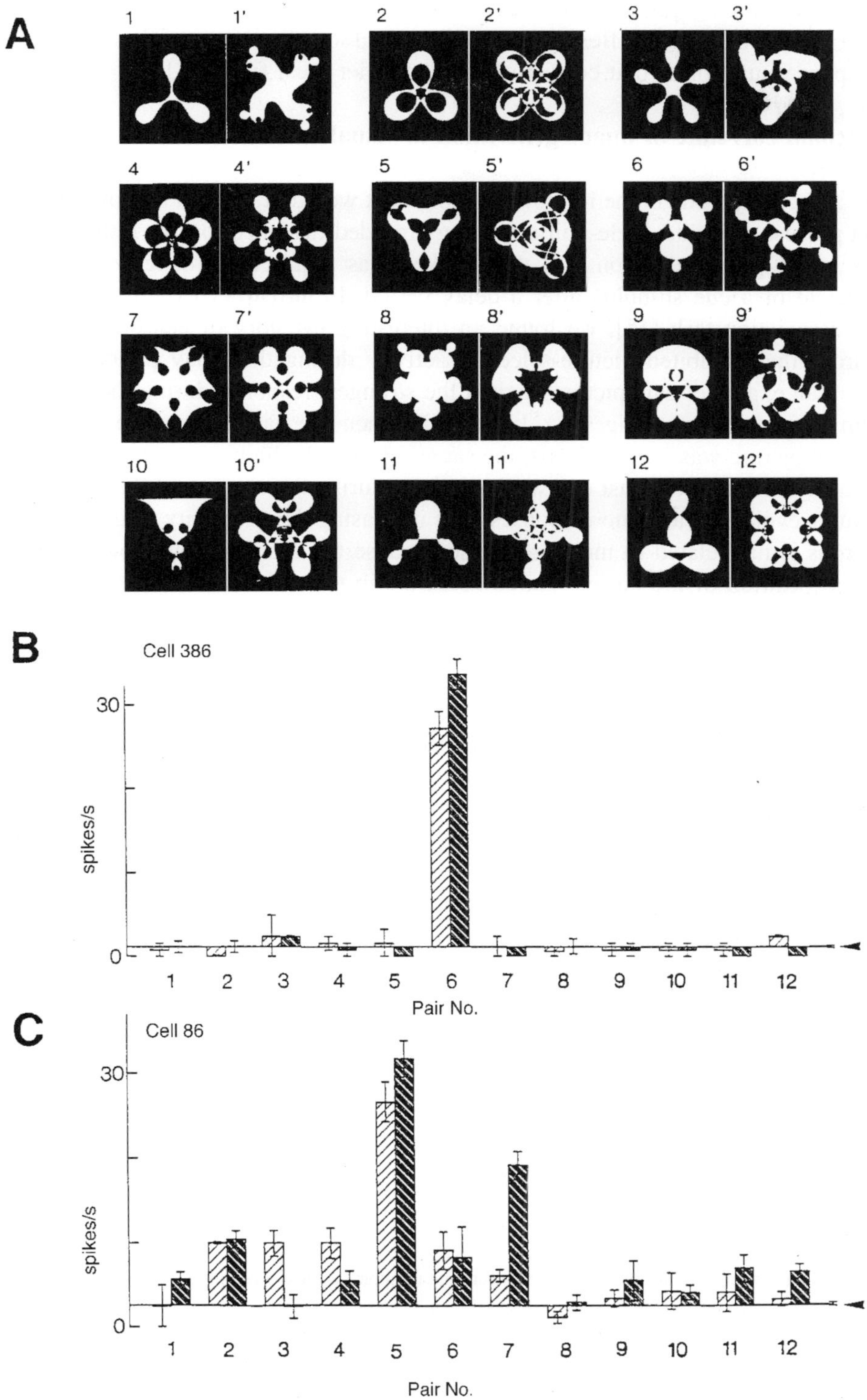
A
1
1'
2
2'
3
3'
4
4'
5
5'
6
6'
7
7'
8
8'
9
9'
10
10'
11
11'
12
12'
B
Cell 386
30
0
spikes/s
1 2 3 4 5 6 7 8 9 10 11 12
Pair No.
C
Cell 86
30
0
spikes/s
1 2 3 4 5 6 7 8 9 10 11 12
Pair No.

association among geometrically different complex forms. It also suggests that the neuronal mechanisms for flexibly creating those associations is crucial in memorizing complex visual forms that consist of many simpler local figures [18].

Neuronal correlate of image-generation in visual association cortex

The associative code in the inferotemporal cortex would be used in the image generation process. Indeed, single-unit recording provided evidence for this [10,12]. Recall that in the pair-association task the monkey was required to recognize the paired associate of a cue stimulus after a delay period. In an earlier experiment with the pair-association task [15], we found an interesting category of neurons (pair-recall neuron) that exhibited picture-selective activity during the delay period. Figure 3 shows an example. One picture elicited the strongest response during the cue period from a single neuron (Fig. 3A). In the trial when the paired associate of this cue-optimal picture was used as a cue, the same cell had the highest tonic activity during the delay period in contrast to a weak response during the cue period (Fig. 3B). Other pictures evoked weal or no responses. The increasing delay activity of the pair-recall neurons is not related to motor response because the monkey could not predict the spatial position of the correct paired associate in the task. This activity is not only picture-selective, but also closely coupled with the paired associate that is not actually seen but retrieved. The discovery of the pair-recall neuron was the first demonstration of the neuronal correlates of visual imagery.

These neural mechanisms were further investigated by developing a novel task, the pair-association with color switch (PACS) task, in which the necessity for image generation and its time were controlled by a color switch in the middle of the delay period [12]. A control task, in which there is no color switch, corresponds to the conventional delayed matching-to-sample (DMS) task where the monkey chooses the

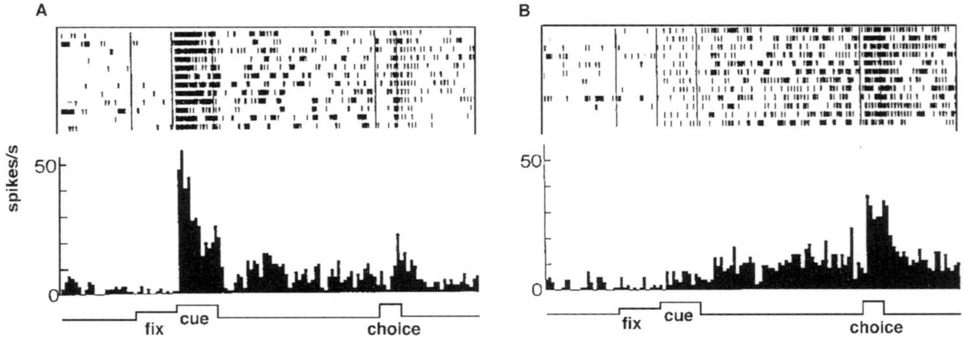

Fig. 3. A neuronal correlate of image-retrieval in the temporal cortex. Responses of a pair-recall neuron, which exhibits picture-selective activity while the subject retrieves the image of the paired associate in the pair association task. **A**: Trials for the cue 12 that elicited the strongest cue response. **B**: Trials for the cue 12′ that elicited the delay response, reflecting retrieval of the paired associate 12. Note the tonic increasing activity during the delay period, which is much stronger than the cue response. Note: fix = fixation period; cue = cue stimulus presentation period; choice = choice stimuli presentation period.

same picture as a cue. In both trials, 12 pairs of Fourier descriptors were used as visual stimuli (G1 and C1-G12 and C12), each pair containing a green picture and a cyan picture (Fig. 4A). The sequence of events in a trial of the PACS or the DMS was as follows (Fig. 4B). When the monkey started to keep pressing a lever in front of the video monitor, a white square was presented at the center of the screen for 1 s (warning). Following the cue presentation of one of 24 pictures for 0.5 s, a square was presented during the delay period. The square's color was the same as the cue's color during the first part of the delay period (delay period 1) for 2 s in the PACS trial or for 5 s in the DMS trial. In the PACS trial, the square's color changed into the color of the paired associate after delay period 1, signaling the initiation of retrieval, and the second part of the delay period (delay period 2) for 3 s started. Delay period 2 was not included in the DMS trial. To balance the visual stimulus conditions in the two trials, a white square was presented for 1 s during the third part of the delay period (delay period 3). After delay period 3, a choice of two stimuli was shown randomly in two of four possible positions (arranged in two rows of two columns). The choice stimuli were the paired associate of the cue (correct) and a distracter (error) in the PACS trial, while the choice stimuli were the same picture as the cue (correct) and a distracter (error) in the DMS trial. The animal obtained a reward for touching the correct picture within 1.2 s. The PACS trials and the DMS trials were randomized. Extracellular discharges of single neurons were recorded in three hemispheres with a glass-insulated tungsten microelectrode.

Figure 5 shows typical responses of a single inferotemporal neuron in PACS task. One picture (G7; light grey – see legend) elicited the strongest response during cue period (Fig. 5, top; here we call this cue-optimal picture the best picture, irrespective of its delay response). The excitatory response was maintained in delay period 1 (Fig. 5, middle). The paired associate (C7; dark grey – see legend) of the best picture (G7) elicited little response during cue period and delay period 1; however, this neuron started to respond just after delay period 2 onset when the square's color changed from the cue's color (C7; dark grey) to that of the paired associate (G7; light grey). The picture-selective activation after the color switch in the PACS task is called the "pair-recall" effect. The pair-recall effect continued from delay period 2 into delay period 3 in which the square's color was the same white in both tasks (Fig. 5, bottom). Thus, the pair-recall effect was not due to the square's color; this was also confirmed by the fact that the pair-recall effect was observed in trials with cue C7 but not in trials where other cyan pictures were used as a cue. In the DMS task, the response selectivity was the same as that of the PACS task during the cue period and delay period 1. However, the color switch of the square had no effect in the DMS task. These results suggest that this delay discharge which is specific to the PACS task was triggered by image generation.

To characterize the pair-recall effect in 15 task-related cells, we collected trials whose cue picture was the best picture's associate. The effect of task on the delay response was significant ($F(1,14) = 19.8$, $p < 0.001$), and there was an interaction between task and period ($F(1,28) = 12.2$, $p < 0.005$). We further analyzed responses among periods in each task, and responses between the tasks in each period. In both

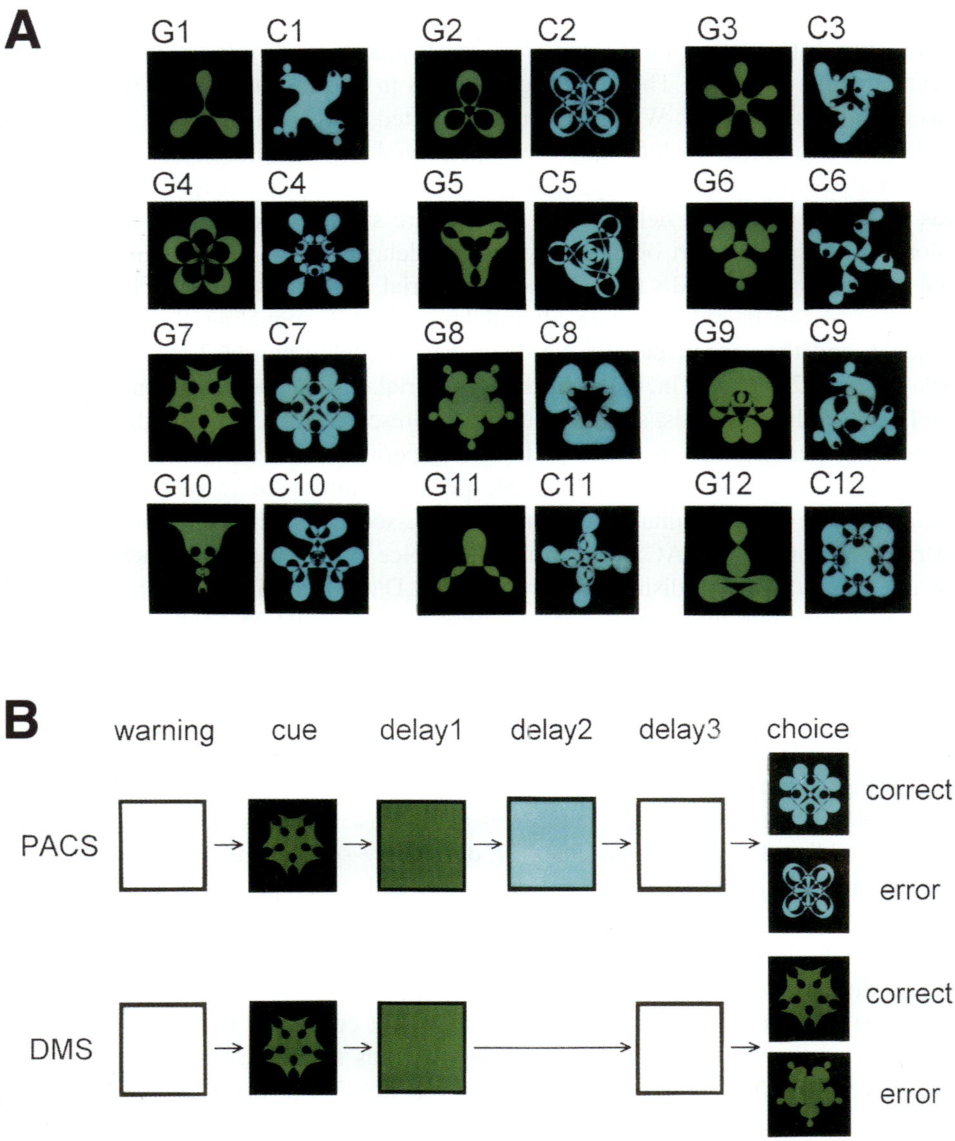

Fig. 4. The pair-association with color switch (PACS) task and an example of neuronal responses in the temporal cortex. **A**: Twelve pairs of colored Fourier descriptors used as stimuli in both PACS task and DMS task. Similar to Fig. 2A, but now stimuli are color-coded for controlling the necessity for image generation and its initiation time. The first pair is picture G1 (green) and picture C1 (cyan), the second pair is G2 and C2, etc. **B**: Sequence of events in a trial of the PACS task or the DMS task. Cue stimuli and squares were presented at the center of a video monitor. Choice stimuli were presented randomly in two of four positions on the video monitor. Warning: white square (1 s in both tasks); cue: one of 24 pictures in A as a cue stimulus (0.5 s); delay period 1: square which has the same color as the cue picture (2 s in the PACS task, 5 s in the DMS task); delay period 2: square which has the same color as the paired associate of the cue picture (3 s in the PACS task); delay period 3: white square (1 s in both tasks); choice: a choice of two stimuli (1.2 s in both tasks), the paired associate of the cue (correct) and one from a different pair (error) in the PACS task, or the same picture as the cue (correct) and one from a different pair (error) in the DMS task.

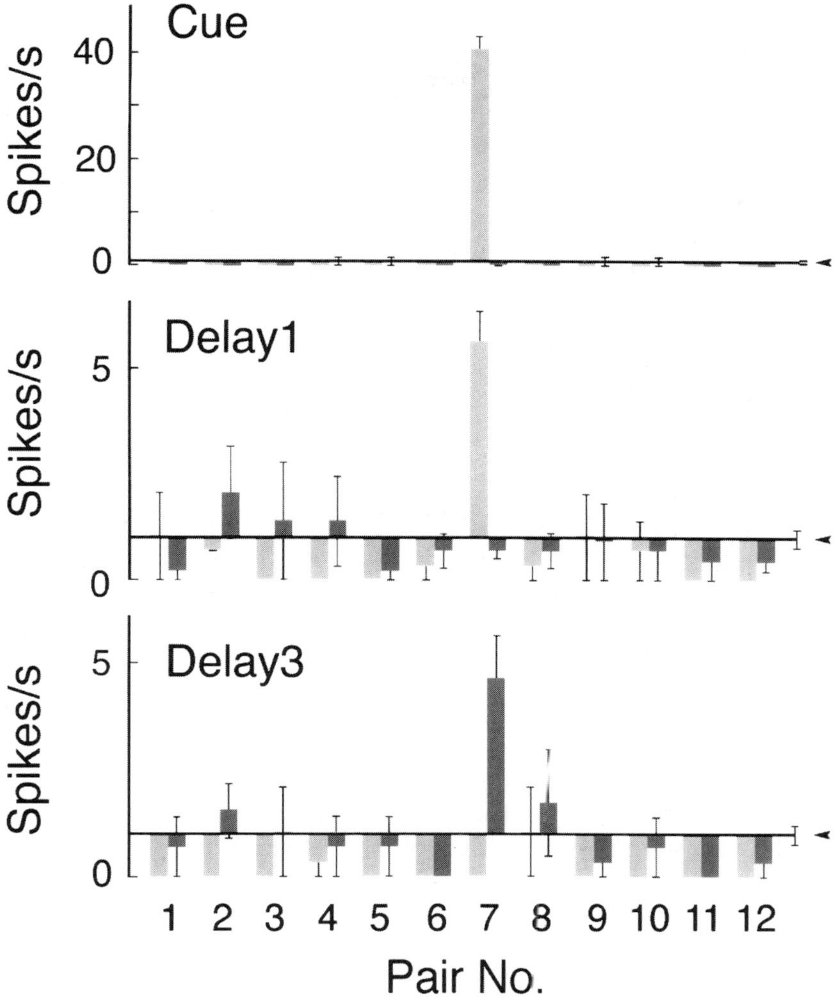

Fig. 5. Image generation triggered by the color switch in the PACS task. Stimulus-selectivity of a neuron in the temporal cortex is shown in different task periods (top: cue period; middle: delay period 1; bottom: delay period 3). Ordinate shows the mean discharge rate in each task period after the presentation of a cue stimulus; cue stimuli are labeled as "pair No." on the abscissa (light grey histogram bar in No. 1: G1; dark grey histogram bar in No. 1: C1, etc.). This neuron selectively responded to the cue stimulus G7 during the cue period. Note that the cue stimulus C7 does not elicit any response during the cue period and delay period 1, but this neuron became active during delay periods 2 and 3.

PACS and DMS tasks, responses in delay period 1 remained equal to warning responses. These delay period 1 responses were thus task independent. In delay period 3 of the PACS task, the responses were significantly stronger than those in delay period 1 ($t = 3.5$, $p < 0.005$), while the delay period 3 responses in the DMS task remained equal to the delay period 1 responses. The delay period 3 responses in the PACS task were significantly stronger than those in the DMS task ($t = 5.2$, $p < 0.001$). These results indicate that the pair-recall effect is triggered by the color switch.

All of the above results support the hypothesis that these neural discharges represent active images of a specific object. However, how this neural firing comes out? Or, what kind of machinery can drive the pair-recall neuron in delay period 2 in spite of null activity in delay period 1? Obviously we cannot assume any homunculus or "quantum microtubule". You may assume an excitatory input from nonvisual association cortex such as the dorsolateral prefrontal cortex. However, the activity in delay period 2 is highly stimulus-selective, and we must remember that the inferotemporal cortex is the highest visual center of the primate brain that has specific visual information of an object. The neural mechanism for the image generation remains to be further identified, but it should involve the pair-recall neurons in the temporal cortex.

Imagery generation requires the backward projection to the primary visual cortex

So far, we described the neuronal origin of visual imagery in the inferotemporal cortex in the case of a visual object, or more generally, the higher-order association areas in the temporal and parietal lobes in the case of images of scenes or living beings. At present, there is general agreement on this issue with neurophysiological and neuroimaging approaches. Contrary to this agreement on high-level visual mechanisms, evidence from neuroimaging appears to both support and contradict the hypothesis that visual imagery is based upon contributions of the early visual areas [14,21]. There are some possible sources of the conflicting observations. First, is the difference in task paradigms, which would result in different requirements for scrutinizing local features of objects or for attention. Second, is the contribution of weak signals near the threshold of detection sensitivity; intersubject averaging, used to enhance signal-to-noise ratios in PET, might mask physiological and morphological variances among subjects. This difficulty might be overcome partially by the use of fMRI [22].

Psychophysical studies have also addressed this problem on the basis of neuronal response characteristics in the primary visual cortex (VI), which are specific to the stimulated eye as well as to orientation and retinal location of the stimulus (Fig. 6). For example, subjects decided whether two line segments were perfectly aligned (a "vernier acuity" task), and accuracy decreased from 80–65% when the subjects formed images compared to when they did not [23]. Moving the image away from the target almost eliminated the effect (location specificity); however, effects observed when the subjects formed images of vertical lines were comparable to those observed when the subjects formed images of horizontal lines (no orientation specificity). It was demonstrated [24] that the McCullough effect (orientation-specific color after effect) could be induced via imagery. However, the imagery effect was transferred from one eye to the other, while the actual McCullough effect is not. Thus, from the

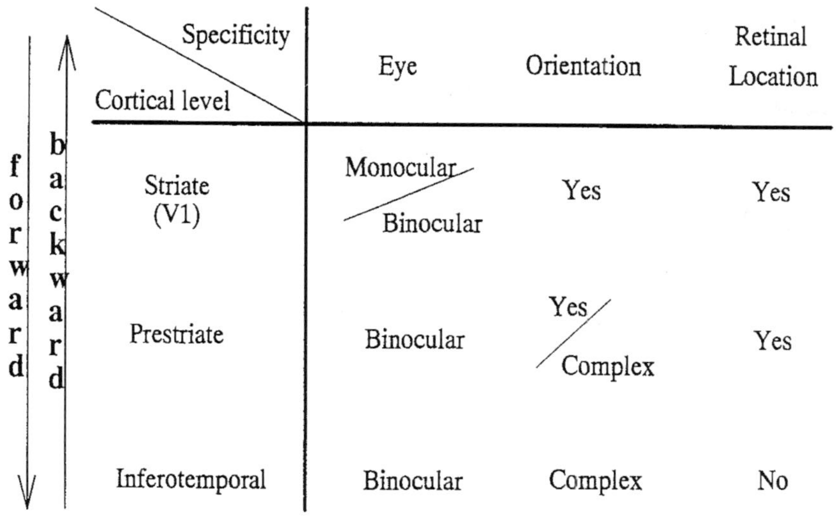

Fig. 6. Psychophysical studies address the problem of brain areas involved in imagery by transfer tests. The test is based on neuronal response specificity in various visual areas. If the imagery experience is specific to the stimulated eye, as well as to orientation and retinal location of the stimulus, it strongly suggests involvement of the striate cortex.

results of these psychophysical experiments, involvement of early visual areas (especially V1) in visual imagery have been unable to be ascertained.

Recently, a new approach was devised [25], which investigated imagery effect upon visual perception in a psychophysical detection paradigm. They compared human subjects' abilities to detect a computer-generated "Gabor target" under two conditions; one was the perceptual condition in which the target was flanked by two peripheral Gabor masks, and the other was the imagery condition in which the subjects imagined the absent masks. Experiments under the perceptual condition showed enhancement of detection of a visual target when the subjects perceived actual visual mask stimuli, revealing the existence of facilitatory interactions between spatial channels. The enhancement under the perceptual condition was specific to the stimulated eye, and orientation, spatial frequency and retinal location of the stimulus [26]. Surprisingly the same facilitatory characteristics were found under the imagery condition, such as monocularity, and orientation and retinal location specificity. This observation strongly suggests that the mental images can be interfaced with perceptual representations at early stages of visual information processing (Fig. 6).

Discussion

We have examined several lines of empirical evidence that support the backward-projection hypothesis of visual imagery. The pair-association with color switch (PACS) task, in which the necessity for image generation and its initiation time were controlled by a color switch, provided particularly compelling evidence for the role of the temporal cortical neurons. The hypothesis also predicted that backward signals from higher order representations would reach topographically organized visual areas when an imagery task requires reconstruction of the detailed local geometry of the image (such as counting the columns of the Pantheon). This prediction was confirmed by an elegant psychophysical experiment with Gabor-target detection task.

However, in those experiments the image was generated only in a simple way, in which a subject recalled a previously seen object or event. We use imagery in more flexible ways. First, we can combine or arrange objects in novel configurations in images. Second, we can also mentally draw patterns that we have never actually seen, or we can visualize novel patterns that are not based on rearranging familiar components. To make these image computations possible, imagery system should contain another subsystem that uses associative memory and that constructs local representations on the basis of a global image by top-down attentional shift [5]. Anatomical localization of the subsystem might be hypothesized in the dorsolateral prefrontal cortex, but evidence is still scanty [5,27,28]. It is noted that, although we believe depictive representation lies in the deepest structure for image generation (as described above), propositional representation may also generate imagery by the use of this subsystem, especially in humans [13]. Characterization and neurophysiological analysis of this subsystem would provide further constraints on the neural basis of imagery and its conscious activation.

References

1. Alain EAC. Systeme des Beaux-Art. Paris 1926.
2. Sartre JP. L'Imaginare. Paris: Gallimard, 1938.
3. Farah MJ. Visual Agnosia. Cambridge: MIT Press, 1990.
4. Tye M. The Imagery Debate. Cambridge: MIT Press, 1991.
5. Kosslyn SM. Image and Brain. Cambridge: MIT Press, 1994.
6. Cooper LA, Shepard RN. Chronometric studies of the rotation of mental images. In: Chase WG (ed) Visual Information Processing. New York: Academic Press, 1973;76–176.
7. Roland PE. Brain Activation. New York: Wiley-Liss, 1993.
8. Miyashita Y. How the brain creates imagery. Science 1995;268:1719–1720.
9. Van Essen DC, Anderson CH, Felleman DJ. Information processing in the primate visual system: An integrated systems perspective. Science 1992;255:419–423.
10. Miyashita Y. Inferior temporal cortex: Where visual perception meets memory. Annu Rev Neurosci 1993;16:245–263.
11. Higuchi S, Miyashita Y. Formation of mnemonic neuronal responses to visual paired associates in inferotemporal cortex is impaired by periphinal and entorhinal lesions. Proc Natl Acad Sci USA 1996;93:739–743.
12. Naya Y, Sakai K, Miyashita Y. Activity of primate inferotemporal neurons related to a sought target in a pair-association task. Proc Natl Acad Sci USA 1996;93:2664–2669.
13. Pylyshyn ZW. The imagery debate: Analogue media versus tacit knowledge. Psychol Rev 1981;87: 16–45.
14. Roland PE, Friberg L. Localization of cortical areas activated by thinking. J Neurophysiol 1985;53: 1219–1243.
15. Sakai K, Miyashita Y. Neural organization for the long-term memory of paired associates. Nature 1991;354:152–155.

16. Mishkin M. A memory system in the monkey. Phil Trans A Soc Lond B 1982;298:85–95.
17. Squire LR. Memory and Brain. London: Oxford University Press, 1987.
18. Miyashita Y, Date A, Okuno H. Configurational encoding of complex visual forms by single neurons of monkey temporal cortex. Neuropsychologia 1993;31:1119–1132.
19. Miyashita Y. Neuronal correlate of visual associative long-term memory in the primate temporal cortex. Nature 1988;335:817–820.
20. Murray EA, Gaffan D, Mishkin M. Neural substrates of visual stimulus-stimulus association in rhesus monkey. J Neurosci 1993;13:4549–4561.
21. Kosslyn SM, Alpert NM, Tompson WL, Maljkovic V, Weise SB, Chabris CF, Hamilton SE, Rauch SL, Buonanno FS. Visual mental imagery activates topographically organized visual cortex: PET investigations. J Cogn Neurosci 1993;5:263–287.
22. Le Bihan D, Turner R, Zeffiro TA, Cuenod CA, Jezzard P, Bonnerot V. Activation of human primary visual cortex during visual recall: A magnetic resonance imaging study. Proc Natl Acad Sci 1993;90: 11802–11805.
23. Craver-Lemley C, Reeves A. Visual imagery selectively reduces vernier acuity. Perception 1987;16: 533–614.
24. Kaufman JH, May JG, Kunen S. Interocular transfer of orientation-contingent color aftereffects with external and internal adaptation. Percept Psychophysics 1981;30:547–551.
25. Ishai A, Sagi D. Common mechanisms of visual imagery and perception. Science 1995;268: 1772–1774.
26. Polat U, Sagi D. Spatial interactions in human vision: From near to far via experience-dependent cascades of connections. Proc Natl Acad Sci 1994;91:1206–1209.
27. Fuster JM. The Prefrontal Cortex. New York: Raven Press, 1989.
28. Miyashita Y, Chang HS. Neuronal correlate of pictorial short-term memory in the primate temporal cortex. Nature 1988;331:68–70.

Integrative and Molecular Approach to Brain Function.
M. Ito and Y. Miyashita, editors.

Functional architecture and synaptic plasticity in the monkey inferior temporal cortex

Ichiro Fujita, Makoto Kato, Yusuke Murayama and Takanori Uka
Department of Cognitive Neuroscience, Osaka University Medical School, Osaka, Japan

Abstract. Area TE of the inferior temporal cortex in the monkey has a columnar organization. Most neurons in the TE respond selectively to particular visual features of objects, and those with similar stimulus preferences are clustered in vertical arrays roughly 0.5 mm in width. Evidence for the columnar organization has been obtained in anesthetized monkeys, but is now extended by experiments in awake monkeys. Columns interact with each other via intra-areal horizontal axons which connect laterally displaced sites in the TE. High-frequency electrical stimulation of horizontal axons evokes long-term potentiation (LTP) of synaptic transmission efficacy mediated by this axon system. In the primary visual cortex, however, an identical stimulus protocol produced long-term depression in the same fiber system. The success of evoking LTP in the TE in vivo will allow us to investigate in future studies whether and how LTP affects stimulus selectivity of TE neurons and the functional columnar organization of this area. The results also suggest that visual cortical areas are diverse not only in their anatomical structures and receptive field properties, but also in their synaptic plasticity.

Key words: columns, horizontal axon, long-term depression, long-term potentiation, primate, visual memory, visual recognition.

Introduction

Cellular/subcellular-level analysis of synaptic plasticity and systems-level analysis of cognitive functions represent two major research fields in neuroscience today, yet there is scant interaction between them. In studies on experimental models of synaptic plasticity such as long-term potentiation (LTP) and long-term depression (LTD), research is mostly directed toward elucidating the subcellular and molecular bases of these phenomena, and rodents are commonly used as subjects (reviewed in [1–4]). The number of studies on the biological significance of LTP and LTD is increasing but still small [4–9]. In contrast, monkeys are a favorite subject for systems-level studies of cognitive functions because of their highly developed perception, recognition and motor control. The nature of synaptic plasticity in the monkey brain has not been explored despite the fact that many aspects of cognitive functions involve learning and memory components.

Several lines of evidence suggest that the neocortex, rather than the hippocampus and its associated structures, is the site for storage of long-term recognition memory in the primate [10–13]. The natural equivalents of LTP and LTD are thought to

Address for correspondence: Ichiro Fujita, Department of Cognitive Neuroscience, Osaka University, Medical School, Suita, Osaka 565, Japan. Tel.: +81-6-879-2600. Fax: +81-6-879-2609. E-mail: ifujita@cogni.med.osaka-u.ac.jp modify neuronal circuits in the brain and underlie certain types of memory (e.g., spatial memory in rodents). If this hypothesis holds for visual recognition memory, LTP or LTD should occur in the monkey visual cortices, particularly in the inferior temporal cortex where long-term memory of visual stimuli is stored and can be reorganized even in adulthood (Fig. 1) [14,15].

A goal of our current project is to bridge the gap between the two research fields discussed above. We started our analysis at an intermediate level of brain organization, i.e., neuronal circuit level, to outline the functional and anatomical architecture of area TE of the monkey inferior temporal cortex. Neurons in this area are arranged into columns according to their selective responsiveness to visual features of objects [16]. In this article, we present new evidence, obtained in awake monkeys, of the columnar organization of the TE [17]. Perpendicular to the columnar array of neurons, intrinsic (i.e., intra-areal) horizontal axons traverse the TE cortex [18]. Horizontal axons can mediate interaction between columns and are suggested to be important for regulating representation in some cortical maps [19,20]. We investigated whether LTP and/or LTD can occur at synapses activated by signals from horizontal axons in the TE.

Columnar organization in the inferior temporal cortex

Most neurons in the TE selectively respond to a particular range of stimuli which constitute visual features of objects [21–23]. Preferred object features vary among neurons, but are not as diverse as the number of the neurons. Two adjacent neurons recorded with a single electrode respond maximally or almost maximally to the same or similar stimuli [16]. The most effective stimuli or suboptimal stimuli, however,

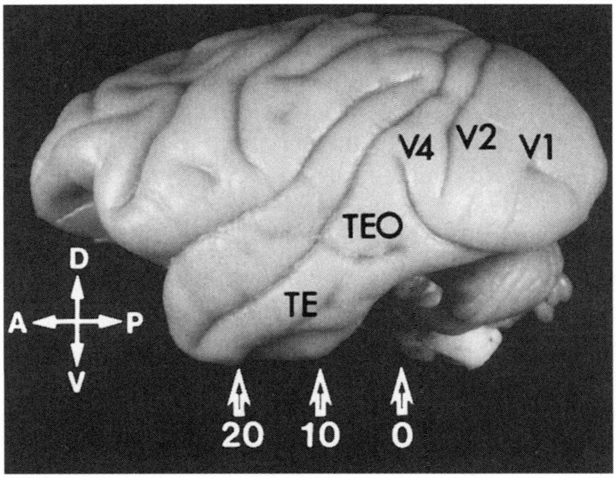

Fig. 1. Five major cortical areas along the visual pathway critically involved in object recognition. V1: the primary visual cortex; V2 and V4: two prestriate cortices; TEO and TE: cytoarchitectonic areas TEO and TE of the inferior temporal cortex. The brain is from a Japanese macaque. Numbers indicate approximate distance (mm) from the external meatus. A: anterior; P: posterior; D: dorsal; V: ventral.

slightly differ between adjacent neurons. The effective stimulus features vary among vertical penetrations 1 mm apart across the cortical surface. Neurons with similar, but not identical, stimulus selectivities are thus clustered locally within the TE [16,24,25].

Neurons with similar selectivities are arranged in columns in the TE: neurons selective for the same or similar stimuli are distributed vertically across most of the cortical layers, but are localized within a tangential extent of 0.4–0.5 mm (Fig. 2) [16]. Preferred stimuli are shared by neurons within a column, although the optimal stimulus or response tuning properties differ among neurons as is already evident from the results of simultaneous recordings of adjacent neurons. Neighboring columns respond to distinct object features. Several columns with similar stimulus selectivities appear to exist, based on the finding that in tangential penetrations, the same stimuli activated two or three clusters of neurons which were separated by a cluster of neurons unresponsive to the stimuli tested or ones selective for other stimuli. The number of columns in one TE is estimated to be 1,300–2,000 [15,16]. The clustering of TE neurons with similar stimulus selectivities has recently been confirmed by optical imaging techniques [26].

Evidence in awake monkeys of the columnar organization of the TE

The evidence summarized in the preceding section was obtained in anesthetized monkeys. The procedure used to determine the stimulus preference of TE neurons in anesthetized monkeys (referred to as the simplification procedure or reduction process in [15] and [27]) is time-consuming and cannot readily be applied to awake monkeys.

In the present experiments we implanted five guide cannulae perpendicular to the cortical surface of the TE. The cannula tips were placed below the dura mater, but above the brain surface (Fig. 3A). Recording electrodes were aimed repeatedly at the same site through these cannulae. We first searched for effective stimuli for neurons

Fig. 2. Columnar organization of the TE. Neurons with shared stimulus selecitivities are grouped together to form a column 0.4–0.5 mm wide.

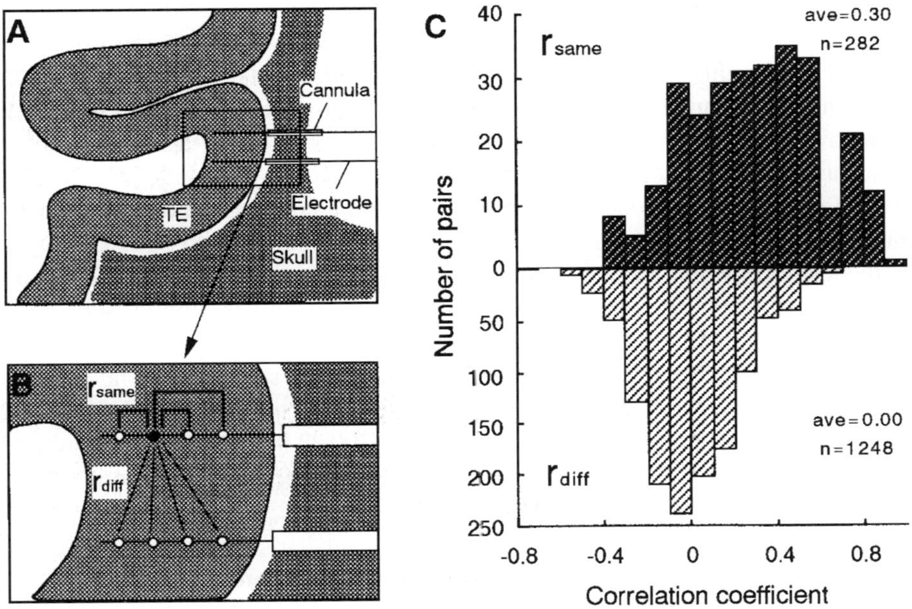

Fig. 3. Evidence in awake monkeys of the columnar organization of the TE. **A**: Experimental setup. Five guide cannulae (two of them are shown in this Figure) were implanted perpendicular to the cortical surface of the TE. Through these cannulae, neuronal recordings were made under both anesthetized and awake conditions. **B**: Correlation coefficients for normalized responses were calculated among pairs of neurons in the same (r_{same}) and different (r_{diff}) penetrations. **C**: Distribution of r_{same} and r_{diff} under awake conditions. Coefficients are significantly larger within the same penetration than between different penetrations, suggesting that neurons with similar response profiles are clustered along vertical penetrations.

along a penetration under anesthetized conditions. For some neurons, critical stimulus features were determined using the simplification procedure. By repeating the same experiments for the five cannulae, penetration-specific effective stimuli were determined. We then devised a stimulus set by combining 40 effective stimuli for neurons along the five penetrations with another 40 stimuli which were arbitrarily chosen. Two to 14 days after the experiments under anesthetized conditions, responses of TE neurons were recorded through the five cannulae under awake conditions. While the monkey fixated a small spot on a color LCD monitor, the predetermined set of 80 stimuli was presented in a pseudorandom order.

We recorded 62 neurons under anesthetized conditions and the same number under awake conditions. For each neuron, responses to the 80 stimuli were averaged over seven to 10 stimulus presentations, and were normalized to its own maximal response. Correlation coefficients were then calculated for normalized response magnitude among pairs of neurons in the same and different penetrations (r_{same} and r_{diff}, respectively; Fig. 3B). The average of r_{same} was 0.30 ± 0.09 whereas r_{diff} was averaged to 0.00 ± 0.05 (Fig. 3C). Twenty-seven percent of r_{same} were larger than 0.5, whereas only 2% of r_{diff} exceeded 0.5. Under anesthetized conditions, the average of r_{same} was smaller (0.14 ± 0.07) than that under awake conditions, but was significantly larger than the average of r_{diff} (0.00 ± 0.03) ($p < 0.001$, t test). These results show a stronger correlation among neurons within a vertical penetration than among neurons in different penetrations under both awake and anesthetized conditions, indicating that neurons in the TE are organized into columns according to their stimulus selectivity.

Horizontal axon system

As in other cortices, vertically running axons across cortical layers and horizontal axons running roughly parallel to the cortical surface are the two major components of intrinsic neuronal connections in the TE [18]. Horizontal axons are present in all layers, and are most numerous in layers 2 and 3. Some of them extend over several millimeters within the TE. These long-range horizontal axons give rise to a few to more than 15 patches of terminal arborization. These patches are columnar in shape in the coronal plane, spanning from layer 1 to layer 3 or sometimes to layer 5, and are 0.5 mm wide on average. Neurons at a given site in the TE are thus preferentially connected to a particular set of neurons which tend to be distributed in columnar clusters, and are not diffusely and randomly connected with other TE neurons.

Most horizontally projecting neurons are pyramidal neurons [28]. The contribution of GABAergic neurons to the horizontal axon system is small and limited to short-distance projections (<1 mm). Horizontal axons are therefore predominantly excitatory in their direct synaptic interactions.

The size of patches of horizontal axon terminal arborization roughly corresponds to the width of columns measured in unit recording experiments [16] and by optical imaging [26]. The spacing of nearby patches is also consistent with the spacing of multiple clusters of neurons with similar stimulus preferences [16]. Relations between patches of horizontal axon terminal arborization and functional columns, however, have not been analyzed in the TE.

Long-term potentiation and depression in the primate neocortex

The evidence that modification in horizontal axons changes a coded representation in the primary visual and motor cortices [19,20] prompted us to investigate whether synaptic transmission mediated by horizontal axons in the TE is modifiable [29].

Extracellular field potentials were recorded in layer 2/3 of the TE, and stimulating electrodes were placed 0.5–1 mm lateral to the recording site to activate horizontal axons. A single electrical stimulus (50 μs duration, 100 μA amplitude, negative monopolar) evoked monophasic negative field potentials with a peak latency of around 5 ms in layer 2/3. The results of analyses of depth profiles of field potentials and current source-density, effects of stimulus position, together with the results of anatomical analyses of horizontal axons discussed above, suggest that this potential largely reflects ensemble excitatory synaptic currents in layer 2/3 (Y. Murayama, I.

Fujita, M. Kato, unpublished observation).

After a high-frequency conditioning stimulus (20–40 pulses at 40–100 Hz, applied every 4 s for 3–5 min, 100–400 μA), the amplitude (Fig. 4) and initial negative slope of field potentials evoked by a single stimulus gradually increased over 50–70 min to a maximum, and this potentiation lasted for more than 3 h [29]. The slow time course of LTP in the TE contrasts with that found in rodent hippocampus where a maximal level of potentiation is induced immediately after a conditioning stimulus. When a new memory is stored into a cortical area, patterned neural inputs are thought to cause changes in synaptic strength at distributed synapses. It has been proposed that the neuronal network accumulates small changes caused by input activity at individual synapses, and that it gradually adjusts the distribution of synaptic strength to find a state where both new and already existing memories can coexist [30,31]. The slow time course of LTP in the TE may reflect the properties required for this gradual adjustment.

When the same experiments were performed in the primary visual cortex (V1), an identical stimulus protocol did not potentiate field potentials in the V1, but instead caused a depression. This depression developed over 5 to 10 min and remained stable until the end of recording. Steady-state depression with a smooth initial development indicates that the depression is a physiological phenomenon and not an experimental artifact due to, for example, damage to the stimulating site. In both areas, field potentials evoked from an unconditioned pathway were not changed in their amplitude and waveform. Thus, both LTP in the TE and LTD in V1 were homosynaptically induced and maintained [29].

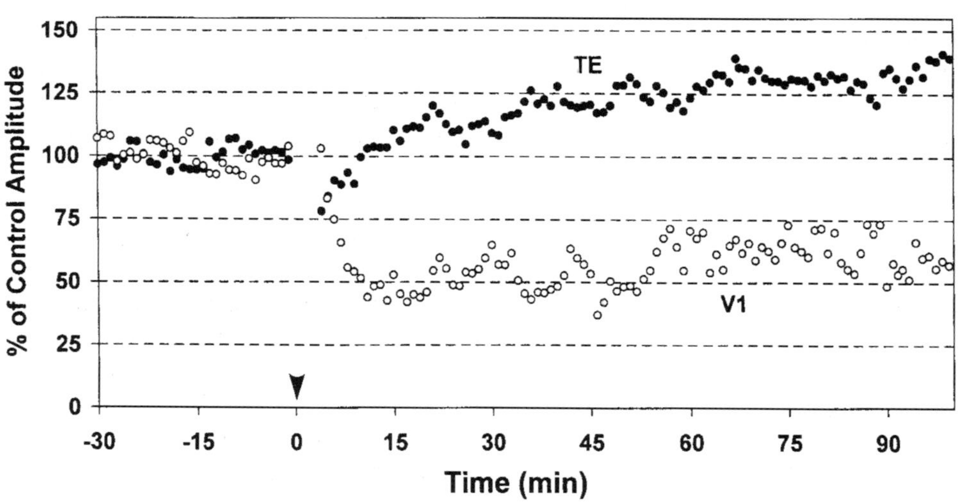

Fig. 4. LTP and LTD in the horizontal axon pathways of the TE and V1. Changes in the amplitude of extracellular field potentials evoked by electrical stimulation of horizontal axons are plotted against time. Arrowhead indicates time of application of a conditioning tetanic stimulus. After the tetanic stimulation, field potentials are potentiated in the TE, whereas those in V1 are depressed.

The present demonstration of LTP and LTD in the TE and V1 represents the first in any part of the monkey brain. It is surprising that, despite their widely proposed role in memory function, LTP and LTD were not previously experimentally shown in monkeys, the animal group with the most developed learning abilities.

Stimulus selectivity of neurons in the columnar organization of the TE can be changed by learning of novel visual stimuli [15,32]. Intrinsic horizontal axons are crucial for maintaining a cortical map in the primary visual cortex [19,33] (but see [34] for another interpretation). The ability of the TE and V1 to undergo changes in functional connections in the horizontal axon system may provide a clue to the mechanism for these changes induced by learning or altered sensory inputs.

The same conditioning stimulus protocol produced contrasting effects in the TE and V1. This suggests that monkey visual cortical areas are diverse in the nature of their synaptic plasticity in addition to other characteristics such as cyto- or chemo-architecture, input and output connections, and functional properties of neurons. The cellular and molecular basis of the difference between the two areas remains to be clarified, although differences in the distributions of some neurochemicals related to synaptic plasticity may partially account for this: the concentrations of protein kinase C (γ subtype) which is implicated in expression of LTP [35,36], and the phosphorylation level of its substrates (protein F1, also known as GAP-43, and a 81 kDa protein) are higher in the TE than in the V1 [37,38].

Conclusions

The confirmation of the columnar organization of the TE of awake monkeys now takes us a step further to accepting that this organization plays a role in object recognition. The success of evoking LTP and LTD in vivo provides us with a unique opportunity to investigate whether and if so how LTP and LTD alter functional properties of cortical neurons through examination of visual responses of neurons before and after expression of LTP or LTD, and opens a way to link systems-level analysis of the primate visual system with studies on LTP and LTD. The primate visual system represents a case where the organization of the system, functional and anatomical architecture of each cortical area, characterization of neuronal properties, and behavioral capacity have been extensively studied. Bringing the paradigm of LTP and LTD into studies on the primate visual system is expected to yield new developments in both research fields.

Acknowledgements

We thank Emiko Sugitani for secretarial assistance in preparation of this article. Parts of our study discussed here were supported by the Uehara Memorial Foundation, and Core Research for Evolutionary Science and Technology (CREST) of Japan Science and Technology Corporation.

References

1. Bliss TVP, Collingridge GL. A synaptic model of memory: long-term potentiation in the hippocampus. Nature 1993;361:31–39.
2. Tsumoto T. Long-term potentiation and long-term depression in the neocortex. Prog Neurobiol 1992; 39:209–228.
3. Linden DJ. Long-term synaptic depression in the mammalian brain. Neuron 1994;12:457–472.
4. Ito M. Long-term depression in the cerebellum. Sem Neurosci 1990;2:381–390.
5. Morris RGM, Andersen E, Lynch GS, Baudry M. Selective impairment of learning and blockade of long-term potentiation by an N-methyl-D-aspartate receptor antagonist, AP5. Nature 1986;319: 774–776.
6. Davis S, Butcher SP, Morris RGM. The NMDA receptor antagonist D-2-amino-5-phosphonopentanoate (D-AP5) impairs spatial learning and LTP in vivo at intracerebral concentrations comparable to those that block LTP in vitro. J Neurosci 1992;12:21–34.
7. Slva AJ, Stevens CF, Tonegawa S, Wang Y. Deficient hippocampal long-term potentiation in α-calcium-calmodulin kinase II mutant mice. Science 1992;257:201–206.
8. Silva AJ, Paylor R, Wehner JM, Tonegawa S. Impaired spatial learning in α-calcium-calmodulin kinase II mutant mice. Science 1992;257:206–211.
9. Rogan MT, LeDoux JE. LTP is accompanied by commensurate enhancement of auditory-evoked responses in a fear conditioning circuit. Neuron 1995;15:127–136.
10. Mishkin M. A memory system in the monkey. Phil Trans R Soc Lond B 1982;298:85–95.
11. Squire LR. Memory and Brain. New York: Oxford University Press, 1987.
12. Sakai K, Miyashita Y. Memory and imagery in the temporal lobe. Curr Opin Neurobiol 1993;3: 166–170.
13. Suzuki W. The anatomy, physiology and functions of the perirhinal cortex. Curr Opin Neurobiol 1996;6:179–186.
14. Miyashita Y. Inferior temporal cortex: where perception meets memory. Ann Rev Neurosci 1993; 16:245–263.
15. Tanaka K. Inferotemporal cortex and object vision. Ann Rev Neurosci 1996;19:109–139.
16. Fujita I, Tanaka K, Ito M, Cheng K. Columns for visual features of objects in monkey inferotemporal cortex. Nature 1992;360:343–346.
17. Kato M, Uka T, Fujita I. Columnar organization in inferior temporal cortex of awake monkey. Soc Neurosci Abstr 1996;22(In press).
18. Fujita I, Fujita T. Intrinsic connections in the macaque inferior temporal cortex. J Comp Neurol 1996; 368:467–486.
19. Gilbert CD, Wiesel TN. Receptive field dynamics in adult primary visual cortex. Nature 1992;356: 150–152.
20. Jacobs KM, Donoghue JP. Reshaping the cortical motor map by unmasking latent intracortical connections. Science 1991;251:944–947.
21. Desimone R, Albright TD, Gross CG, Bruce C. Stimulus-selective properties of inferior temporal neurons in the macaque. J Neurosci 1984;4:2051–2062.
22. Tanaka K, Saito H, Fukada Y, Moriya M. Coding visual images of objects in the inferotemporal cortex of the macaque monkey. J Neurophysiol 1991;66:170–189.
23. Baylis GC, Rolls ET, Leonard CM. Functional subdivisions of the temporal lobe neocortex. J Neurosci 1987;7:330–342.
24. Gochin PM, Miller EK, Gross CG, Gerstein GL. Functional interactions among neurons in inferior temporal cortex of the awake macaque. Exp Brain Res 1991;84:505–516.
25. Gawne TJ, Richmond BJ. How independent are the messages carried by adjacent inferior temporal cortical neurons? J Neurosci 1993;13:2758–2771.
26. Wang G, Tanaka K, Tanifuji M. Optical imaging of functional organization in the monkey inferotemporal cortex. Science 1996;272:1665–1668.
27. Fujita I. Columns in the inferotemporal cortex: machinery for visual representation of objects.

Biomed Res 1993;14(S4):21–27.

28. Fujita I, Tanigawa H, Ojima H, Kato M. Neurons mediating horizontal interactions in the monkey inferior temporal cortex. Soc Neurosci Abstr 1995;21:660.
29. Murayama Y, Fujita I, Kato M. Long-term potentiation and depression in the monkey visual cortex. Soc Neurosci Abstr 1996;22(In press).
30. Aroniadou VA, Teyler TJ. Induction of NMDA receptor-independent long-term potentiation (LTP) in visual cortex of adult rats. Brain Res 1992;584:169–173.
31. McClelland JL, McNaughton BL, O'Reilly RC. Why there are complementary learning systems in the hippocampus and neocortex: insights from the successes and failures of connectionist models of learning and memory. Psychol Rev 102:419–457.
32. Kobatake E, Tanaka K, Wang G, Tamori Y. Effects of adult learning on the stimulus selectivity of cells in the inferotemporal cortex. Soc Neurosci Abstr 1993;19:975.
33. Heinen SJ, Skavenski AA. Recovery of visual responses in foveal V1 neurons following bilateral foveal lesions in adult monkey. Exp Brain Res 1991;83:670–674.
34. Garraghty PE, Kaas JH. Dynamic features of sensory and motor maps. Curr Opin Neurobiol 1992; 2:522–527.
35. Malenka RC, Kauer JA, Perkel DJ, Mauk MD, Kelly PT, Nicoll RA, Waxham MN. An essential role for postsynaptic calmodulin and protein kinase activity in long-term potentiation. Nature 1989;340: 554–557.
36. Malinow R, Schulman H, Tsien RW. Inhibition of postsynaptic PKC or CaMKII blocks induction but not expression of LTP. Science 1989;245:862–866.
37. Nelson RB, Friedman DP, O'Neill JB, Mishkin M, Routtenberg A. Gradients of protein kinase C substrate phosphorylation in primate visual system peak in visual memory storage areas. Brain Res 1987;416:387–392.
38. Huang FL, Yoshida Y, Nakabayashi H, Friedman DP, Ungerleider LG, Young III WS, Huang KP. Type I protein kinase C isozyme in the visual-information-processing pathway of monkey brain. J Cell Biochem 1989;39:401–410.

Integrative and Molecular Approach to Brain Function.
M. Ito and Y. Miyashita, editors.

Plasticity of early visual learning

Charles D. Gilbert
The Rockefeller University, New York, New York, USA

Key words: cortical plasticity, perceptual learning, primary visual cortex, retinal lesions.

The mechanisms of learning and memory must be explored at many levels, from the molecular to the integrative. While there has been an explosion in work on the mechanisms of synaptic plasticity, it has been a greater challenge to understand how information is encoded by neuronal ensembles and by alterations in synaptic weights of specific circuits. The linkage between the molecular, cellular, integrative and cognitive aspects of learning may become more accessible by the use of primary sensory cortex as the model system. This possibility has been opened by a radical change in our understanding of the early stages of sensory processing. Rather than these stages having cells with fixed functional properties and cortical areas with fixed functional architectures, a growing body of evidence indicates that experience-dependent mutability of function occurs in primary sensory cortex throughout life. On the perceptual level, rather than thinking of learning in terms of storing the most complex objects, it has long been known that even the discrimination of simple attributes can undergo substantial improvement with practice, and this improvement has characteristics suggestive of the involvement of early stages sensory processing.

Cortical plasticity and recovery following lesions

For many years it was believed that the functional properties of cortical cells were fixed after the end of the critical period, a period of experience-dependent plasticity that lasted for the first few months of life. This plasticity was associated with an ability to change the extent of axonal fields in the projection from thalamus to cortex. The experiments on critical period dealt with the ability to shift the balance of input from the two eyes impinging on cortical cells [1]. Based on these experiments it had been widely assumed that primary sensory cortex represented a preprocessor, filtering the input from the sensory periphery in a fixed and stereotyped manner, independent of context or experience. While some measure of plasticity had to exist in adulthood, this was believed to be a specialized property of higher order cortical areas, such as

Address for correspondence: Charles D. Gilbert MD, PhD, The Rockefeller University, 1230 York Avenue, New York, NY 10021, USA. Tel.: +1-212-327-7670. Fax: +1-212-327-7844. E-mail: gilbert@rockvax.rockefeller.edu the inferotemporal cortex, which is known to play a role in object recognition and visual associations [2].

The traditional view was turned around when investigators looked at the central effects of peripheral lesions. In the somatosensory system, for example, digit amputation initially removes input from the area of cortex representing that digit. Over time the cortex is remapped, receiving input instead from the digits adjacent to the amputated one [3–5]. Motor nerve lesion also leads to reorganization of motor cortex [6]. Similarly, in the auditory system, localized cochlear lesions lead to a remapping of the tonotopic map on the surface of primary auditory cortex [7]. In the original experiments it had been supposed that the reorganization was cortical in origin, but it was not known whether some of the reorganization may have occurred at subcortical stages. Work in the laboratory of Wall had shown, in fact, that peripheral nerve section could lead to reorganization in the spinal cord itself [8], which probably was attributable to the existence of long-range connections of somatosensory afferents within the spinal cord [9].

We attempted to determine whether a similar form of plasticity could exist in the visual cortex. We were particularly intrigued by the possibility that known circuit elements within the cortex, the long-range horizontal connections, would be ideally suited for mediating this reorganization. To deafferent functionally a restricted area of primary visual cortex we made focal binocular retinal lesions at homologous retinal locations. As expected, this procedure initially inactivated the area of cortex representing the lesioned part of the retina. Over several months, however, the cells in this region regained functioning visual input, only now their receptive fields had shifted to positions outside of the lesioned part of the retina [10,11]. Similar experiments involving combinations of retinal lesions and enucleation also led to comparable effects [12–14].

To determine the source of the reorganization, we recorded at various stages along the visual pathway. We found that when the area of cortex had reorganized, there was still a large silent area in the lateral geniculate nucleus, showing the mechanism of recovery to be cortically based [11,15]. The spatial extent of the cortical reorganization was roughly 6–8 mm, matching that of the long-range horizontal connections. Long-range horizontal connections are a plexus of connections formed by cortical pyramidal cells running for considerable distances parallel to the cortical surface. The targets of these connections integrate information over an area of cortex representing a part of the visual field much larger than their own receptive fields. Because of their visuotopic extent, they can provide visual input into the cortical scotoma from cells outside the scotoma [16,17]. The thalamocortical connections, although also a potential source of reorganization, are much more limited in their lateral extent. We found that they did not change their overall extent with the functional reorganization, indicating that they could not account for the observed changes. The long-range horizontal connections intrinsic to primary visual cortex were, therefore, the most likely source of the reorganization. We have not ruled out, however, the potential involvement of feedback connections from higher order cortical areas. The observation that the pattern of orientation columns in the reorganized cortical region was similar to that seen before lesioning was supportive of the involvement of the intrinsic horizontal connections, since they run between columns of similar orientation preference [18].

Anatomical studies showed that the reorganization involved a sprouting of axon collaterals and synaptogenesis. The horizontal connections were labeled by focal injections of an anterograde label, biocytin, in the area of cortex surrounding the deafferented region. This is the area that was presumed to provide input to the inactivated region via the horizontal collaterals. We found that the density of the horizontal plexus had increased by almost a factor of two, and was even more pronounced in the superficial cortical layers [19]. Because of the precise topography of the reorganization, this system is ideal for exploring the molecular basis of cortical plasticity. We have found an increase in the expression of a number of factors that may account for the collateral sprouting, including neurotrophins and their receptors, as well as substances that may be markers for synaptic density, such as synapsins (Obata and Gilbert, unpublished results). The remarkable conclusion that emerges from these studies is that formation of new connections, a feature traditionally associated with developing cortex, can occur throughout life. Furthermore, they demonstrate quite clearly that plasticity is not limited to cortical areas involved in processing information of the highest order complexity, but is a pervasive property of entire sensory pathways, beginning with primary sensory cortex.

Short-term cortical plasticity and visual experience

Another developing theme of cortical plasticity is the possibility of very rapid changes in receptive field properties resulting from short-term exposure to patterned stimuli. In the visual system, this was first evident again in the retinal lesion experiments. If one records from cells in the cortical scotoma within minutes after the lesion is made, there is quite a remarkable enlargement in receptive field size and a shift in receptive field position. This is seen for cells whose receptive fields were located initially just within the boundary of the area subsequently covered by the lesion [11,15].

The possibility for such short-term changes, particularly when the lesion often destroyed only the photoreceptor layer, opened the possibility for observing changes in receptive field structure without making a lesion at all. We explored this by using what we call an "artificial scotoma", a large area of patterned stimulation within which there is an occluder or mask. The pattern used could be a set of drifting lines or twinkling dots. We could explore the effect of such a pattern on individual cortical neurons by first isolating a cell, measuring its receptive field, and then placing the receptive field within the occluder. Generally, the occluder was sized to approximately 3–5 times the diameter of the receptive field. After a few minutes of exposure to the pattern, the receptive field was again mapped. We found a large increase in receptive field size – up to a tripling in receptive field diameter. When the occluder is removed and the cell receives visual stimuli within the receptive field, the field collapses back down to its original size. In this way receptive fields can be expanded and contracted numerous times [20–22]. It presents to sensory neurophysiology a problem equivalent to the Heisenberg uncertainty principle in physics: anything you do to measure the receptive field may change it.

The perceptual consequences of this form of plasticity are several. One potential role is in perceptual fill-in, which in turn may reflect the attempt of the visual system to link contours and textures belonging to a common surface, and to segregate surfaces belonging to different objects. In the phenomenon of perceptual fill-in, the masked area appears to fill in with the surrounding color or texture [23–26]. By expanding receptive fields within the occluder, the stimulus outside the occluder begins to activate cells that are line labels for positions inside the occluder, leading to the percept that the surrounding stimulus is located inside the occluder. A gradual increase in activation of cells in V4 with fields in an artificial scotoma has also been reported, with a time course of activation similar to that of perceptual fill-in [27]. The proposed shift in stimulus position towards the center of an artificial scotoma was demonstrated quantitatively using a three-line bisection task, where human subjects were asked to judge whether the central of three lines is located nearer one or the other outer lines. Lines located near the boundary of the scotoma are perceived as being shifted towards its center [28].

The mechanisms mediated the receptive field expansion may once again involve the long-range horizontal connections within primary visual cortex. They allow information from one part of the visual field to propagate to other parts of the visual field. By strengthening these connections, the input from outside the classical receptive field can be boosted from a subthreshold modulatory influence to a suprathreshold driving influence. The change in effective connectivity between cells undergoing the receptive field expansion was demonstrated with cross-correlation analysis. This measure of the strength of connection between cells, and the strength of common input to them, shows up to a 3-fold increase under conditions of receptive field expansion [21]. The specific synaptic mechanisms have yet to be elucidated. We have shown that the horizontal connections are capable of use-dependent increase in strength [29], but it is also possible that the changes could be mediated by a decrease in inhibition [30]. The horizontal connections formed by cortical pyramidal cells contact other excitatory cells as well as inhibitory interneurons [31]. Any potentiation of their influence could be mediated either by potentiation of the excitatory connections or an unmasking from a sustained reduction of inhibition.

Cortical plasticity and learning

The plasticity observed after lesions of the central nervous system is likely to reflect a more fundamental and normal process of cortical reorganization occurring throughout life. It is likely to reflect a continuing process of calibration and normalization of visual processing, and could provide insight into the mechanisms of learning itself. Of the two major forms of long-term memory, declarative and nondeclarative, the former is thought to involve primarily the medial temporal structures of the brain. Of the nondeclarative forms of memory, one in particular, known as priming or implicit memory, involves the neocortex. Implicit learning involves an improvement in performance with practice without conscious awareness, and without error feedback, but simply as a result of performing a discrimination task after a number of trials. It operates on various time scales, ranging from seconds to weeks.

Learning in the sensory systems has been shown to apply to nearly every attribute, including visual acuity, hue discrimination, velocity estimation, acoustic pitch discrimination, and two-point somatosensory acuity (for review see [32]). The specificity of the effect indicates an involvement of early stages in sensory pathways. Improvement in visual performance with repeated trials has been observed for a number of visual submodalities, including acuity, stereoacuity, texture, motion and orientation, and probably can be seen with any sensory attribute. Hyperacuity, defined as a spatial threshold that is smaller than the grain of the sensory mosaic (often by an order of magnitude or more), shows improvement with training. This has been demonstrated with various stimuli [33,34]. Some experiments show improvement over days, others over just a few tens of trials. One of the most robust forms of perceptual learning is seen in stereoacuity [35]. Texture discrimination shows improvement, as evidenced by the ability to determine, within a set period of time, the orientation of an array of oriented lines in a background array of differently oriented lines [36].

Perceptual learning has been studied for well over a 100 years, but its cellular basis is poorly understood. Learning in the somatosensory system has been found to lead to cortical recruitment, an increase in the area of cortex representing the digit exposed to the training stimulus [37]. In the auditory system as well, aversive conditioning associated with a particular tone shifts the receptive fields of cells in primary auditory cortex towards the trained frequency [38] and training on a frequency discrimination task has been shown to expand the representation of that frequency in primary auditory cortex [39]. In the visual system, the properties of perceptual learning are also suggestive of the involvement of early stages in the visual pathway. To determine the physiological basis of perceptual learning, we first needed to devise a stimulus that would allow us to do parallel experiments in human psychophysics and in studies of the response properties of cells in alert primates [40]. The visual stimulus we chose is a three-line bisection task, involving determining the central point between two parallel lines. By performing this task repeatedly, several hundred trials a day, over a number of weeks, there is a consistent, substantial improvement over all subjects tested. The specificity of this learning indicates a potential involvement of V1: it is specific to the trained position in visual space; it is specific to the trained orientation, showing no transfer when the triad of lines is rotated to the orthogonal orientation. This points towards the involvement of V1, since there the retinotopic organization and orientation specificity is the most refined. There is an additional specificity, however, that suggests the mechanism does not simply involve cortical recruitment, the enlargement of the representation of the trained position in visual space. There is an additional involvement of higher order influences, since the task does not transfer to other tasks in the same position in the visual field and, therefore, involving the same area of cortex as the trained task. Thus, one sees no improvement in an orientation discrimination task or in a vernier acuity task, even though the latter involves discriminating the position of the same line in the same position and of the same orientation as in the trained task, but relative to a different line. The dependence on behavioral context and on the spatial configuration of the visual stimulus suggests that the cortical changes might be gated by influences from higher order cortical areas. We plan to explore the receptive field and architectural changes that correlate with the perceptual improvement.

Summary

The substrate for learning and memory may be observed in changes in gene expression, neocortical circuitry, receptive field properties and functional architecture. These changes may be associated with different forms of learning in different cortical areas, but characteristics of perceptual learning suggest an involvement of the earliest stages of cortical sensory processing. The plasticity of primary visual cortex represents a new view of the algorithm used by cortex in processing information. Rather than being fixed in their functional properties, cells are active filters, changing according to context and according to previous experience. The mutability of functional properties allows these early stages of sensory processing to integrate information over large parts of visual space, linking contours and surfaces, calibrating the system to visual attributes, and improving performance in sensory discriminations.

Acknowledgements

This work has been supported by grants from the National Eye Institute (EY07968), National Science Foundation (IBN 92-21854) and the McKnight Foundation.

References

1. Hubel DH, Wiesel TN. The period of susceptibility to the physiological effects of unilateral eye closure in kittens. J Physiol 1970;206:419–436.
2. Miyashita Y, Ohbayashi M, Kameyama M, Naya Y, Yoshida M. Origin of visual imagery: role of temporal cortex and top-down signal. This volume, 133–145.
3. Rasmusson DD. Reorganization of raccoon somatosensory cortex following removal of the fifth digit. J Comp Neurol 1982;205:313–326.
4. Merzenich MM, Nelson RJ, Stryker MP, Cynader MS, Schoppmann A, Zook JM. Somatosensory cortical map changes following digital amputation in adult monkeys. J Comp Neurol 1984;224: 591–605.
5. Calford MB, Tweedale R. Immediate and chronic changes in responses of somatosensory cortex in adult flying-fox after digit amputation. Nature 1988;332:446–448.
6. Sanes JN, Suner S, Lando JF, Donoghue JP. Rapid reorganization of adult rat motor cortex somatic representation patterns of motor nerve injury. Proc Natl Acad Sci 1988;85:2003–2007.
7. Robertson D, Irvine D. Plasticity of frequency organization in auditory cortex of guinea pigs with partial unilateral deafness. J Comp Neurol 1989;282:456–471.
8. Wall JT, Cusick CG. Cutaneous responsiveness in primary somatosensory (S-I) hindpaw cortex before and after partial hindpaw deafferentation in adult rats. J Neurosci 1984;6:1499–1515.
9. Wall PD, Werman R. The physiology and anatomy of long ranging afferent fibres within the spinal

cord. J Physiol 1976;255:321–334.
10. Gilbert CD, Hirsch JA, Wiesel TN. Lateral interactions in visual cortex. Cold Spring Harbor Symp Quant Biol 1990;55:663–677.
11. Gilbert CD, Wiesel TN. Receptive field dynamics in adult primary visual cortex. Nature 1992;56: 150–152.
12. Kaas JH, Krubitzer LA, Chino YM, Langston AL, Polley EH, Blair N. Reorganization of retinotopic cortical maps in adult mammals after lesions of the retina. Science 1990;248:229–231.
13. Heinen SJ, Skavenski AA. Recovery of visual responses in foveal V1 neurons following bilateral foveal lesions in adult monkey. Exp Brain Res 1991;83:670–674.
14. Chino YM, Kaas JH, Smith III EL, Langston AL, Cheng H. Rapid reorganization of cortical maps in adult cats following restricted deafferentation in retina. Vision Res 1992;32:789–796.
15. Darian-Smith C, Gilbert CD. Topographic reorganization in the striate cortex of the adult cat and monkey is cortically mediated. J Neurosci 1995;15:1631–1647.
16. Gilbert CD, Wiesel TN. Morphology and intracortical projections of functionally identified neurons in cat visual cortex. Nature 1979;280:120–125.
17. Gilbert CD, Wiesel TN. Clustered intrinsic connections in cat visual cortex. J Neurosci 1983;3: 1116–1133.
18. Das A, Gilbert CD. Cortical point spreads in normal and reorganized primary visual cortex reflect long range intracortical connections. Nature 1995;375:780–784.
19. Darian-Smith C, Gilbert CD. Axonal sprouting accompanies functional reorganization in adult cat striate cortex. Nature 1994;368:737–740.
20. Pettet MW, Gilbert CD. Dynamic changes in receptive field size in cat primary visual cortex. Proc Natl Acad Sci 1992;89:8366–8370.
21. Das A, Gilbert CD. Receptive field expansion in adult visual cortex is linked to dynamic changes in strength of intrinsic cortical connections. J Neurophysiol 1995;74:779–792.
22. Gilbert CD, Das A, Ito M, Kapadia MK, Westheimer G. Spatial integration and cortical dynamics. Proc Natl Acad Sci 1996;93:615–622.
23. Yarbus AL. The perception of an image fixed with respect to the retina. Biophysics 1957;2:683–690.
24. Krauskopf J. Heterochromatic stabilized images: a classroom demonstration. Am J Psychol 1961;80: 632–637.
25. Crane HD, Piantanida TP. On seeing reddish-green and yellowish-blue. Science 1983;221:1078–1079.
26. Ramachandran VS, Gregory TL. Perceptual filling in of artificially induced scotomas in human vision. Nature 1991;350:699–702.
27. DeWeerd P, Gattas R, Desimone R, Ungerleider LG. Responses of cells in monkey visual cortex during perceptual filling-in of an artificial scotoma. Nature 1995;377:731–734.
28. Kapadia MK, Gilbert CD, Westheimer G. A quantitative measure for short-term cortical plasticity in human vision. J Neurosci 1994;14:451–457.
29. Hirsch J, Gilbert CD. Long-term changes in synaptic strength along specific intrinsic pathways in the cat's visual cortex. J Physiol 1993;461:247–262.
30. Xing J, Gerstein GL. Simulation of dynamic receptive fields in primary visual cortex. Vision Res 1994;34:1901–1911.
31. McGuire BA, Gilbert CD, Rivlin P, Wiesel TN. Targets of horizontal connections in Macaque primary visual cortex. J Comp Neurol 1991;305:370–392.
32. Gilbert CD. 1994 Early perceptual learning. Proc Natl Acad Sci 1994;91:1195–1197.
33. McKee SP, Westheimer G. Improvement in vernier acuity with practice. Percept Psychophysics 1978; 24:258–262.
34. Poggio T, Fahle M, Edelman S. Fast perceptual learning in visual hyperacuity. Science 1992;256: 1018–1021.
35. Westheimer G, Truong T. Target crowding in foveal and peripheral stereoacuity. Am J Optom Physiol Optics 1988;65:395–399.
36. Karni A, Sagi D. Where practice makes perfect in texture discrimination: evidence for primary visual cortex plasticity. Proc Natl Acad Sci 1992;88:4966–4970.

37. Recanzone GH, Merzenich MM, Jenkins WM, Grajski KA, Dinse HR. Topographic reorganization of the hand respresentation in cortical area 3b of owl monkeys trained in a frequency-discrimination task. J Neurophysiol 1992;67:1071–1091.
38. Bakin JS, Weinberger NM. Classical conditioning induces CS-specific receptive field plasticity in the auditory cortex of the guinea pig. Brain Res 1990;536:271–286.
39. Recanzone GH, Schreiner CE, Merzenich MM. Plasticity in the frequency representation of primary auditory cortex following discrimination training in adult owl monkeys. J Neurosci 1993;13:87–103.
40. Crist RE, Kapadia MK, Westheimer G, Gilbert CD. Perceptual learning of spatial localization: Specificity for position, oreintation and context. Soc Neurosci Abstr 1996;(In press).

Integrative and Molecular Approach to Brain Function.
M. Ito and Y. Miyashita, editors.

Development of cortical output projections: axon guidance, target recognition, and plasticity

Dennis D.M. O'Leary, Martin Bastmeyer, Maryellen Daston, Susan E. Koester, Linda Richards and Kathleen Yee

Molecular Neurobiology Laboratory, The Salk Institute, La Jolla, California, USA

Abstract. This article summarizes work done in our laboratory on the development of the major output projection of the mammalian cerebral cortex, which arises from layer 5. Layer 5 subcortically projecting neurons in the mammalian cortex innervate multiple targets in the brainstem and spinal cord, which in the adult differs upon the area of cortex in which the neuron resides. Early in development, these neurons innervate a similar set of targets, and the area-specific pattern of projections emerges later through a process of selective axon elimination. This elimination process is plastic, and can be modified by altering the neuron's location in the developing cortex.

Layer 5 neurons innervate their targets exclusively by collateral branches that form de novo along the axon shaft, well behind the advancing growth cone. Thus, in this system, the process of target recognition is mediated by the axon's shaft rather than its growth cone. Evidence obtained from in vitro experiments indicates that the formation and directed growth of these collateral branches is promoted by an unidentified diffusible chemotropic activity released by targets of cortical layer 5 axons. Recent evidence suggests that the neurotrophin, NT-3, and the polysialic acid modification of NCAM may have selective, but important roles in the development of these layer 5 projections.

Key words: axon guidance, collateral branching, corticospinal axons, netrin, neurotrophin-3, spinal cord, target recognition, time-lapse video microscopy, trkC.

Introduction

Generation of a properly functioning nervous system is dependent on the development of precise connections between appropriate sets of neurons. As a consequence, a fundamental issue in neurobiology is defining the mechanisms by which neurons recognize and innervate their targets, and the molecules that control these phenomena. We have been studying the development of the major efferent projection of the mammalian cortex, that from layer 5 neurons to the brainstem and spinal cord, to gain a better understanding of the mechanisms and molecules involved. Layer 5 is the major output layer of the cortex, being the only cortical source of projections to several targets in the brainstem and the spinal cord [1]. This projection is the sole means by which we mediate conscious control over bodily movements; in addition, it has other important functions, such as contributing to the processing of sensory information and to the maintenance of body posture.

Address for correspondence: Dennis D.M. O'Leary, MNL-O, The Salk Institute, 10010 N. Torrey Pines Rd, La Jolla, CA 92037, USA. Tel.: +1-619-453-4100 (ext. 1415). Fax: +1-619-558-6207. E-mail: dennis_oleary@qm.salk.edu

Pioneering and early development of cortical efferent projections

The internal capsule and corpus callosum are the two major efferent axonal pathways of the mammalian neocortex. In developing rats, the internal capsule is pioneered on embryonic day (E) 14 [2], but the corpus callosum is not pioneered until E17 [3] (Fig. 1). The first cortical axons to grow through the internal capsule, the pathway from cortex to its subcortical targets, are extended by subplate neurons, which are the earliest generated neurons in the neocortex [2]. In contrast, the first axons to project through the other major efferent pathway of the cortex, the midline corpus callosum, which connects the two cortical hemispheres, originate from a discrete group of neurons in medial (presumptive cingulate) cortex [3]. Although these early callosal cells are among the earliest generated neurons in this cortical region, based on morphological criteria they appear not to be subplate neurons but instead to be deep-layer cortical plate neurons of cingulate cortex. Subplate neurons not only do not pioneer the corpus callosum, but never even project through it in significant numbers. We conclude that the two major efferent pathways from cortex, the corpus callosum and the internal capsule, are pioneered by developmentally and spatially distinct populations of early-generated cortical neurons.

Using the antibody TuJ1, a neuron-specific marker directed against acetylated β-tubulin [4], and DiI as an anterograde and retrograde axonal tracer [5,6], we have found that the timing difference in the development of these two major cortical efferent pathways is due to an early polarization of cortical axon outgrowth [7]. From E12, when the first cortical neurons are generated, through E16, essentially all axon extension is directed laterally toward the internal capsule. The extension of cortical

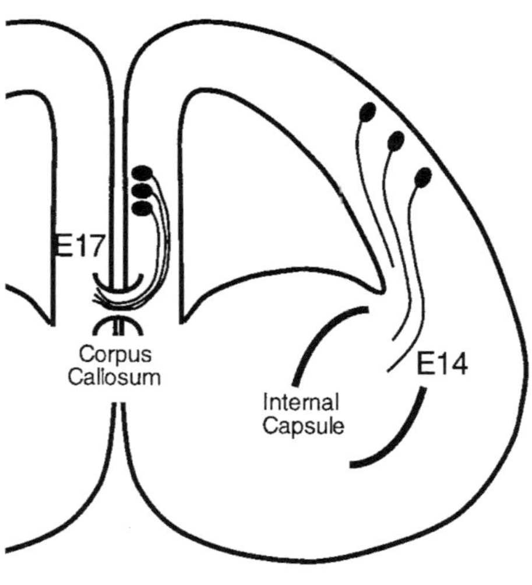

Fig. 1. The two efferent pathways from cortex are pioneered by distinct populations of early generated neurons at different times in development. The subcortical projection through the internal capsule is pioneered by neocortical subplate neurons. These axons first arrive in the internal capsule on E14. The callosal projection is pioneered by a small group of cortical plate neurons in cingulate cortex. The axons of these cells first cross the midline on E17.

axons medially toward the corpus callosum is not evident until E17. Since callosal and subcortical projection neurons are cogenerated, these findings indicate that callosal neurons are delayed in axon extension toward the midline. The possible causes for the delay in axon extension medially are illustrated in Fig. 2. This delay in medially directed axon extension is the basis of the difference in the time of formation of the two major efferent pathways from cortex.

We have studied the mechanisms by which this early polarity of cortical axon extension may be mediated and have investigated the possibility that soluble molecules, present in intermediate target regions, may confer axonal polarity in either a chemorepellent or chemoattractive manner. To investigate this hypothesis, explants of E15 rat cortex were cocultured with regions of their prospective intermediate targets, either explants taken from the nascent internal capsule, to look for chemoattractant activity, or from the medial dorsal telencephalon, to look for a chemorepellent activity (Fig. 3). When cultured with the developing internal capsule, axons from the cortical explant grew in a directed manner towards the internal capsule explant indicating that the nascent internal capsule does produce a soluble chemoattractive activity for cortical axons [7]. However, when cortex was cocultured with explants of medial dorsal telencephalon, the cortical neurites grew out in a symmetrical, nondirected manner, indicating that these axons were neither chemo-repulsed nor attracted, at least at this early stage of development.

Recently, soluble chemoattraction has been studied in the developing spinal cord, where dorsally located commissural axons, are attracted towards the floor plate, a ventral midline structure. We examined whether the ventrolaterally directed axon extension that we observed in the cortex could be mimicked by molecules expressed from the floor plate in coculture and by Netrin-1, a chemoattractant expressed by the floor plate and by the ventral forebrain [8]. We have found that cortical axons were attracted toward both the floor plate and 293T cells transfected with a Netrin-1 expression construct. It is, therefore, possible that the mechanism which directs early cortical axons involves the release of a chemoattractive molecule by cells in the nascent internal capsule, and that this molecule may be Netrin-1.

Targeting potentials of layer 5 projection neurons in the developing neocortex

Layer 5 is the only source of cortical projections to the spinal cord and a number of

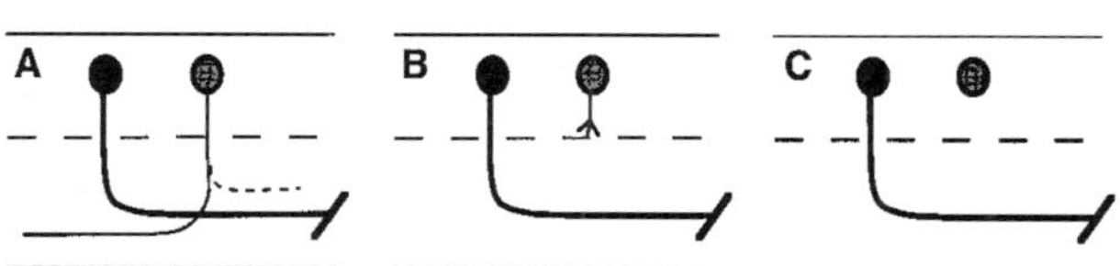

Fig. 2. Schematic of possible causes for delay in medial axon extension. **A:** Both subcortical (filled circle) and callosal (hatched circle) neurons initially extend axons laterally. Later callosal neurons eliminate their laterally directed axons (dotted line) and then extend axons medially. **B:** Both projection neuron types extend axons at the same time but callosal axons "wait" before they make their medially directed turn in the intermediate zone. **C:** Callosal neurons do not extend axons until later in development, days after subcortically projecting neurons have extended their axons.

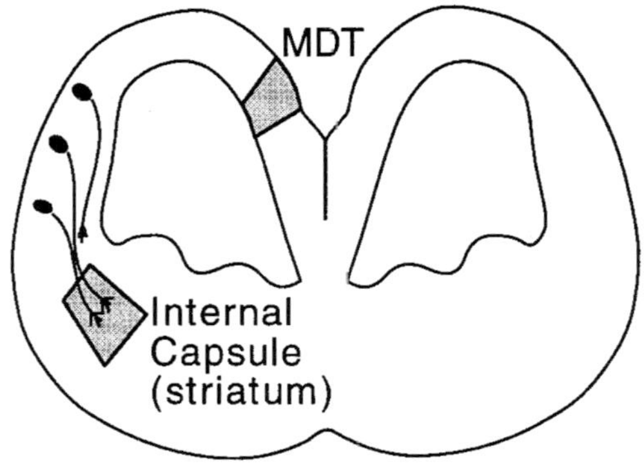

Fig. 3. Schematic representation of coculture experiments of cortex with intermediate targets. **A:** Coronal section through an E15 rat brain showing the regions where explants of potential intermediate targets were dissected, the nascent internal capsule (which develops within the striatum) and medial dorsal telencephalon (MDT). As described in the text, cortical explants were cultured alone or cocultured in collagen gels with test explants of either internal capsule or MDT, or with floor plate or agar cubes containing transfected 293T cells.

midbrain nuclei, including the superior colliculus and deep mesencephalic nuclei, such as the red nucleus, and hindbrain nuclei, including the basilar pons, inferior olive, and dorsal column nuclei (see review [1]). In the adult, layer 5 neurons that project to these targets are limited to specific regions of the cortex that differs from target to target [9,10]. It has been known for several years that the limited distribution of layer 5 neurons that project to the spinal cord or the superior colliculus in adult rats arises from earlier widespread distributions by the process of selective collateral pruning [9]. Only recently has the complexity of this pruning process become apparent. With the introduction of DiI as an anterograde axon tracer, it was found in rats that during development of layer 5 projections to all brainstem targets, collateral branches are extended from primary axons that go on to form the corticospinal projection, and, as a population, layer 5 neurons in motor and visual cortex initially develop a comparable set of collateral branches [11].

Within the framework of these DiI findings, a reinterpretation of the earlier retrograde labeling studies showing a widespread distribution of corticospinal neurons during development is that layer 5 neurons in all areas of the developing neocortex initially extend a spinally directed primary axon. Subsequent retrograde labeling studies confirm this interpretation and show that primary spinal axons of neurons from all areas of the developing neocortex give rise to the major collateral projections to brainstem targets. Therefore, as illustrated in Fig. 4, the functionally appropriate, area-specific patterns of layer 5 subcortical projections characteristic of the adult rat do not develop from the outset; instead, they are pruned from a common projection pattern developed in late fetal and early postnatal life by layer 5 neurons across the entire neocortex. These findings suggest that subcortically projecting layer 5 neurons in diverse areas of the developing neocortex have similar potentials in axon targeting

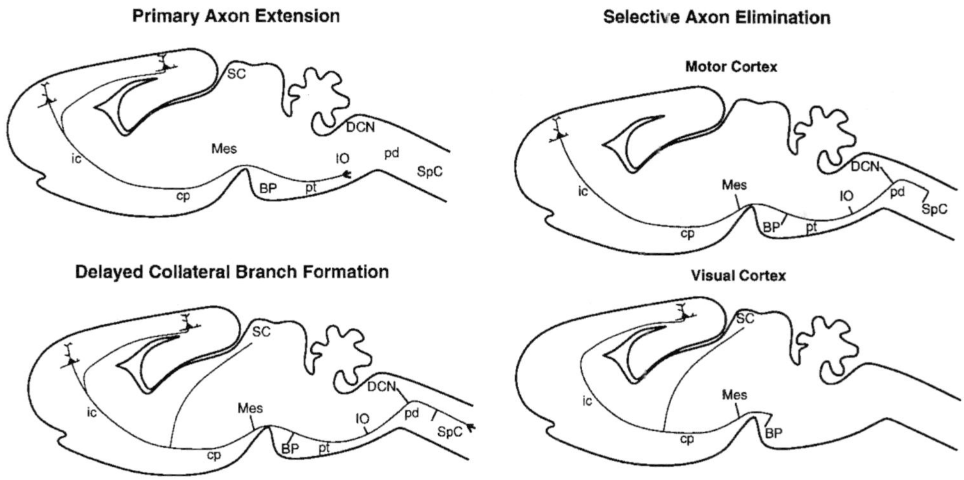

Fig. 4. Development of area-specific subcortical projections of layer 5 neurons in rat neocortex. This process, illustrated in schematics of a sagittal view, occurs in three phases: 1) Primary Axon Extension: Layer 5 neurons extend a primary axon out of cortex toward the spinal cord (SpC) bypassing their subcortical targets. 2) Delayed Collateral Branch Formation: Subcortical targets are later contacted exclusively by axon collaterals that develop by a delayed extension of collateral branches interstitially along the spinally directed primary axon. As a population, L5 neurons in all areas of the neocortex develop a common set of branches. 3) Selective Axon Elimination: As illustrated for visual and motor cortex, specific collateral branches or segments of the primary axon are selectively eliminated to generate the mature projections functionally appropriate for the area of neocortex in which the L5 neuron is located. Abbreviations: BP, basilar pons; cp, cerebral peduncle; DCN, dorsal column nuclei; ic, internal capsule; IO, inferior olive; Mes, mesencephalic component; pd, pyramidal decussation; pt, pyramidal tract; SC, superior colliculus. Adapted from O'Leary and Koester, 1993.

and respond in a qualitatively similar way to cues that govern their collateral branching, although they will ultimately have distinct projection patterns in the adult [9,12].

Although the collateral branches of a developing layer 5 axon form independent of whether the collateral projection is functionally appropriate for the cortical area in which the neuron is located, the subsequent remodeling of the immature axonal projection pattern is tightly correlated with the neuron's areal location, suggesting a causal relationship. Findings from heterotopic cortical transplant experiments support this suggestion [13,14]. Late fetal cortex transplanted heterotopically into newborn rats establish subcortical projections characteristic of their new cortical location. Motor cortical neurons transplanted to visual cortex initially extend a spinal axon, but lose it while maintaining a projection to the superior colliculus; visual cortical neurons transplanted to motor cortex permanently retain spinal axons. Thus, the pruning of the initial common projection pattern is not a fixed property of layer 5 neurons, but is highly plastic and is dependent on their location in the developing cortex.

Dynamics of target recognition by interstitial axon branching along developing cortical axons

During development, layer 5 projections to each of these targets are established by corticospinal neurons through a novel mechanism. In this system, the axon shaft, rather than the primary growth cone, is responsible for the process of target recognition and innervation. DiI labeling studies indicate that layer 5 axons extend out of the cortex along a spinally directed pathway, grow past their targets, and seemingly ignore them as they continue to elongate through the corticospinal tract. The brainstem and spinal targets are later contacted exclusively by axon collaterals that develop by a "delayed interstitial branching" of the parent layer 5 axon [11,15,16]. These branches form well behind the advancing growth cone, and grow directly to their appropriate target.

To establish definitively that the axon shaft rather than the axonal growth cone is responsible for target recognition in this system, and to characterize the dynamics of interstitial branch formation, we have studied this process in an in vivo like setting using slice cultures from neonatal mice containing the entire pathway of corticospinal axons [16]. DiI-labeled corticospinal axons were imaged using time-lapse video microscopy of their pathway overlying the basilar pons, their major hindbrain target. The axon shaft millimeters behind the growth cone exhibits several dynamic behaviors, including the de novo formation of varicosities and filopodia-like extensions, and a behavior that we term "pulsation" which is characterized by a variable thickening and thinning of short segments of the axon. An individual axon can have multiple sites of branching activity, with many of the branches being transient. These dynamic behaviors occur along the portion of the axon shaft overlying the basilar pons but not just caudal to it. Once the collaterals extend into the pontine neuropil, they branch further in the neuropil, while the parent axon becomes quiescent. Thus, the branching activity is spatially restricted to specific portions of the axon, as well as being temporally restricted to a relatively brief time window. These findings provide definitive evidence that collateral branches form de novo along corticospinal axons and establish that the process of target recognition in this system is a property of the axon shaft rather than the leading growth cone.

Influence of a target-derived diffusible activity on layer 5 cortical axon targeting

The results summarized above demonstrate the importance of delayed interstitial collateral branching in the development of the projections of layer 5 neurons to the brainstem and spinal cord. These findings of stereotyped branching patterns suggest that targeting information is localized to specific sites along the axonal pathway. However, the identity of the signals which induce branch formation and provide directional cues to guide the growing collaterals are unknown. To date, most of the work on this issue has focused on cortical innervation of the basilar pons, a major brainstem target of layer 5 axons. Several lines of evidence indicate that the basilar pons controls the development of its own innervation by secreting a diffusible factor(s) that influences corticospinal axons. This was demonstrated with in vitro studies in which cortical explants were cocultured with explants of pons or a control tissue taken from a brain region not innervated by layer 5 axons [17]. The explants were embedded in a three-dimensional collagen matrix, a method which prevents direct contact between the tissues but allows for the graded diffusion of soluble molecules [18–20]. Cortical axons and their collateral branches showed a dramatic tendency to grow towards the pons explant but not the control explants (Fig. 5). Further, layer 5 axons were preferentially influenced by the diffusible activity [17]. Other studies using the same in vitro assay have indicated that the spinal grey [21,22] also exerts a tropic influence on cortical axons.

Coculture experiments in three-dimensional collagen gels, using both "delayed" cocultures (Fig. 6) and time-lapse video microscopy of "isochronic" cocultures (as in Fig. 5), have shown that the pons-derived activity also enhances the formation of collateral branches along the shaft of cortical axons [23]. We found that pontine explants enhanced the de novo initiation of branches along cortical axons in a target-directed manner, also enhanced the stable bifurcation of growth cones, and prompted directional changes by growth cone turning and collateral extension. These effects were distance-dependent and mimicked by pons-conditioned medium. This evidence indicates that the pons activity promotes branch initiation interstitially along cortical axons, a novel property for a chemoattractant, and provides a directional cue for their growth. These findings suggest that the pons chemoattractant serves as a diffusible target-recognition molecule.

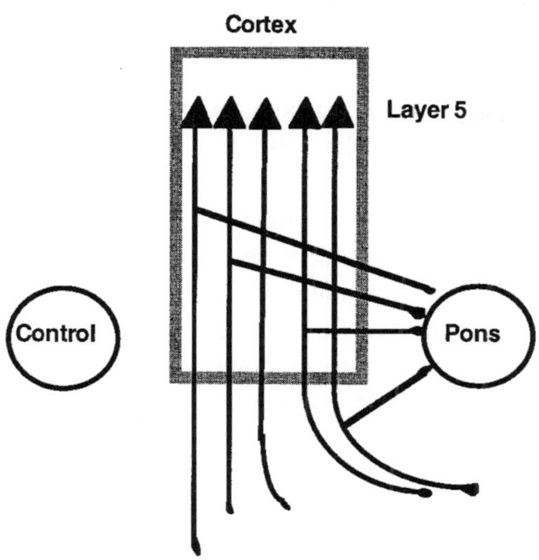

Fig. 5. "Isochronic" three-dimensional collagen gel coculture experiments. P0-P1 explants of cortex (Cortex), basilar pons (Pons) and control tissue (Control) were cocultured at a distance in three-dimensional collagen gels for up to 48 h. The pial surface of the cortical explant is to the top, the ventricular surface is to the bottom. Triangles represent layer 5 neurons; thin lines are axons and collaterals. The growth of cortical axons and collateral branches were directed preferentially toward explants of the pontine target tissue. Layer 5 neurons were found to be the principal responding population of cortical neurons.

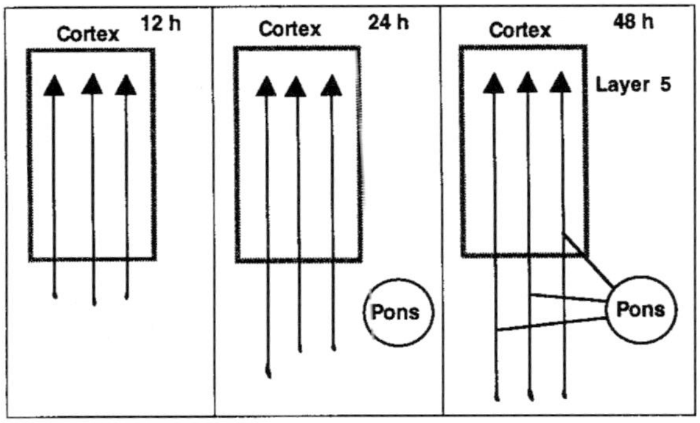

Fig. 6. "Delayed" coculture experiments in three-dimensional collagen gels. Left panel: Cortical explants (Cortex) were cultured alone for an initial period. Middle panel: After 24 h, an explant of basilar pons (Pons) was placed off to the side of the axons that had extended "inferiorly" from the ventricular surface of the cortical explant. Right panel: After an additional 24 h, collateral branches had formed and were directed toward the pontine explant. The branches were preferentially extended by layer 5 axons (black triangles). Branch formation was not stimulated when control tissue was used as the delayed explant.

Experimental manipulations in intact animals also suggest that target-derived cues control the branching and directed growth of cortical axon collaterals in vivo [15,24]. The most compelling of these comes from x-irradiation experiments. X-irradiation of fetal rats during the time of generation and migration of pontine neurons results in the formation of ectopic aggregations of basilar pontine neurons and a greatly reduced basilar pons; corticospinal axons form collateral branches at aberrant locations coincident with the ectopic pontine aggregates, but branches fail to form at the usual location of the basilar pons. Thus, the basilar pons, rather than local cues in the axon pathway, appears to control the location of cortical axon branching. In other experiments, subsets of corticospinal axons were prevented from invading the basilar pons; in these cases, the remaining cortical axons extend collaterals in a directed manner to regions of the basilar pons deprived of cortical input, a behavior consistent with a response to a diffusible chemoattractant emanating from these regions. Taken together, these in vitro and in vivo findings suggest that a diffusible, target-derived chemotropic molecule(s) controls target recognition in the developing corticopontine system by initiating the formation of collateral branches from corticospinal axons and directing their growth into the target.

Interstitial branching of cortical axons related to the local environment of their pathway

In a parallel study, we have explored the possibility that contact-dependent mechanisms might, in addition to a diffusible factor, influence the formation of the corticopontine projection. We examined, in neonatal rats, the region of the axon tract overlying the basilar pons, where branches form, to determine if any features of the substrate correlate spatially or temporally with branch formation [25]. We have used morphological techniques to study three aspects of the neural environment that have been implicated in axon guidance in other systems: extracellular matrix molecules (ECM), glial cells and dendrites. We found that the ECM molecules chondroitin sulfate proteoglycan (CSPG) and cytotactin/tenascin are excluded from the axonal tract. Within the axon tract their distributions do not differ between regions at which branches do or do not form. These results suggest that these ECM molecules do not influence branch formation. Immature astrocytes (positive for vimentin and nestin) and mature and immature oligodendrocytes (O4- and RIP-positive) are present throughout the axon tract. However, they are not enriched in the region of branch formation and the orientation of these cells does not correspond to the direction of axon collateral ingrowth.

Radial glia and mature (GFAP-positive) astrocytes are excluded from the cerebral peduncle in the region over the basilar pons, but not in surrounding parts of the axon tract. Since they are not in contact with corticospinal axons in the region of branch formation they are not likely to play a role in this process. However, dendrites are present at high density in the cerebral peduncle over the basilar pons. Many arise from neurons of the basilar pons. Dendritic processes are oriented parallel to the direction of axon branches. At the light and electron microscopic level, dendrites often appear to be closely apposed to axon branches. Therefore, axonal-dendritic contact or proximity could be a mediator of branch-formation and/or extension in the development of the corticopontine projection.

NT-3 may influence the development of layer 5 projections

The neurotrophins are principally thought of as agents that promote neuronal survival and differentiation; in particular, the trophic effects of NGF on sympathetic and sensory neurons have been well characterized [26]. However, there is limited evidence for possible tropic influences of neurotrophins, most of which concerns NGF. Intracerebral injections of NGF result in the growth of sympathetic fibers into the spinal cord and up the neuraxis towards the site of injection [27]. Direct experimental support for a tropic effect of NGF comes from the in vitro work of Gundersen and Barrett [28,29]. They demonstrated that the growth cones of dorsal root ganglion neurons will turn toward a localized source of NGF. However, it is clear that NGF does not act as a chemoattractant for peripheral sensory neurons in vivo since the synthesis of NGF and its receptor begins after sensory axons have reached their target [30].

Systemic administration of NGF has been shown to cause a nonspecific "sprouting" of cutaneous sensory fibers in rats [31]. Similarly, systemic NT-3 administration is reported to enhance the "sprouting" of corticospinal axons in the developing rat spinal cord, and, when adult rats are cotreated with systemic NT-3 and antibodies that neutralize the myelin-associated proteins that inhibit axonal growth, the sprouting and regeneration of corticospinal axons following injury is also enhanced [32]. The parallel between the effects of NGF on sensory axons and NT-3 on corticospinal axons suggests that NT-3 may act in vivo to promote the targeting of corticospinal axons. This possibility prompted us to initiate studies to assess this hypothesis. Two pieces of data are prerequisites if NT-3 is to have a role in the development of layer 5 projections to their subcortical targets. One, that NT-3 is expressed in subcortical targets of layer 5 axons during development, and two, that corticospinal neurons express trkC, the preferred receptor for NT-3.

The developmental expression of NT-3 mRNA has been detected in Northern blots as early as E13 in the rat brain, with a prominent rise in the perinatal period, and then a gradual decrease to diminished adult levels [33,34]. However, since details on the expression in the early postnatal brainstem and spinal cord at the cellular level are lacking, we have done in situ hybridization on sections of neonatal rat brainstem and spinal cord using antisense cRNA radiolabeled probes for NT-3 mRNA [35]. Several of the brainstem targets of layer 5 axons contain cells that show high expression of NT-3, including the dorsal column nuclei and inferior olive (but not the basilar pons). In addition, another major target, the superior colliculus, also contains many highly expressing cells. Thus, at an age when layer 5 axons begin to form their collateral branches that will innervate these targets, the target cells are expressing NT-3 mRNA. In early postnatal spinal cord, in addition to the previously reported NT-3-expressing motorneurons in the ventral horn [36], numerous cells scattered in the intermediate grey of the cervical region also highly express NT-3. Again, this corresponds to the cortical innervation pattern of the spinal cord: layer 5 axons densely innervate the intermediate grey and to a lesser extent the ventral horn.

In situ hybridization studies show that in the developing cortex trkC expression is strongest during the first postnatal week and then gradually declines [37–39]. In adult rats, corticospinal neurons have been shown to express trkC and NT-3 has been shown to support their survival following axotomy [40]. To examine whether corticospinal neurons express trkC during development, we have used retrograde tracing combined with immunohistochemistry to show that corticospinal neurons in neonatal rats express trkC [35]. Further, we find that trkC protein is distributed along the corticospinal tract in neonatal rats, suggesting that it is present on corticospinal axons at the time that they recognize and innervate the inferior olive, dorsal column nuclei and intermediate gray of the spinal cord – their targets that express NT-3. These findings suggest that NT-3 could influence the development of these layer 5 projections, and may act to control target innervation by regulating the formation and directed growth of collateral branches of layer 5 axons.

Spatially restricted increase in polysialic acid enhances target recognition and innervation by corticospinal axons

We have done a series of experiments in rats to determine whether target recognition in the development of corticospinal projections is influenced by a posttranslational addition of a polysialic acid (PSA) moiety to the neural cell adhesion molecule (N-CAM) [41]. It has been shown in several systems that cell-cell interactions can be modulated by the PSA moiety that is uniquely associated with N-CAM. High levels of PSA have been shown to attenuate interactions between cell surface or cell substrate ligands [42]. PSA levels are temporally and spatially regulated during development so that high levels of PSA are often found in regions where axons reach decision points and diverge from major axon tracts to innervate their specific targets. For example, studies in chicks of the innervation of limb muscles by motor axons have implicated PSA-modified NCAM in the process of axon sorting in the limb plexus and peripheral nerve defasciculation [43–45].

As a prelude to studying the role of PSA in target recognition and collateral branching by developing corticospinal axons, we examined the distribution of PSA and NCAM over the relevant developmental stages [41]. Whereas NCAM is distributed along the entire length of corticospinal axons as they grow from the cortex to the spinal cord, PSA-modified NCAM only becomes evident later in development, and is restricted to the spinal segment of corticospinal axons at the time they begin to form collateral branches in the spinal cord (Fig. 7). This unique spatiotemporal

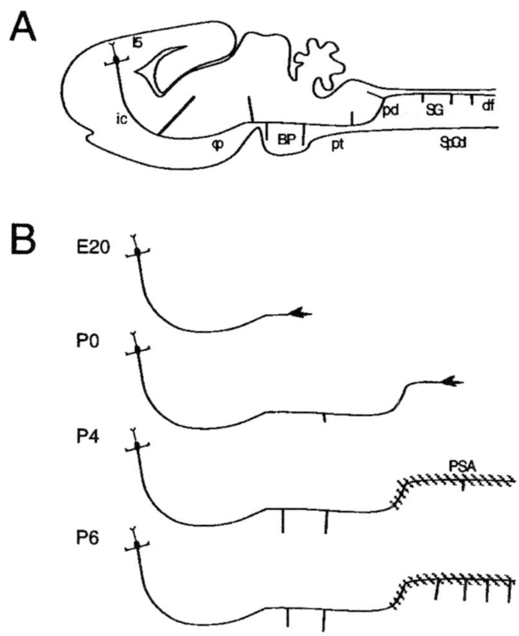

Fig. 7. PSA expression is largely restricted to the spinal segment of corticospinal axons. **A:** Sagittal view of an early postnatal rat brain illustrating the pathway of corticospinal axons, which originate from cortical layer 5 (L5) neurons, and the location of collateral branches to some of their major subcortical targets. **B:** Developmental sequence of PSA expression on corticospinal axons. Corticospinal axons reach the basilar pons at E20 and form the first collateral branches to the pons at P0. In the spinal cord collateral branch formation starts at P4. NCAM is present along the entire length of corticospinal axons at all ages, but PSA (cross-hatching) is found only on the distal portion of these axons and only at later stages of their development. E20-P0: PSA is not detected on corticospinal axons as they grow from cortex to the spinal cord. P4-P6: PSA is upregulated on a restricted distal portion of corticospinal axons starting near the pyramidal decussation. The spatiotemporal dynamics of PSA localization on corticospinal axons correlates with their formation of collateral branches to the spinal grey. Abbreviations: BP, basilar pons; cp, cerebral peduncle; df dorsal funiculus; ic, internal capsule; L5, layer 5; pd, pyramidal decussation; pt, pyramidal tract; SG, spinal grey; SpCd, spinal cord.

distribution of PSA suggests its involvement in corticospinal axon branching. To test this suggestion, PSA was removed in vivo by a single injection in neonatal rats of endoneuraminidase N, an enzyme that selectively removes PSA without affecting NCAM. This treatment did not interfere with the growth or pathfinding of corticospinal axons. However, PSA removal delayed the onset of collateral branching by corticospinal axons within the spinal cord, and later diminished the magnitude of branching. These findings suggest that PSA influences the process of target recognition and innervation by corticospinal axons within the spinal cord. Although PSA itself is not an inducer of axon branching, it does enhance the process of interstitial collateral branching, perhaps by lessening physical interactions with neighboring axons and permitting more effective interaction with external signals that promote branching [41].

Acknowledgements

The work from the O'Leary lab summarized in this chapter has been supported by NIH grants R01 NS31558 (D.O'L.) and P01 NS31249 (D.O'L.) and SCRF grant 1459 (K.Y.). We gratefully acknowledge and thank past and present lab members and collaborators who have contributed in many ways to the work summarized here, especially Juan De Carlos, Chris Heffner, David Litwack, Laura Lopez-Mascaraque, Heiko Possel, Makoto Sato, Brent Stanfield, Toshio Terashima, Rebecca Tuttle and Paul Yates. We also thank Douglass Borngasser for help with making the figures.

References

1. Jones EG. Laminar distributions of cortical efferent cells. In: Peters A, Jones EG (eds) Cerebral Cortex. New York: Plenum Press, 1984;521–553.
2. De Carlos JA, O'Leary DDM. Growth and targeting of subplate axons and establishment of major cortical pathways. J Neurosci 1992;12:1194–1211.
3. Koester SE, O'Leary DDM. Axons of early generated neurons in cingulate cortex pioneer the corpus callosum. J Neurosci 1994;14:6608–6620.
4. Moody SA, Quigg MS, Frankfurter A. Development of the peripheral trigeminal system in the chick revealed by an isotype-specific anti-beta-tubulin monoclonal antibody. J Comp Neurol 1989;279: 567–580.
5. Godement P, Vanselow J, Thanos S, Bonheoffer F. A study in developing nervous system with a new method of staining neurones and their processes in fixed tissue. Development 1987;101:697–713.
6. Honig MG, Hume RI. DiI and DiO: versatile fluorescent dyes for neuronal labelling and pathway tracing. Trends Neurosci 1989;12:333.
7. Richards LJ, Koester SE, Tuttle R, O'Leary DDM. Directed growth of early cortical axons may be controlled by a chemoattractant released by an intermediate target. 1996 (Submitted).
8. Kennedy TE, Serefini T, de la Torre JR, Tessier-Lavigne M. Netrins are diffusible chemotropic factors for commissural axons in the embryonic spinal cord. Cell 1994;78:425–435.
9. Stanfield BB, O'Leary DDM, Fricks C. Selective elimination in early postnatal development restricts cortical distribution of rat pyramidal neurons. Nature 1982;298:371–373.
10. Bates CA, Killackey HP. The emergence of a discretely distributed pattern of corticospinal projection neurons. Dev Brain Res 1984;13:265–273.
11. O'Leary DDM, Terashima T. Cortical axons branch to multiple targets by interstitial axon budding: implications for target recognition and "waiting periods". Neuron 1988;1:910.

12. Terashima T, O'Leary DDM. Quantitative analysis of the cortical distribution of pyramidal tract neurons in young and adult rats. Anat Rec 1988;220:96A.
13. O'Leary DDM, Stanfield BB. Selective elimination of axons extended by developing cortical neurons is dependent on regional locale. Experiments ultilizing fetal cortical transplants. J Neurosci 1989;9:3776–3795.
14. Stanfield BB, O'Leary DDM. Fetal occipital cortical neurons transplanted to the rostral cortex can extend and maintain a pyramidal tract axon. Nature 1985b;313:135.
15. O'Leary DDM, Bicknese AR, De Carlos JA, Heffner CD, Koester SE, Kutka LJ, Terashima T. Target selection by cortical axons: alternative mechanisms to establish axonal connections in the developing brain. CSHS Quant Biol 1990;55:453–468.
16. Bastmeyer M, O'Leary DDM. Dynamics of target recognition by interstitial axon branching along developing cortical axons. J Neurosci 1996;2016:1450–1459.
17. Heffner CD, Lumsden AGS, O'Leary DDM. Target control of collateral extension and directional axon growth in the mammalian brain. Science 1990;247:217–220.
18. Lumsden AGS, Davies AM. Earliest sensory nerve fibers are guided to peripheral targets by attractants other than nerve growth factor. Nature (London) 1983;306:786–788.
19. Lumsden AGS, Davies AM. Chemotropic effect of specific target epithelium in the developing mammalian nervous system. Nature (London) 1986;323:538–539.
20. Tessier-Lavigne M, Placzek M, Lumsden AGS, Dodd J, Jessell TM. Chemotropic guidance of developing axons in the mammalian central nervous system. Nature 1988;336:775–777.
21. Joosten EA, Gispen WH, Bar PR. Tropism and corticospinal target selection in the rat. J Neurosci 1994;59:33–41.
22. Joosten EA, van der Ven PF, Hooiveld MH, ten Donkelaar HJ. Induction of corticospinal target finding by release of a diffusible chemotropic factor in cervical grey matter. Neurosci Lett 1991;128:25–28.
23. Sato M, Lopez-Mascaraque L, Heffner CD, O'Leary DDM. Action of a diffusible target-derived chemoattractant on cortical axon branch induction and directed growth. Neuron 1994;13:791–803.
24. O'Leary DDM, Heffner CD, Kutka L, Lopez-Mascaraque L, Missias A, Reinoso BS. A target-derived chemoattractant controls the development of the corticopontine projection by a novel mechanism of axon targeting. Development 1991;(Suppl 2):123–130.
25. Bastmeyer M, Daston MM, Possel H, O'Leary DDM. Interstitial branching of cortical axons related to the local environment of their pathway. 1996 (Submitted).
26. Snider WD. Functions of the neurotrophins during nervous system development: what the knockouts are teaching us. Cell 1994;77:627–638.
27. Levi-Montalcini R. Developmental neurobiology and the natural history of nerve growth factor. Ann Rev Neurosci 1982;5:341–362.
28. Gundersen RW, Barrett JN. Neuronal chemotaxis: chick dorsal-root axons turn toward high concentrations of nerve growth factors. Science 1979;206:1079–1080.
29. Gundersen RW, Barrett JN. Characterization of the turning response of dorsal root neurites toward nerve growth factor. J Cell Biol 1980;87:546–554.
30. Davies AM, Bandtlow C, Heumann R, Korsching S, Rohrer H, Thoenen H. Timing and site of nerve growth factor synthesis in developing skin in relation to innervation and expression of the receptor. Nature 1987;326:353–358.
31. Diamond J, Holmes M, Coughlin M. Endogenous NGF and nerve impulses regulate the collateral sprouting of sensory axons in the skin of the adult rat. J Neurosci 1992;12:1452–1466.
32. Schnell L, Schneider R, Kolbeck R, Barde Y-A, Schwab ME. Neurotrophin-3 enhances sprouting of corticospinal tract during development and after adult spinal cord lesion. Nature 1994;367:70–173.
33. Ernfors P, Ibanez CF, Olson L, Persson H. Molecular cloning and neurotrophic activities of a protein with structural similarities to nerve growth factor: developmental and topographical expression in the brain. PNAS 1990;87:5454–5458.
34. Maisonpierre PC, Belluscio L, Friedman B, Alderson RF, Wiegand SJ, Furth ME, Lindsay RM, Yancopoulos GD. NT-3, BDNF, and NGF in the developing rat nervous system: parallel as well as

reciprocal patterns of expression. Neuron 1990;5:501–509.

35. Yee KT, O'Leary DDM. NT-3 and trkC expression patterns suggest their role in the development of corticospinal axon projections to the brainstem and spinal cord. Soc Neurosci Abstr 1996;(In press).
36. Schecterson LC, Bothwell M. Novel roles for neurotrophins are suggested by BDNF and NT-3 mRNA expression in developing neurons. Neuron 1992;9:449–463.
37. Lamballe F, Smeyne RJ, Barbacid M. Developmental expression of trkC, the neurotrophin-3 receptor, in the mammalian nervous system. J Neurosci 1994;14:14–28.
38. Ringstedt T, Lagercrantz H, Persson H. Expression of members of the trk family in the developing postnatal rat brain. Dev Brain Res 1993;72:119–131.
39. Tessarollo L, Tsoulfas P, Martin-Zanca D, Gilbert DJ, Jenkins NA, Copeland NG, Parada LF. trkC, a receptor for neurotrophin-3, is widely expressed in the developing nervous system and in non-neuronal tissues. Development 1993;118:463–475.
40. Giehl KM, Tetzlaff W. BDNF and NT-3, but not NGF, prevent axotomy-induced death of rat corticospinal neurons in vivo. Eur J Neurosci 1996;8:1167–1175.
41. Daston MM, Bastmeyer M, Rutishauser U, O'Leary DDM. Spatially restricted increase in polysialic acid enhances corticospinal axon branching related to target recognition and branching. J Neurosci 1996;16:5488–5497.
42. Rutishauser U. Polysialic acid as a regulator of cell interactions. In: Margolis RU, Margolis RK (eds) Neurobiology of Glycoconjugates. New York: Plenum Press, 1989;367–382.
43. Landmesser L, Dahm L, Tang J, Rutishauser U. Polysialic acid as a regulator of intramuscular nerve branching during embryonic development. Neuron 1990;4:655–667.
44. Tang J, Landmesser L, Rutishauser U. Polysialic acid influences specific pathfinding by avian motoneurons. Neuron 1992;8:1031–1044.
45. Tang J, Rutishauser U, Landmesser L. Polysialic acid regulates growth cone behavior during sorting of motor axons in the plexus region. Neuron 1994;13:405–414.

Mechanisms of memory as the basis of behaviour

Integrative and Molecular Approach to Brain Function.
M. Ito and Y. Miyashita, editors.

Cellular organization of the hippocampal formation

T. Kosaka, T. Fukuda, N. Fujise and Y. Aika

The Third Department of Anatomy, Faculty of Medicine, Kyushu University, Higashi-ku, Fukuoka, Japan

Abstract. In the present study we show characteristics of two groups of neurons presumably relating to the coactivation of the neuronal population in the rodent hippocampus

Parvalbumin (PV)-containing neurons, a subpopulation of GABAergic nonprincipal neurons, in the hippocampus are mainly basket and axo-axonic cells, making synapses on somata and axon initial segments of principal cells. They are intensely immunoreactive for GluR4 and display some specific surface and/or extracellular substances characterized by several monoclonal antibodies and lectins on their somata and proximal dendrites, suggesting that they might have some particular microenvironment. They make numerous synaptic contacts with one another, as well as gap junctions, both of which are considered to be important in generating synchronized interneuronal bursts.

Mossy cells, the principal neurons in the dentate hilus, receive excitatory synapses from dentate granule cells and, in turn, extend long-associational/commissural axonal branches which make excitatory synapses on proximal dendrites of granule cells, as well as dendrites of GABAergic neurons. Thus, mossy cells are regarded as a part of "a disynaptic associational network between granule cells". However, in the dorsal part of the mouse dentate gyrus, there are fewer mossy cells, whereas in the ventral part numerous mossy cells are clustered, indicating a possible dorsoventral difference in the neuronal circuit.

Key words: calcium binding protein, GABA, gap junction, interneuron, mossy cell, synchronization.

Many studies have revealed that some subsets of neurons contain multiple neuroactive substances and specific calcium binding proteins. Similarly, several surface markers and particular subunits of various receptors have been reported to be expressed on particular subsets of neurons. These chemical substances are regarded as chemical markers characterizing some subpopulations of neurons. Neurons in the hippocampal formation might be divided into two groups, principal neurons and nonprincipal neurons. Principal neurons in each area are generally regarded to be glutamatergic and rather homogeneous, whereas the vast majority of hippocampal nonprincipal cells are assumed to be GABAergic and extensively heterogeneous. These principal and nonprincipal neurons are characterized by various chemical markers (Table 1), as well as their structural features. In the present study we show characteristics of two groups of neurons presumably relating to the coactivation of the neuronal population in the mouse and rat hippocampal formation: 1) parvalbumin (PV)-containing neurons, a particular subpopulation of GABAergic neurons, and 2) the mossy cells, the principal neurons in the hilus of the dentate gyrus (DG).

Address for correspondence: T. Kosaka, The Third Department of Anatomy, Faculty of Medicine, Kyushu University, Higashi-ku, Fukuoka 812-82, Japan. Fax: +81-9-2-633-6774. E-mail: kosaka@a3rd.med.kyushu-u.ac.jp

Table 1. Chemical markers of hippocampal neurons.

I) Hippocampal principal neurons	II-A) Hippocampal GABAergic nonprincipal neurons	II-B) Hippocampal non-GABAergic nonprincipal neurons
Granule cells; glutamate dynorphin calbindin Hilar mossy cells; glutamate calretinin (in mouse) CA3 pyramidal cells; glutamate CA2 pyramidal cells; glutamate markers for microenvironments (lectins, HNK-1, etc.) CA1 pyramidal cells; glutamate calbindin	A. Calcium binding proteins parvalbumin calbindin calretinin B. Neuroactive substances peptides CCK somatostatin VIP neuropeptide Y nitric oxide (NOS) C. Markers for microenvironments lectins; VVA, SBA, monoclonal Abs; HNK-1 proteoglycans; 473, 3B3 tenasin D. Receptors and channels glutamate receptors; metabotrophic AMPA GR1,4 $GABA_A$ receptors muscarine Ach receptor peptide receptors steroid hormone receptors estrogen glucocorticoid KV3.1b	Long spiny multipolar cells glutamate calretinin (in rat) Cajal-Retzius cells glutamate calretinin

Characteristics of PV-containing GABAergic neurons

As we previously proposed, PV-containing GABAergic neurons are a separate subpopulation of GABAergic neurons from those containing other neuroactive substances [1], although, in the rat hippocampus, about 14% of PV-containing neurons also contain the somatostatin-like immunoreactivity (our unpublished observations). They are also different from GABAergic neurons containing calbindin and calretinin [2]. Thus, the PV neuron group appears to be chemically a distinct subpopulation of GABAergic neurons. Although PV neurons appear to be somewhat heterogeneous in minor respects, the characteristics of most of PV neurons in the hippocampus, especially in the CA1 region are summarized as follows [1,3–10]. They are so-called GABAergic nonpyramidal cells whose somata are located in the strata pyramidale and oriens in the Ammon's horn and in the stratum granulosum and subgranular zone in the DG (Figs. 1–3). The numerical density of PV neurons in the CA1 region of the rat dorsal hippocampus is $1.1 \times 10^3/mm^3$, which is 3.2% of all neurons [3]. PV positive synaptic terminals are also mainly located in the strata pyramidale and granulosum (Fig. 2), i.e., the layers in which PV positive cell somata themselves, as well as principal cell somata, are located; in these layers they make symmetrical synapses on principal cell somata, axon initial segments and proximal dendrites, as well as nonpyramidal cell somata and dendrites, including PV positive ones [4]. These observations indicate that PV neurons in the hippocampus might be basket cells and axo-axonic cells, which was confirmed by recent combined physiology-immunocytochemistry studies [11]. Combined physiology-immunocytochemistry using hippocampal slice preparations revealed that physiologically identified fast-spiking or repetitively firing neurons are PV positive [6], although a recent in vivo recording showed a fast-spiking neuron containing calbindin but not PV [11]. Almost all PV neurons in the hippocampus are intensely GluR4 positive (Figs. 3 and 4), whereas the majority of other types of neurons, both principal and nonprincipal cells, except some somatostatin neurons (Fig. 3), some NOS containing neurons and a few calbindin neurons (our unpublished observations), do not express the GluR4 immunoreactivity so intensely. Dendrites of PV neurons are frequently covered by numerous axon terminals making asymmetrical synapses, where the intense GluR4 immunoreactivity is located (Fig. 4). The majority of PV neurons are only faintly immunoreactive for GAD 65 [12]. PV neurons are also reported to express a specific subunit of K-channels, KV 3.1b, which is assumed to be deeply related with the physiological characteristics of fast-spiking neurons [14,15]. In addition, some surface markers [1,7,9,16] listed in Table 1 also characterize a major part of PV neurons (about 80% of hippocampal PV neurons are HNK-1 positive), indicating some specific microenvironments of these neurons (Fig. 5). However, these surface markers are not necessarily specific to PV neurons and they also label CA2 pyramidal cells, as well as some PV negative nonprincipal cells [7]. Substances labeled with these markers are supposed to be highly negative-charged [17], and thus assumed to determine the microenvironments of these neurons (such as ionic environments of immediate vicinity of synaptic sites) which might deeply influence the physiological characteristics of these neurons, although their exact functional roles remain to be determined.

Our previous electron microscopic, as well as combined Golgi-electron microscopic analyses revealed the presence of gap junctions between nonprincipal presumed GABAergic cells [18–20]. Later we confirmed gap junctions on PV neurons [5], indicating that PV neurons can communicate with each other via electrical coupling. In addition, our recent analysis [12,13] clearly revealed that the majority of PV neurons, especially those located in the stratum pyramidale, receive numerous GABAergic synaptic contacts on their somata and proximal dendrites (Fig. 2), whereas PV negative GABAergic neurons usually receive a far smaller number of GABAergic synapses. Furthermore, the majority of these GABAergic terminals on PV neurons are also PV positive and most of them persist even after the fimbria-fornix transection, indicating that PV positive GABAergic neurons in the hippocampus make mutual synaptic connections.

Recently, inhibitory neurons are considered to play an important role in synchronizing the activities of principal neurons [15,21–23]. GABAergic interneurons themselves are reported to be able to become synchronized via at least two different mechanisms; one is GABA-mediated synaptic interactions through their recurrent collaterals and the other is a mechanism probably involving electrotonic coupling [23–25]. Furthermore, Whittington et al. [15] proposed that networks of GABAergic inhibitory neurons can generate 40-Hz oscillations in pyramidal cells, which have been implicated in cognitive function. Mutual interaction through reciprocal $GABA_A$

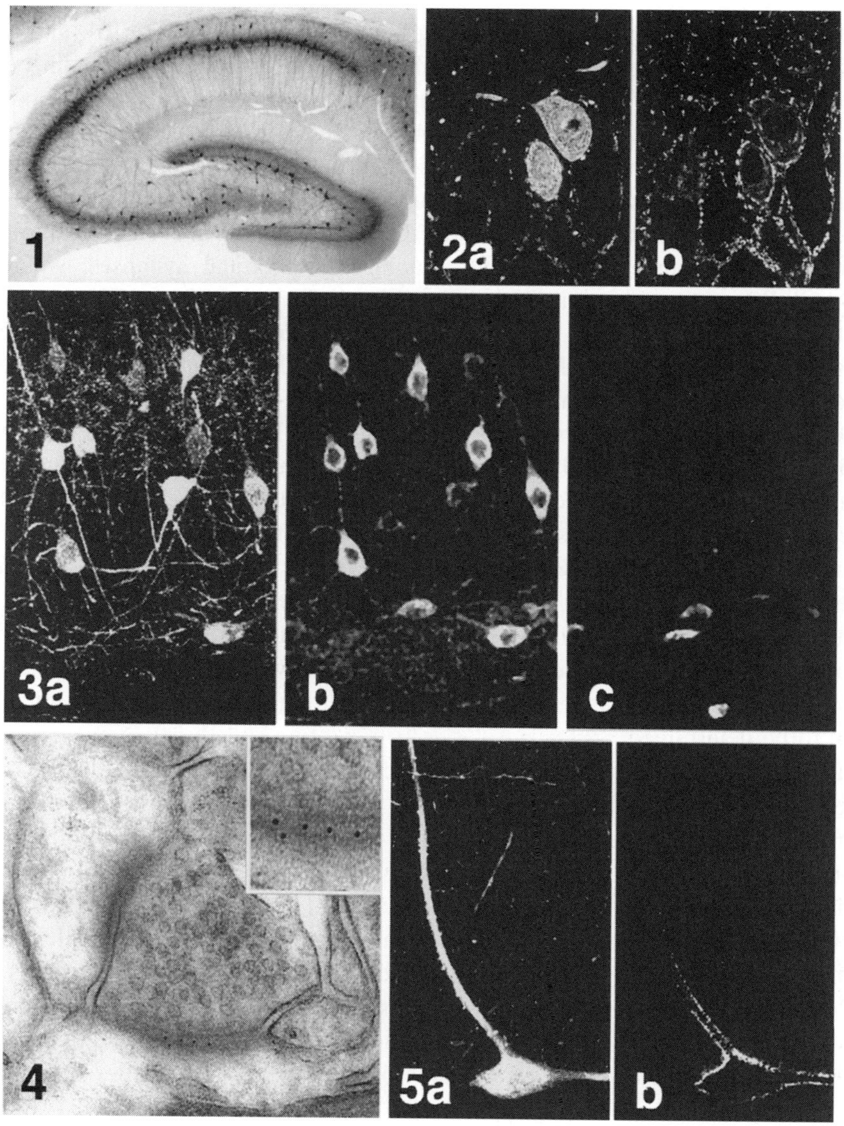

Fig. 1. Light micrograph of the rat dorsal hippocampus immunostained with anti-PV.

Fig. 2. Immunofluorescent confocal laser scanning light microscopic (CLSM) images of the CA1 stratum pyramidale double stained with **a** anti-PV and **b** anti-GAD, showing many GABAergic terminals on PV positive somata as well as on pyramidal cell somata.

Fig. 3. Immunofluorescent CLSM images of the CA1 strata pyramidale and oriens triple stained with **a** anti-PV, **b** anti-GluR4, and **c** antisomatostatin. All PV neurons are also intensely or intermediately GluR4 positive. One somatostatin positive neuron is also intensely GluR4 positive.

Fig. 4. Electron micrograph showing the postembedding colloidal gold immunocytochemistry for GluR4 applied on a freeze-substituted Lowicryl-embedded ultrathin section. Gold particles are located at the synaptic site between an axon terminal and a nonpyramidal cell dendrite. The same axon terminal also makes a synapse on a spine-like profile of a presumed pyramidal cell dendrite, where no gold particles are observed.

Fig. 5. Immunofluorescent CLSM images of the CA1 stratum oriens double stained with **a** anti-PV and **b** anti-HNK-1.

---

synapses is assumed to be sufficient to maintain the oscillatory synchrony of GABAergic inhibitory neuron population, although gaps junctions, if present, are assumed to enhance synchronization. Thus, so far as we know, among GABAergic inhibitory neuron groups, PV neurons appear to correspond most well to those neurons showing synchronized bursts. Furthermore, PV neurons are reported to be activated by pyramidal cells through a single release site [26], which is theoretically assumed to be due to possible active dendritic conductances in these cells [27]. Traub [28] also proposed that these active dendritic conductances are an important factor in the generation of synchronized population bursts of inhibitory neurons via gap junctions. Thus, PV neurons in hippocampus and presumably in other cortical regions are a very distinctive subpopulation of GABAergic neurons in their structural, chemical and physiological features (Fig. 6), and are assumed to participate in the interneuronal networks imposing an oscillatory synaptic input onto the principal cells [21].

Mossy cells in the hilus; dorsoventral difference in their distribution

Mossy cells are located in the hilus of the DG and receive excitatory synapses from mossy terminals of granule cells on the characteristic thorny excrescences of their proximal dendrites (Figs. 9 and 10). They, in turn, extend long-associational and commissural axonal branches in the inner molecular layer of the DG and make excitatory synapses on proximal dendrites of granule cells, as well as dendrites of GABAergic neurons which are also assumed to make synapses on granule cells [29–31]. Recently Buckmaster and Schwartzkroin [30] proposed the "granule cell association" hypothesis of mossy cell function, i.e., mossy cells integrate inputs from granule cells and distribute that information to distant granule cells and, thus, mossy

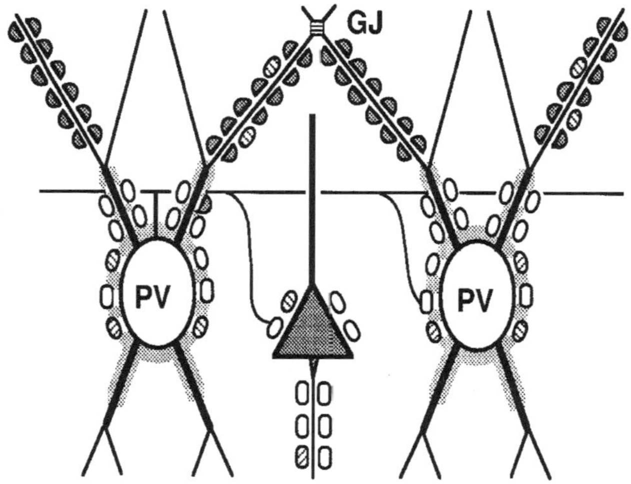

Fig. 6. Scheme showing characteristics of PV positive neurons. Cell body and proximal dendrites are covered with some specific substances characterized by several monoclonal antibodies and lectins. PV neurons also express GluR4 and KV3.1b intensely. Many PV positive (○) and PV negative (⦸) GABAergic terminals make symmetrical synapses on somata and dendrites of PV neurons. Many synaptic boutons making asymmetrical synapses (●) cover dendrite of PV neurons. PV neurons extend axonal branches in the principal cell layer and make synapses on somata, initial segments and proximal dendrites of principal cells, as well as nonprincipal cells. PV neurons make mutual synaptic contacts and gap junctions (GJ).

cells support the coactivation of spatially dispersed granule cells. Thus, mossy cells are regarded as "a part of a disynaptic associational network between granule cells" [30] which "may share functional capabilities analogous to the monosynaptic associational network in the CA3 field" [30]. Interestingly our recent observations in the mouse hippocampal formation [32] indicate that mossy cells show a significant difference in their distribution and chemical nature between dorsal and ventral parts.

In the hilus of the mouse DG, the calretinin (CR) immunoreactivity displays a drastic dorsoventral difference [32]; in the dorsal hilus a few CR-IR cells are scattered, whereas almost the entire area of the ventral hilus appears to be occupied by the CR immunoreactivity (Figs. 7 and 8). In this CR-IR hilar region, numerous large CR-IR multipolar cells are clustered. They bear grape-like clusters of small CR-IR spheroidal or irregularly shaped elements at their proximal dendrites (Fig. 11), resembling the thorny excrescences of Golgi-impregnated mossy cells (Fig. 10); thus, these large CR-IR hilar cells are assumed to be mossy cells. We also confirmed this identification as mossy cells with several methods (Fujise et al. In preparation): 1) mossy cells identified by intracellular labeling were CR-IR, 2) electron microscopic analysis of CR-IR elements clearly revealed structural features of mossy cells, 3) detailed three-dimensional analysis of CR-IR processes with a CLSM revealed the characteristic thorny excrescences of mossy cells, and 4) anterograde and retrograde

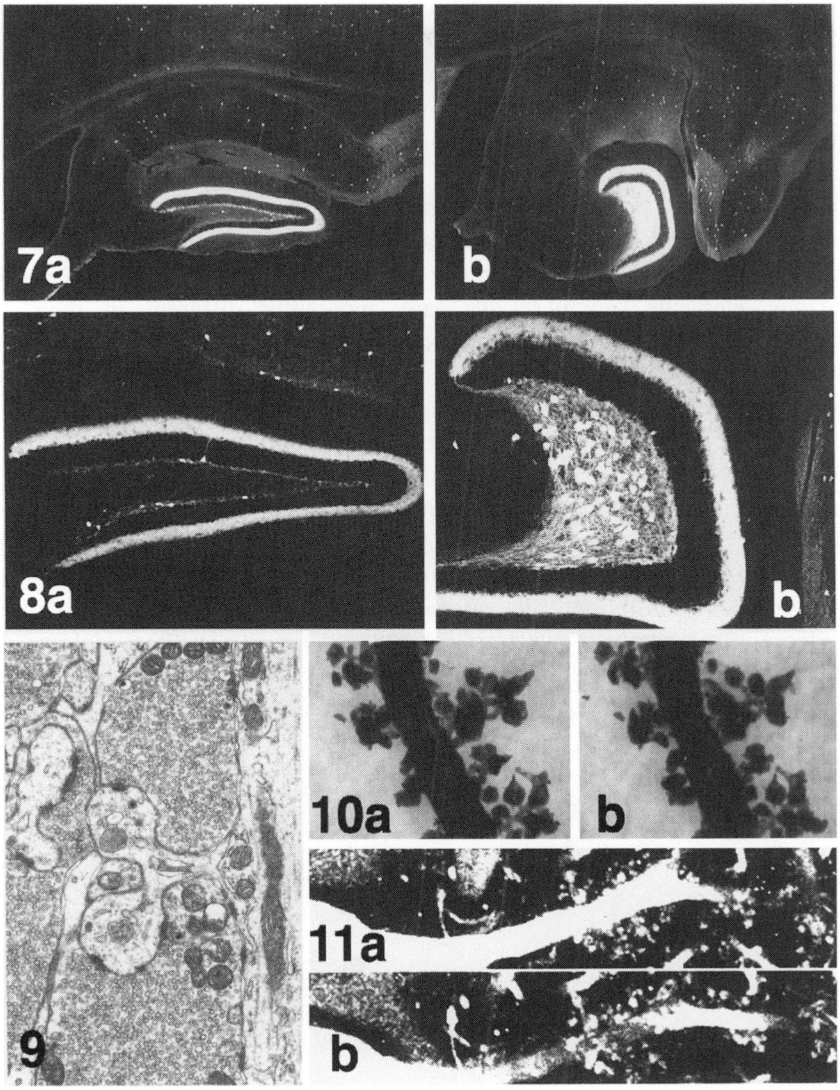

Fig. 7. Photomicrographs of **a** dorsal and **b** ventral parts of hippocampal formations of the mouse, fluorescent-immunostained with anti-CR.

Fig. 8. Immunofluorescent CLSM images, showing the distribution of the CR-IR elements in the **a** dorsal and **b** ventral parts of the mouse dentate gyrus.

Fig. 9. Electron micrograph showing synapses on a thorny excrescence from mossy terminals.

Fig. 10. Stereo pair of high voltage electron micrographs showing Golgi-impregnated thorny excrescences of a mossy cell dendrite.

Fig. 11. Immunofluorescent serial CLSM images, showing the thorny excrescences of a CR-IR mossy cell of the mouse dentate gyrus.

tracer experiments, as well as degeneration experiments indicated that the CR-IR band at the supragranular zone, the target zone of mossy cell axon terminals, originated from ventrally located large CR-IR hilar cells. CR-IR mossy cells increase in number near the caudal pole of the horizontally extending dorsal hippocampus and make a distinct cellular column in the ventral hippocampus [32]. However, we must consider the presence of faintly CR-positive or CR-negative mossy cells in the dorsal DG, which we confirmed by their GluR2/3 immunoreactivity. Even when we take these CR negative mossy cells into consideration, mossy cells are concentrated in the ventral part of the DG but less in the dorsal part of the DG. Furthermore, the presynaptic elements, granule cells, are apparently different in their axonal arborization in the hilus between the dorsal and ventral parts (Fig. 12). These observations lead us to speculate that there might be a prominent difference between the dorsal and ventral parts of the DG in their neuronal organization (Fig. 13), i.e., granule cells in the dorsal part make a relatively small number of synaptic contacts on fewer mossy cells but receive numerous inputs from ventrally located mossy cells, whereas those in the ventral part make numerous synapses on mossy cells and in turn receive numerous inputs from the latter. In addition, granule cells in both regions might receive synaptic contacts from inhibitory neurons which are supposed to receive synapses from mossy cells, as well as granule cells. Although further details

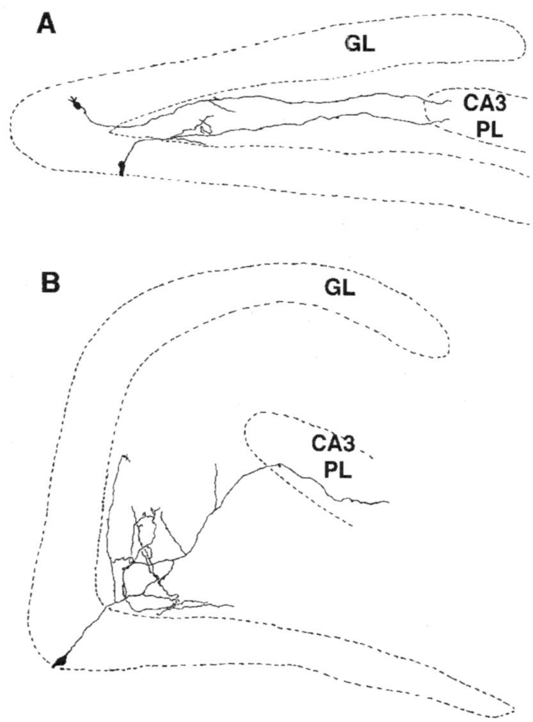

Fig. 12. Camera lucida drawings of axonal branches of Golgi-impregnated granule cells in the **A** dorsal and **B** ventral dentate gyrus. GL, stratum granulosum. CA3 PL, stratum pyramidale in the CA3 region.

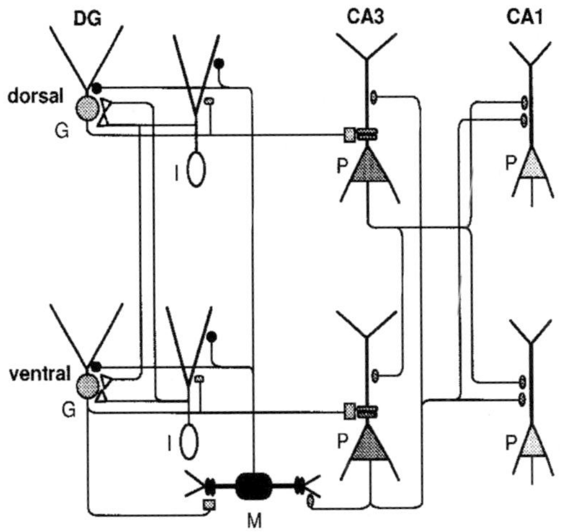

Fig. 13. Scheme showing the possible dorsoventral difference in a neuronal circuit. Although a few mossy cells are present in the dorsal hilus, they are not shown in this scheme to emphasize the dorsoventral difference. G, granule cell. I, dentate interneuron. M, mossy cell. P, pyramidal cell.

of this possible difference in the neuronal organization between the dorsal and ventral parts of the DG remain to be revealed, the possible "disynaptic associational network between granule cells" [30] in which mossy cells are assumed to participate might be somewhat different dorsoventrally.

Acknowledgements

The authors are grateful to Dr Heizmann for his kind supply of anti-PV. We thank Ms Suematsu for her secretarial assistance. This work was partially supported by Grants-in-Aid for Scientific Research from the Japanese Ministry of Education, Science and Culture and grants from Uehara Memorial Foundation and Mitsubishi Foundation.

References

1. Kosaka T, Isogai K, Barnstable CJ, Heizmann CW. Monoclonal antibody HNK-1 selectively stains a subpopulation of GABAergic neurons containing the calcium binding protein parvalbumin in the rat cerebral cortex. Exp Brain Res 1990;82:566–574.
2. Miettinen R, Gulyás AI, Baimbridge KG, Jacobowitz DM, Freund TF. Calretinin is present in non-pyramidal cells of the rat hippocampus. II. Co-existence with other calcium binding proteins and GABA. Neuroscience 1992;48:29–43.
3. Aika Y, Ren JQ, Kosaka K, Kosaka T. Quantitative analysis of GABA-like-immunoreactive and parvalbumin-containing neurons in the CA1 region of the rat hippocampus using a stereological method, the disector. Exp Brain Res 1994;99:267–276.
4. Katsumaru H, Kosaka T, Heizmann CW, Hama K. Immunocytochemical study of GABAergic neurons containing the calcium-binding protein parvalbumin in the rat hippocampus. Exp Brain Res 1988;72:347–362.

5. Katsumaru H, Kosaka T, Heizmann CW, Hama K. Gap junctions on GABAergic neurons containing the calcium-binding protein parvalbumin in the rat hippocampus (CA1 region). Exp Brain Res 1988;72:363–370.
6. Kawaguchi Y, Katsumaru H, Kosaka T, Heizmann CW, Hama K. Fast spiking cells in the rat hippocampus (CA1 region) contain the calcium-binding protein parvalbumin. Brain Res 1987;416: 369–374.
7. Kosaka T, Heizmann CW. Selective staining of a population of parvalbumin-containing GABAergic neurons in the rat cerebral cortex by lectins with specific affinity for terminal N-acetylgalactosamine. Brain Res 1989;483:158–163.
8. Kosaka T, Heizmann CW, Barnstable CJ. Monoclonal antibody VC1.1 selectively stains a population of GABAergic neurons containing a calcium binding protein parvalbumin in the rat cerebral cortex. Exp Brain Res 1989;78;43–50.
9. Kosaka T, Heizmann CW, Fujita SC. Monoclonal antibody 473 selectively stains a subpopulation of GABAergic neurons containing the calcium-binding protein parvalbumin in the rat cerebral cortex. Exp Brain Res 1992;89:109–114.
10. Kosaka T, Katsumaru H, Hama K, Wu JY, Heizmann CW. GABAergic neurons containing the Ca^{2+}-binding protein parvalbumin in the rat hippocampus and dentate gyrus. Brain Res 1987;419:119–130.
11. Sik A, Penttonen M, Ylinen A, Buzsaki G. Hippocampal CA1 interneurons: an in vivo intracellular labeling study. J Neurosci 1995;15:6651–6665.
12. Fukuda T, Kosaka T. Parvalbumin-containing GABAergic neurons in the mouse hippocampus receive more numerous GABAergic inputs on their somata than parvalbumin-negative GABAergic neurons. Soc Neurosci Abst 1995;176:6.
13. Fukuda T, Aika Y, Heizman CW, Kosaka T. Dense GABAergic input on somata of parvalbumin-immunoreactive GABAergic neurons in the hippocampus of the mouse. Neurosci Res 1996;26:(In press).
14. Du J, Weiser M, Rudy B, McBain CJ. Developmental expression and functional characterization of the potassium-channel subunit Kv3.1b in parvalbumin-containing interneurons of the rat hippocampus. J Neurosci 1996;16:506–518.
15. Whittington MA, Traub RD, Jefferys JGR. Synchronized oscillations in interneuron networks driven by metabotropic glutamate receptor activation. Nature 1995;373:612–615.
16. Celio M. Perineuronal nets of extracellular matrix around parvalbumin-containing neurons of the hippocampus. Hippocampus 1993;3:55–60.
17. Brückner G, Brauer K, Härtig W, Wolff JR, Rickmann MJ, Derouiche A, Delpech B, Girard N, Oertel W, Reichenbach A. Perineural nets provide a polyanionic, glia-associated form of microenvironment around certain neurons in many parts of the rat brain. Glia 1993;8:183–200.
18. Kosaka T. Gap junctions between non-pyramidal cell dendrites in the rat hippocampus (CA1 and CA3 regions). Brain Res 1983;271:157–161.
19. Kosaka T. Neuronal gap junctions in the polymorph layer of the rat dentate gyrus. Brain Res 1983;277:347–351.
20. Kosaka T, Hama K. Gap junctions between non-pyramidal cell dendrites in the rat hippocampus (CA1 and CA3 regions): a combined Golgi-electron microscopy study. J Comp Neurol 1985;231: 150–161.
21. Buzsáki G, Chrobak JJ. Temporal structure in spatially organized neuronal ensembles: a role for interneuronal networks. Curr Opin Neurobiol 1995;5:504–510.
22. Cobb SR, Buhl EH, Halasy K, Paulsen O, Somogyi P. Synchronization of neuronal activity in hippocampus by individual GABAergic interneurons. Nature 1995;378:75–78.
23. Ylinen A, Bragin A, Nádasdy Z, Jandó G, Szabó I, Sik A, Buzsáki G. Sharp wave-associated high-frequency oscillation (200 Hz) in the intact hippocampus: network and intracellular mechanisms. J Neurosci 1995;15:30–46.
24. Michelson HB, Wong RKS. Excitatory synaptic responses mediated by $GABA_A$ receptors in the hippocampus. Science 1991;253:1420–1423.
25. Michelson HB, Wong RKS. Synchronization of inhibitory neurons in the guinea-pig hippocampus

in vitro. J Physiol 1994;477:35–45.

26. Gulyás AI, Miles R, Sik A, Tóth K, Tamamaki N, Freund T. Hippocampal pyramidal cells excite inhibitory neurons through a single release site. Nature 1993;366:683–687.
27. Traub RD, Miles R. Pyramidal cell-to-inhibitory cell spike transduction explicable by active dendritic conductances in inhibitory cell. J Comput Neurosci 1995;2:291–298.
28. Traub RD. Model of synchronized population bursts in electrically coupled interneurons containing active dendritic conductances. J Comput Neurosci 1995;2:283–289.
29. Buckmaster PS, Strowbridge BW, Kunkel DD, Schmiege DL, Schwartzkroin PA. Mossy cell axonal projections to the dentate gyrus molecular layer in the rat hippocampal slice. Hippocampus 1992;2:349–362.
30. Buckmaster PS, Schwartzkroin PA. Hippocampal mossy cell function: a speculative view. Hippocampus 1994;4:393–402.
31. Scharfman HE. Electrophysiological evidence that dentate hilar mossy cells are excitatory and innervate granule cells and interneurons. J Neurophysiol 1995;74:179–194.
32. Liu Y, Fujise N, Kosaka T. Distribution of calretinin immuncreactivity in the mouse dentate gyrus: I. General description. Exp Brain Res 1996;108:389–403.
33. Weiser M, Bueno E, Sekirnjak C, Martone ME, Baker H, Hillman D, Chen S, Thornhill W, Ellisman M, Rudy B. The potassium channel subunit KV3.1b is localized to somatic and axonal membranes of specific populations of CNS neurons. J Neurosci 1995;15:4298–4314.

Integrative and Molecular Approach to Brain Function.
M. Ito and Y. Miyashita, editors.

Cortical and hippocampal contributions to declarative memory

H. Eichenbaum[1], G. Schoenbaum[2], B. Young[3] and M. Bunsey[4]

[1]*Laboratory of Cognitive Neurobiology, Department of Psychology, Boston University;* [2]*Department of Psychology, University of North Carolina;* [3]*Department of Psychology, University of Otago; and* [4]*Department of Psychology, Kent State University, USA*

Abstract. In humans declarative or explicit memory is supported by the hippocampus and related structures of the medial temporal lobe working in concert with the cerebral cortex. This presentation reviews our progress in developing an animal model for studies of cortical-hippocampal interactions in memory processing, based on close associations between the olfactory and hippocampal systems. Our findings support the view that the cerebral cortex maintains various forms of memory representation, and that hippocampal structures extend the persistence and mediate the organization of these codings. Specifically, the cerebral cortex encodes representations of specific stimuli, as well as their significance and stimulus associations. The parahippocampal region mediates the convergence and extended persistence of cortical codings. The hippocampus itself is critical to the organization cortical representations in terms of relationships among items in memory and in the "flexible", inferential memory expression that is the hallmark of declarative memory.

Key words: cerebral cortex, entorhinal cortex, olfactory learning, parahippocampal region, perirhinal cortex, rats.

Introduction

It is widely accepted that the hippocampus and closely related structures of the medial temporal lobe interact with the cerebral cortex in support of the persistence and organization of memories [1–4]. This account is guided by a simple conception of cortical-hippocampal pathways in which there are three main serially and bidirectionally connected components: the cortex, the parahippocampal region, and the hippocampus itself (Fig. 1A). The beginning and end point of this system involves several tertiary "association" cortical areas whose outputs converge on the parahippocampal region [5–7]. This region, composed of the interconnected perirhinal, parahippocampal (or postrhinal), and entorhinal cortices, surround the hippocampus in both rats and monkeys and merge different sources of cortical input bound for the hippocampus. Within the hippocampus itself there are several stages of serial and parallel processing and the outcome of this integration is sent back to the parahippocampal region, which in turn sends its main outputs to the same tertiary cortical areas that provided the source of input.

Address for correspondence: Howard Eichenbaum, Laboratory of Cognitive Neurobiology, Department of Psychology, 2 Cummington Street, Boston University, Boston, MA 02215, USA.

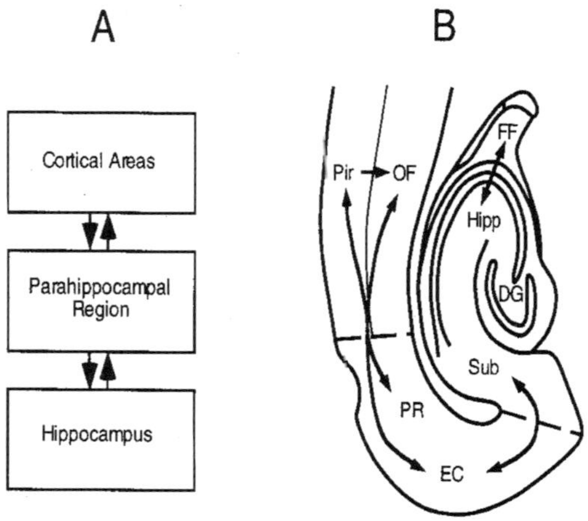

Fig. 1. **A**: Simple schematic diagram of cortical-hippocampal connections. **B**: Outline of a horizontal rat brain section illustrating the locations and flow of information between components of the hippocampus, parahippocampal region, and adjacent cortical areas. DG = dentate gyrus; EC = entorhinal cortex; FF = fimbria-fornix; Hipp = hippocampus proper; OF = orbitofrontal cortex; Pir = perirhinal cortex; PR = perirhinal cortex; Sub = subiculum.

Here we review our own recent developments of a model system for exploring the nature of cortical and hippocampal memory representation and the specific roles played by hippocampal structures in memory formation. Our explorations focus on the set of interconnected olfactory-cortical and hippocampal structures that are fully evolved in rodents (Fig. 1B). Odor information initially processed by the olfactory bulb is sent directly to the piriform cortex and closely interconnected orbital prefrontal cortex [8,9]. Both of these cortical areas, as well as the olfactory bulb, project heavily to the perirhinal and entorhinal components of the parahippocampal region, which then provides the primary source of olfactory sensory information to the hippocampus itself. In the return pathway, the outputs of hippocampal processing involve direct projections from the parahippocampal region to both the piriform and orbital prefrontal cortices. In addition, physiological studies indicate close interactions in the synchrony of processing between olfactory and hippocampal areas [10,11]. In sum then, olfactory-hippocampal pathways provide a straightforward and relatively simple instantiation of the general model of cortical-hippocampal interconnections. Our efforts in exploring the functional mechanisms of this system involve a combination of neuropsychological studies on the effects of selective ablation of these areas, and electrophysiological studies that characterize the coding properties of single neurons in each area during memory performance.

Representations of odor memories and odor associations in the olfactory prefrontal cortex

Damage to the orbitofrontal cortex severely impairs olfactory learning and memory

[12,13]. Guided by this finding, and by our model of olfactory-hippocampal interactions, we developed a complex odor discrimination task that permits simultaneous examination of several aspects of odor coding in the prefrontal area [14,15]. The discrimination task involved a sequence of trials in which the rat was signaled by a light that it could initiate an odor presentation by poking its nose into a stimulus sampling port. When the odor was an assigned as a rewarded (S+) stimulus an additional nose poke into a separate water port resulted in water delivery, and no reward was given regardless of the response for other (S–) odors. On each trial one of eight different odor stimuli (identified here by number and assigned reward values) was presented, half of which were associated with a sweetened water reward and the other half not rewarded. We examined the coding of odor identity by comparing neural activity during odor sampling across the eight odors. In addition we assessed whether cortical odor codings reflect the significance of items by comparing responses across the group of S+ and S– stimuli.

We found that many neurons in the orbitofrontal cortex and piriform cortex encode both odor identity and odor valence. Although a few cells fired only to one odor, most were coarsely tuned as reflected by different levels of activation across the eight stimuli (Fig. 2A). In addition, most cells fired more strongly either to all the S+ odors or to all the S– odors. For example in Fig. 2B the response was stronger to all S+ odors, and strongest to the third S+. An equivalent proportion of cells showed the opposite pattern, firing more strongly to all S– odors. The example shown in Fig. 2C shows larger responses to all S– odors, with the greatest response to the second S–. This pattern of findings indicates, not surprisingly, that the olfactory prefrontal area encodes odor identity. In addition, assigned stimulus valences are also very prominently represented in the cortex, indicating that cortical odor representations are indeed "tuned" or "biased" in association with reward contingencies.

In addition we sought to determine whether neural activity in the sensory cortex would reflect a simple kind of learned stimulus association, specifically an acquired association between two odors where presentation of one of them predicts the subsequent occurrence of the other [16]. Assuming that rats can learn a predictable relationship between odor stimuli, a neural reflection of the acquired association might be observed in the ability of presentation of the first odor to evoke or modify the neural representation of the second. We trained rats to associate sequentially presented odors by including specific exceptions to the otherwise random stimulus presentation order. For example, odors 5+ and 6– always were followed by odor 7+;

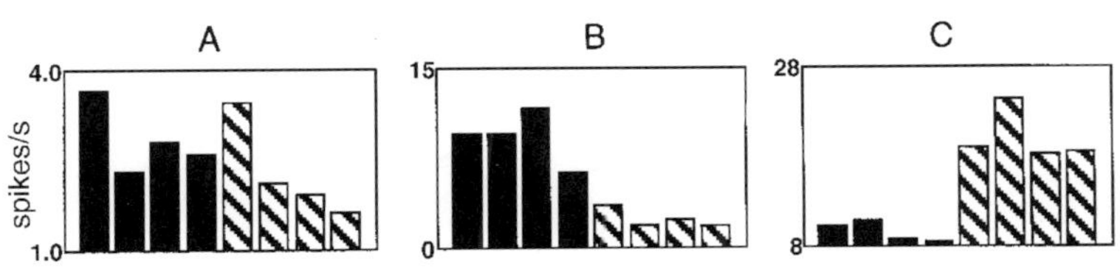

Fig. 2. Odor responses of olfactory cortex neurons. The dark bars represent neural responses to S+ odors and striped bars represent responses to S– odors.

as a control, odor 7+ also sometimes followed other odors unpredicted. Rats learned these incidental predictabilities in the odor sequence, as reflected in more rapid trial initiation when an S+ was predicted and slowed initiation of trials where S– odors were predicted.

We found two kinds of neural correlates of the learned odor associations. First, we identified olfactory cortex cells that fired in anticipation of a predicted odor. In the example shown in Fig. 3A the cell fired maximally prior to as well as during the presentation of odor 1+ when it was predicted. By contrast there was almost no response when the odor was not predicted by its associate. Second, we observed cells that fired differently in response to an odor depending on whether the preceding odor predicted its occurrence. For example, in Fig. 3B the response to odor 1+ was much reduced when it was predicted than when it was unexpected. This pattern of results indicates that learned stimulus associations are indeed established within the olfactory cortex.

In sum, our characterization of the olfactory prefrontal cortex indicated that neurons in this area encode aspects of stimulus memories that should occur independent of hippocampal function, specifically the tuning and biasing of odor representations, as well as aspects of memory that may depend on hippocampal function, such as odor-odor associations.

Critical role of the parahippocampal region in extending the persistence of item memories

In monkeys, several studies have shown that damage to the hippocampal region results in a selective impairment in object recognition memory, as well as in several related object memory tasks that involve rapid acquisition and extended retention demands [17,18]. Our studies focused on which hippocampal system structures are critical to these features of memory in a simple recognition test known as delayed nonmatch to sample [19,20]. These experiments employed two strategies, neuropsychological studies that determine the effects of damage to different hippocampal structures on performance of the delayed nonmatch to sample task, and electrophysiological studies that characterize the firing properties of neurons in different components of the hippocampal region in animals performing this task. In both lines of study we exploited rats' superb olfactory learning and memory in the development of a variant of the delayed nonmatch to sample task that employed odor cues and involved a stimulus presentation protocol suitable for characterizing neural firing

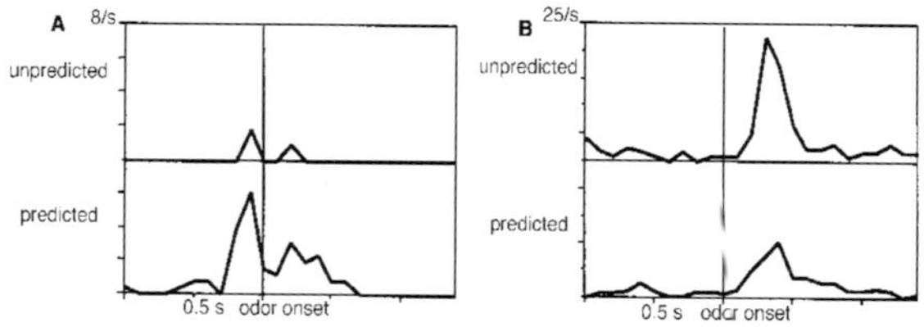

Fig. 3. Neural correlates of odor-odor associations in the olfactory cortex. See text for description.

patterns to single memory stimuli as well as behavioral responses in accordance with the nonmatch memory contingency [21,22]. In our version of the task, called continuous delayed nonmatch to sample (cDNM), one of a large set of odors is presented on each trial with the contingency that an appropriate response to the odor is rewarded only if that odor was different from (i.e., a nonmatch with) the immediately preceding odor. Rats were trained initially with a very brief interval between odor presentations. Subsequently the interstimulus interval was manipulated to vary the retention delay, allowing an assessment of the persistence of memory as in the earlier studies on monkeys.

In an initial experiment we compared the effects of selective ablation of the parahippocampal region vs. fornix transection that selectively disrupts hippocampal function [21]. Normal rats acquired the task within approximately 150 trials and neither lesion affected the acquisition rate. In subsequent tests across memory delays, intact rats showed 90% or better retention at the shortest delay and their memory performance gradually declined as the retention interval was increased. Rats with damage to the parahippocampal region also showed good retention at the shortest delay, but their performance declined abnormally rapidly showing a severe deficit within 1 min. By contrast, rats with fornix lesions performed identically with normal rats both during task acquisition and across delays, showing both intact performance at the short delay and the same gradual memory decay as normal rats. These findings indicate that neither the parahippocampal region or hippocampus is critical to the perception of the odor cues, nor is either component of the system necessary for acquisition of the nonmatch rule or short-term memory for odors. However, the parahippocampal region is critical to mediating a memory representation that persists beyond immediate memory in rats, as it is in monkeys. Furthermore, we infer from these results that the parahippocampal region, through its direct reciprocal connections with the cortex, is sufficient to mediate the persistence of single items memories independent of hippocampal processing.

In complementary electrophysiological studies we have also examined the response properties of neurons in the parahippocampal region and in the hippocampus of rats performing the odor-guided cDNM task [23]. In our studies on the parahippocampal region we examined the firing patterns of cells in these areas during both the period when rats were sampling the odor cues, in order to assess the extent to which odors were selectively encoded, and during the memory delay, to determine the capacity of these areas to maintain an odor memory representation. For the latter we focused on the trial period at the end of the memory delay, i.e., immediately preceding the initiation of trials succeeding a particular odor presentation. At this time the overt behavior of the animal is consistent (it is approaching the stimulus port), allowing us to determine if neural activity varies with the identity of the preceding just before the recognition judgment must be made.

We found that a substantial proportion of cells in each of the subdivisions of the parahippocampal region and subiculum encode the identity of the odor cues during the odor sampling period (Fig. 4A). Some of these cells fired selectively or differentially to odors at odor onset and ceased firing when odor sampling was

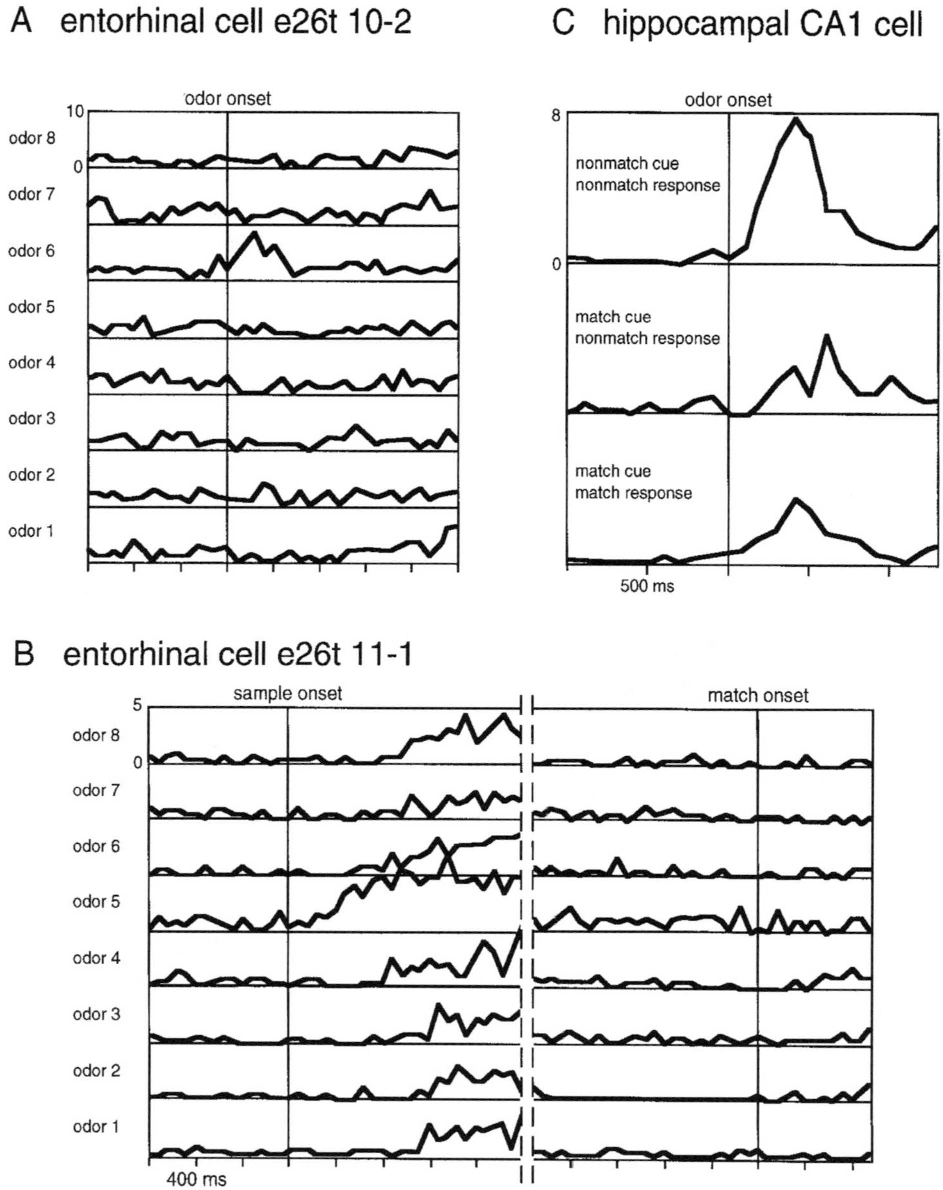

Fig. 4. Neural activity in the parahippocampal region and hippocampus associated with delayed nonmatch to sample performance. **A**: Responses of a cell during the 2 s following the onset of different odors. **B**: Activity of another cell during the cue sampling period and during the last 2 s of that delay period, i.e., prior to the onset of the subsequent odor-matching cue. **C**: Responses of a hippocampal CA1 neuron selectively to nonmatch stimuli and not match stimuli regardless of whether the rat makes the correct response.

concluded, much as one would expect of a sensory neuron. In addition, other cells showed striking odor-specific activity during odor sampling and then throughout the delay period, such as the example shown in Fig. 4B. Indeed, substantial numbers of cells in each of the subdivisions of the parahippocampal region demonstrated selective or differential firing patterns that were apparent in our systematic analysis of firing patterns at the end of the delay period.

To assess odor memory representations in the hippocampus itself we recorded from CA1 pyramidal neurons of rats performing the cDNM task [22]. A large proportion of hippocampal cells was active associated with virtually every identifiable behavioral event in the task. A substantial subset of these cells was selectively active during the odor stimulus sampling period, and the activity of some of these hippocampal cells reflected the "match" or "nonmatch" relationship critical to performance in this task (Fig. 4C). By contrast with cells in the parahippocampal region, no hippocampal cells fired associated with the sampling of a particular odor or with particular combinations of odors that composed specific matching comparisons. Rather, hippocampal cellular activity reflected all comparisons with the same outcome. We interpreted this finding as entirely consistent with the results of our lesion studies. It appears that the hippocampus itself is not involved in the encoding and storage of representations for specific items in this task. Instead it appears the hippocampus encodes abstract relations among cues, as it does in many other test situations. Thus, for example, the match/mismatch relationship reflected in the activity of hippocampal cells in animals performing the cDNM task can be viewed as analogous to the spatial relations among cues reflected in the firing patterns of hippocampal place cells when rats explore open fields [24]. Indeed across a broad range of tasks hippocampal cells encode all manner of relations among salient stimuli as well as behavioral actions [25–27].

The hippocampus, memory for stimulus relations, and "flexible" memory expression

The above-described findings on hippocampal neuronal firing patterns, among a substantial body of other neuropsychological and electrophysiological evidence, led us to suggest that the hippocampus is critical for mediating the organization of memories according to the relevant relationships among remembered stimuli [27–30]. In addition we have proposed that such a relational representation supports the ability to express knowledge about indirectly related items in the memory organization in a variety of ways outside repetition of learning events. We have now assessed the role of the hippocampus in several different memory tasks, each of which require the animal to learn an organization of relationships among odor cues, and each of which require the animal to use its memory representations "flexibly", i.e., in situations that challenge the animal to employ the learned memory organization outside repetition of the learning trials. One such task that exploits rats superb memory capacities with odors will be described here.

Paired associate learning is a memory task that requires the encoding of specific relations among cues. The verbal paired associate task has been exceedingly useful in understanding cognitive aspects of associative learning in humans and is often applied in the assessment of human amnesia. It seemed to us that a paired associate task adapted for animals was the simplest paradigm that we could exploit for neurobiological studies on the learning of new associations between distinct and neutral stimulus events. The paired associate task as typically used for humans involves presentation of a list of arbitrarily paired words followed by testing in which the subject is cued with the first item of each pair and must recall the second item.

For rats, we designed an analogous task using odor stimuli and a recognition format that required subjects to distinguish in each of a series of trials the appropriate associate for a sample odor [31]. Exploiting rodents' natural foraging strategies that employ olfactory cues, animals were trained with stimuli that consisted of distinctive odors added to a mixture of ground rat chow and sand through which they dug to obtain buried cereal rewards. On each paired associate trial one of two sample odors initially presented was followed by two choice odors, each assigned as the "associate" of one of the samples and baited only when preceded by that sample (Fig. 5A,B). Following training on two sets of overlapping odor-odor associations, subsequent probe tests were used to characterize the extent to which learned representations supported two forms of flexible memory expression, transitivity, the ability to judge inferentially across stimulus pairs that share a common element, and symmetry, the ability to associate paired elements presented in the reverse of training order.

Intact rats learned paired associates rapidly and hippocampal damage did not affect acquisition rate on either of the two training sets (Fig. 5C). Intact rats also showed strong transitivity across the sets with a preference of approximately 2:1 in favor of choice items indirectly associated with the presented sample (Fig. 5D). By contrast, rats with selective hippocampal lesions were severely impaired in that they showed no evidence of transitivity. In the symmetry test, intact rats again showed the appropriate preference of approximately 3:1 in the direction of the symmetrical association (Fig. 5E). By contrast, rats with hippocampal lesions again were severely impaired, showing no significant capacity for symmetry. These findings indicate that some form of representation that supports stimulus-stimulus learning is intact in rats with selective hippocampal damage, as previously reported for rats and monkeys with hippocampal region damage [32–35] as well as human amnesics [36,37]. However, this form of representation is "hyperspecific", i.e., can only be expressed within the confined context of reproduction of the learning event [38]. Only a hippocampally mediated representation can support the flexible expression of associations among items within a larger organization in various and even novel situations.

Conclusions

Combining the findings across the studies presented here provides a preliminary picture of the interactive processes that mediate declarative memory. The cerebral cortex is the site of long-term memory representations for individual items. Cortical representations can be "biased" or "tuned" to support performances in ways that

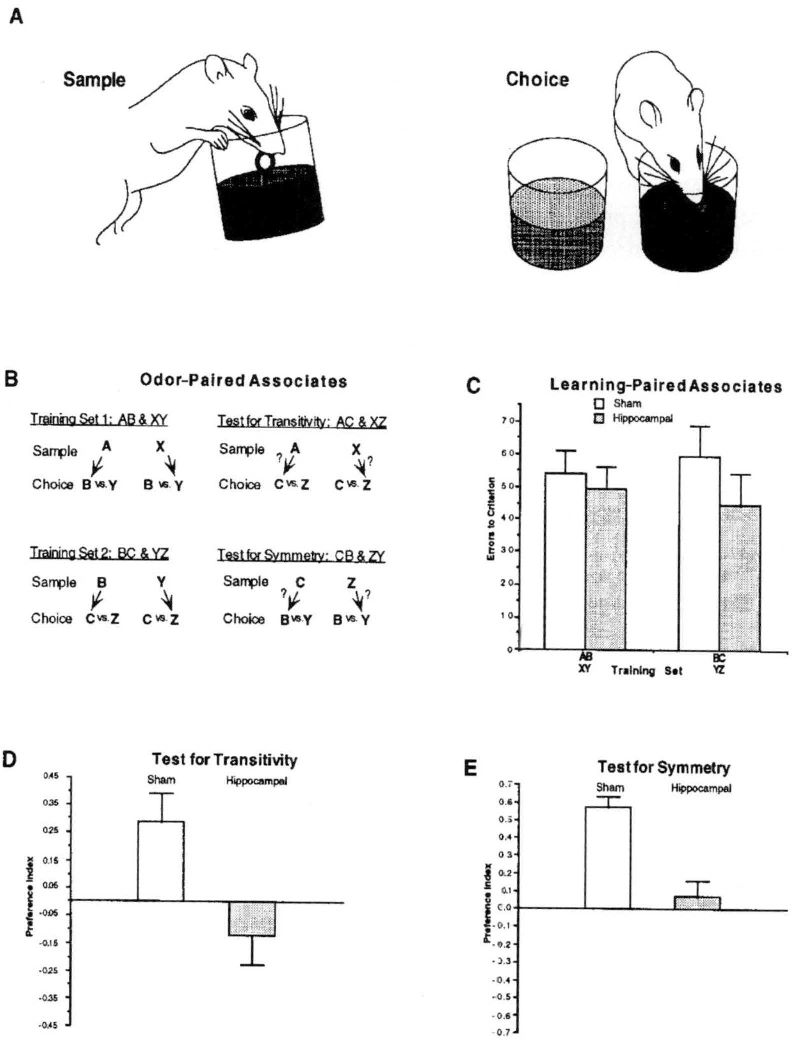

Fig. 5. Inferential expression of odor-odor associations in sham-operated rats and rats with selective neurotoxic lesions of the hippocampus. **A**: Illustrations of the rat and stimulus cups during sample and choice phases of the task. **B**: Schematic diagram of paired associate training and probe testing. Letters represent odor stimulus items; arrows without question marks indicate trained pairings, whereas arrows with question marks indicate expected transitive and symmetrical choices. Rats are first trained on two overlapping sets of paired associates (left). Then (right) they are tested for inferential expression in two ways. In the test for transitivity, they are presented with one of two sample cues from the first training set and then required to select between the choice cues from the second set, based on the shared associates of these items. In the test for symmetry or "reversibility" of the associations, they are presented with one of two choice cues from the second set and required to select the appropriate sample cue from that set. **C**: Errors to criterion on acquisition of the two sets of paired associates for sham-operated and hippocampal subjects. **D**: Preference indices on the test for transitive inference. For these probe trials a preference score was calculated as (X–Y)/(X+Y), where X and Y were the digging times in the transitive and alternate choices, respectively. **E**: Preference indices on the test for symmetrical expression.

likely do not require hippocampal mediation, and such representations can be associated with one another in ways that may well depend on and reflect the influence of hippocampal processing. It appears the parahippocampal region contributes to memory by "buffering" specific cortical representations that can be accessed and manipulated by the hippocampus over prolonged periods. The hippocampus represents the critical relations among the items held by the parahippocampal region, and presumably has access to the much larger organization of item representations in cortical association areas via the parahippocampal region.

We propose that large-scale relational organizations come about through multiple iterations of cortical input to the parahippocampal region and temporary storage there, followed by hippocampal relational processing that influences interconnections among representations in the parahippocampal region and then in the cortex. Over extended time periods, new experiences that bear on the established organization reactivate established representations as well as add new ones, and these are processed together by this hippocampal circuit to weave the new information into the established relational network. Precisely because this network is so extensive and systematically interconnected, access to items via novel routes and in novel experiences is not only possible but also occurs continuously as we express memories to guide almost every aspect of daily life.

Acknowledgements

This work was supported by NIMH, NIA, and ONR.

References

1. McClelland JL, McNaughton BL, O'Reilly RC. Why there are complementary learning systems in the hippocampus and neocortex: Insights from successes and failures of connectionist models of learning and memory. Psychol Rev 1995;102:419–457.
2. Halgren E. In: Butters N, Squire LR (eds) Neuropsychology of Memory. New York: Guilford, 1984;165–182.
3. Rolls E. In: Byrne JH, Berry WO (eds) Neural Models of Plasticity: Theoretical and Empirical Approaches. New York: Academic Press, 1989.
4. Squire LR, Cohen NJ, Nadel L. In: Weingartner H, Parker E (eds) Memory Consolidation. Hillsdale: Erlbaum, 1984;185–210.
5. Amaral DG, Witter MP. The three-dimensional organization of the hippocampal formation: A review of anatomical data. Neuroscience 1989;31:571–591.
6. Burwell RD, Witter MP, Amaral DG. Perirhinal and postrhinal cortices in the rat: A review of the neuroanatomical literature and comparison with findings from the monkey brain. Hippo Campus 1995;5:390–408.
7. Witter MP, Groenewegen HJ, Lopes da Silva FH, Lohman AHM. Functional organization of the extrinsic and intrinsic circuitry of the parahippocampal region. Prog Neurobiol 1989;33:161–254.
8. Kosel KC, Van Hoesen GW, West JR. J Comp Neurol 1981;198:467–482.
9. Price JL, Carmichael ST, Carnes K, Clugnet MC, Kuroda M, Ray JP. In: Davis JL, Eichenbaum H (eds) Olfaction: A Model System for Computational Neuroscience. Cambridge: MIT Press, 1991;101–120.
10. Macrides F, Eichenbaum H, Forbes WB. Temporal relationship between sniffing and limbic theta rhythm during odor discrimination reversal learning. J Neurosci 1982;2:1705–1717.

11. Otto T, Eichenbaum H, Wiener SI, Wible CG. Learning-related patterns of CA1 spike trains parallel stimulation parameters optimal for inducing hippocampal long-term potentiation. Hippo Campus 1991;1:181–192.
12. Eichenbaum H, Eckmann KW, Shedlack KJ. Thalamocortical mechanisms in odor guided behavior: I. The effects of lesions on the mediodorsal nucleus and frontal cortex on olfactory discrimination in the rat. Brain Behav Evol 1980;17:255–275.
13. Eichenbaum H, Clegg RA, Feeley A. A re-examination of functional subdivisions of the rodent prefrontal cortex. Exp Neurol 1983;79:434–451.
14. Schoenbaum G, Eichenbaum H. Information coding in the rodent prefrontal cortex. I. Single neuron activity in the orbitofrontal cortex compared with that in the piriform cortex. J Neuro Physiol 1995;74:733–750.
15. Schoenbaum G, Eichenbaum H. Information coding in the rodent prefrontal cortex. II. Ensemble activity in the orbitofrontal cortex. J Neuro Physiol 1995;74:751–762.
16. Sakai K, Miyashita Y. Neural organization for the long term memory of paired associates. Nature 1991;354:152–155.
17. Mishkin M. A memory system in the monkey. Phil Trans R Soc Lond 1982;B209:85–95.
18. Zola-Morgan S, Squire LR. Medial temporal lesions in monkeys impair memory on a variety of tasks sensitive to human amnesia. Behav Neurosci 1985;99:22–34.
19. Gaffan D. Recognition impaired and association intact in the memory of monkeys after transection of the fornix. J Comp Physiol Psychol 1974;86:1100–1109.
20. Mishkin M, Delacour J. An analysis of short-term visual memory in the monkey. J Exp Psychol: Animal Behav Proc 1975;1:326–334.
21. Otto T, Eichenbaum H. Neuronal activity in the hippocampus during delayed non-match to sample performance in rats: Evidence for hippocampal processing in recognition memory. Hippo Campus 1992;2:323–334.
22. Otto T, Eichenbaum H. Complementary roles of orbital prefrontal cortex and the perirhinal-entorhinal cortices in an odor-guided delayed non-matching to sample task. Behav Neurosci 1992;106:763–776.
23. Young BJ, Otto T, Fox GD, Eichenbaum H. Neuronal activity in the parahippocampal region of rats performing a delayed non-match to sample task. Soc Neurosci Abstr 1995;21:943.
24. O'Keefe JA. A review of hippocampal place cells. Prog Neurobiol 1979;13:419–439.
25. Deadwyler SA, Bunn T, Hampson RE. Hippocampal ensemble activity during spatial delayed-nonmatch-to-sample performance in rats. J Neurosci 1995;16:354–372.
26. Eichenbaum H. Thinking about brain cell assemblies. Science 1993;261:993–994.
27. Eichenbaum, H. Is the rodent hippocampus just for 'place'? Curr Opin Neurobiol 1996;6:187–195.
28. Eichenbaum H, Otto T, Cohen NJ. The hippocampus – What does it do? Behav Neural Biol 1992;57:2–36.
29. Eichenbaum H, Otto T, Cohen NJ. Two functional components of the hippocampal memory system. Brain Behav Sci 1994;17:449–518.
30. Cohen NJ, Eichenbaum H. Memory, Amnesia, and the Hippocampal System. Cambridge: MIT Press, 1993.
31. Bunsey M, Eichenbaum H. Conservation of hippocampal memory function in rats and humans. Nature 1996;379:255–257.
32. Eichenbaum H, Bunsey M. On the binding of associations in memory: Clues from studies on the role of the hippocampal region in paired-associate learning. Curr Dir Psychol Sci 1995;4:19–23.
33. Cho YH, Kesner RP Relational object association learning in rats with hippocampal lesions. Behav Brain Res 1995;67:91–98.
34. Murray EA, Gaffan D, Mishkin M. Neural substrates of visual stimulus-stimulus association in rhesus monkeys. J Neurosci 1993;13:4549–4561.
35. Saunders RC, Weiskrantz L. The effects of fornix transection and combined fornix transection, mammillary body lesions and hippocampal ablations on object pair association memory in the rhesus monkey. Behav Brain Res 1989;35:85–94.
36. Musen G, Squire LR. On the implicit learning of novel associations by amnesic patients and normal

subjects. Neuropsychol 1993;7:119–135.

37. Moscovitch M. Memory and working with memory: Evaluation of a component process model comparisons with other models. In: Schacter DL, Tulving E (eds) Memory Systems 1994. Cambridge: MIT Press, 1994;269–310.
38. Eichenbaum H, Cohen NJ, Otto T, Wible CG. A snapshot without the album. Brain Res Rev 1992;16:209–215.

Integrative and Molecular Approach to Brain Function.
M. Ito and Y. Miyashita, editors.

Neural mechanisms of episodic memory in the monkey septo-hippocampal system

Taketoshi Ono, Hisao Nishijo, Ryoi Tamura and Teruko Uwano
Department of Physiology, Faculty of Medicine, Toyama Medical and Pharmaceutical University, Toyama, Japan

Abstract. We analyzed the activity of the septo (SP)-hippocampal (HF: it includes the hippocampus proper, the dentate gyrus, and the subicular complex) neurons in monkeys to investigate the characteristics of spatial and nonspatial information processing in the SP-HF system, which are neurophysiological bases of episodic memory. First, the activity of the HF neurons was tested with the two different stimulus-response association tasks (Go/Nogo tasks with symmetrical reinforcement). In a place-dependent, conditional, stimulus-response association (PCR) task, the monkey's location was the condition upon which the stimulus (object)-behavioral response association depended. In a place-independent, simple, conditional stimulus-response association (SCR) task, the object-behavioral response contingency was independent of the monkey's location. The results indicated that some HF neurons preferentially responded in the PCR task, while none of the HF neurons responded preferentially in the SCR task. The characteristics of the HF neuronal response patterns suggest that HF neurons encode both stimulus percept and attributes, such as the place where the stimulus is presented and the conditional relations imposed in the task. Furthermore, the selectivity to external stimuli and the neurophysiological characteristics of the HF neurons suggest the dichotomy of the monkey HF neurons which might be functionally comparable to that of the rat-complex spike cell vs. θ cell. Second, the SP neurons were recorded during performance of a place-dependent asymmetrical Go/Nogo (PGN) task in which a bar press was reinforced only if each object was seen in one corresponding place, and during performance of a place-independent asymmetrical and symmetrical Go/Nogo tasks. The results indicated that some SP neurons responded differentially to the monkey's location, and responded to the objects only in the PGN task. Activity of other SP neurons was related to the reinforcement contingency of an object regardless of the tasks. These results suggest that the SP are involved in integrating spatial information, conditional place-object association, and reinforcement contingency. Taken together, these results suggest that SP-HF neurons extract relevant information from a given context, and relationally encode this diverse information based on spatial context.

Key words: episode, hippocampus, memory, place, septum.

Background

Since Scoville and Milner [1] first reported hippocampal (HF: it includes the hippocampus proper, the dentate gyrus, and the subicular complex) relation to amnesia, extensive research has described the role in memory; the HF is important in formation of declarative memory [2–5] and working memory [6], important in the processing of cognitive mapping [7], and relational [8–10] or configural [11]

Address for correspondence: Dr Taketoshi Ono, Department of Physiology, Faculty of Medicine, Toyama Medical and Pharmaceutical University, Sugitani 2630, Toyama 930-01, Japan. Tel.: +81-764-34-2281. Fax: +81-764-34-5013. E-mail: onotake@ms.toyama-mpu.ac.jp association. Although the theoretical bases of those studies were different, these theories all postulated the role of the HF in association with or relation to multiple factors. Anatomically, the HF receives highly integrated diverse information to form declarative or episodic memory from all association cortices [12–14]. One way to enable this postulated function of the HF (i.e., encoding of association with or relation to multiple factors) is for the HF to work as a memory index by which neuronal activation in various neocortical areas is organized as a pattern, and information of this neocortical activation pattern is encoded in the HF [15,16].

There is some earlier evidence to suggest that the HF encodes multiple factors. First, rat HF neurons were often reported to respond to various combinations of sensory stimulus and place [17–20]. In monkeys, some HF neurons responded only when a stimulus was presented in a particular position [21,22], or when a monkey saw an object from a particular direction [22,23], or when the monkey was in a particular location [24–26]. These previous results indicated that a population of HF neurons could differentially encode information about a given object with reference to the spatial location of the object or the animal itself, with or without reference to some biological meaning of the object.

Septal nuclei (SP) and the HF have been suggested to act as one functional unit by behavioral [27,28] and anatomical [29] results. Especially, one major efferent pathway of the HF enters the lateral SP [29,30]. Long-term potentiation, regarded as a cellular mechanism of learning [31,32], was reported in these HF-SP synapses [33]. Taken together, SP neurons could receive spatially organized information such as place-object relation from the HF, and this information might be further processed during learning in the SP. In the present study, to investigate associative encoding of spatial and nonspatial information in the SP-HF system, activity of the HF and the SP neurons in monkeys was analyzed during performance of place-dependent conditional object discrimination tasks in which place was the condition upon which reward contingency or behavioral requirements of the object depended.

Methods

Four male *Macaca fuscata* weighing between 3.8 and 6.0 kg were used. A monkey was restrained painlessly in a stereotaxic apparatus, by a surgically fixed head holder, and sat in a chair in a cab, which was set in a 2.5 × 2.5 m space located within a 5 × 6 m experimental shielded room (Fig. 1A), as previously described [24,25,34]. The room contained several visual landmarks, such as a stereomicroscope, a refrigerator, a table, a rack, and some experimental devices; the monkey faced the right side of Fig. 1A. The cab was moved horizontally from one place to another by computer controlled electric motors. The front, left, and right walls of the cab were half-mirrored, and the rear wall was a steel plate equipped with two lamps (cab lamps) and two speakers on the inside. The landmarks in the room could be seen from inside the cab, through the half-mirrored walls, only when the cab lamps were extinguished. The upper part of the front wall also contained a double half-mirrored shutter (15 × 15 cm) just above an operant responding bar. There was an illuminated stage for

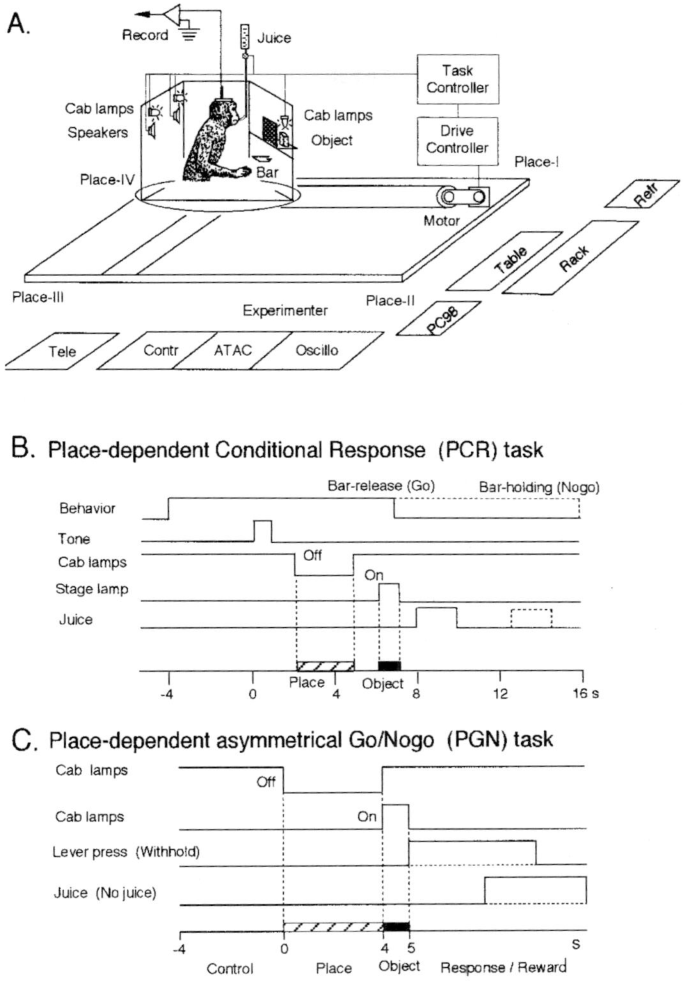

Fig. 1. Schemata of the experimental setup (A) and paradigms of the place-dependent object discrimination tasks in monkeys (B, C). **A**: Movable monkey cab in experimental shielded room (5 × 6 m). The monkey sat in a chair placed in a 0.7 m long × 0.8 m wide × 0.85 m high cab. The cab was moved to one of four places (places-I, II, III and IV) in 2.5 × 2.5 m space. Front, left and right walls of the cab were half-mirrored. The rear wall, made of steel, was equipped with two cab lamps and speakers. A half-mirrored shutter (15 × 15 cm) was in the upper half of the cab front wall just above an operant bar. A front lamp behind the shutter illuminated an object in the bay. Abbreviations: Tele, telemeter; Contr, controller; ATAC, microcomputer ATAC 3700; Oscillo, oscilloscope; PC98, microcomputer; Refr, refrigerator. **B**: Time sequences of the place-dependent conditional response (PCR) task. When the object indicated Go behavior based on place-object relations, the monkey was required to release the bar (Go trial, solid line); while, Nogo required the monkey to hold the bar (Nogo trial, dashed line). **C**: Time sequences of place-dependent Go/Nogo (PGN) task. During 4 s cab-lamp-off, monkey could see room environment through half-mirrored walls (place phase). While the front lamp was on for 1 s, a given object was visible behind the shutter (object phase). If the object meant a reward, a drop of juice was dispensed from a spout after keeping the bar pressed for 3 s (Go trial). If the object meant a nonreward, juice was never available and pressing the bar produced a loud coarse buzzer sound (Nogo trial).

object presentation behind the shutter; an object on the stage was visible to the monkey, through the shutter, when a cab lamp was on. Juice, controlled by an electromagnetic valve, was delivered from a tube that projected through the rear wall.

With this movable cab, activity of the HF neurons was analyzed during performance of the place-dependent conditional symmetrical stimulus-response association (PCR) task and its control task [35], and activity of the SP neurons, during place-dependent conditional asymmetrical Go/Nogo (PGN) task and its control tasks [36].

PCR and its control tasks

In the PCR task, the monkey's location was the context which indicated the behavioral attributes (Go or Nogo responses) of the objects. The control task was a place-independent, simple, conditional stimulus-response association (SCR) task, in which each object indicated Go or Nogo responses, regardless of the monkey's location. In the PCR task (Fig. 1B), each trial began when the monkey pressed and held a response bar. After holding the bar for 4–6 s (control phase), a pure tone was presented to alert the monkey. After a 1-s delay following the tone, the inside cab lamps were turned off for 3 s (place phase). During the place phase, the monkey could see outside the cab through the half-mirrored cab walls. The inside cab lamps were turned on again, and after another 1-s delay, an object (a red or blue block) behind the half-mirrored shutter was illuminated for 1 s (object phase). When the object indicated a Go response, based on the monkey's location, the monkey was required to release the response bar within 2 s (Go trial). If the monkey responded correctly, a juice reward was presented after another 1-s delay. The time course of a trial with a Nogo response (Nogo trial) was similar, but the monkey was required to keep holding the bar for 5 s.

The behavioral attributes of the objects depended upon the locations of the cab. When the cab was located at place I, a red block required the monkey to release the bar (Go trial), and a blue block required it to hold the bar (Nogo trial) to obtain a juice reward. When the cab was located at place II, the behavioral responses indicated by the two objects were reversed; a red block required holding the bar (Nogo trial) and a blue block required releasing of the bar (Go trial) to obtain a reward. In the SCR task, two pairs of familiar objects were associated with Go/Nogo responses, as in the PCR task, but the red object always required Go while the blue object always required Nogo, to obtain a reward. In some neurons, the reversed version of the SCR task, in which the object-response association was reversed, was also tested. In the SCR task, the inner cab lamps were turned on throughout the trials so the monkey could not see outside the cab to visually identify its location. Because of this, there was no place phase before the object phase in the SCR task; the time course was the same as that of the PCR task, except that there was no place phase (3 s) and no delay (1 s) between the place and object phase.

Place-dependent conditional asymmetrical Go/Nogo task

In the same set-up as noted above, monkeys were trained in the following three tasks: place-dependent Go/Nogo (PGN), place-independent asymmetrical Go/Nogo (AGN), and place-independent symmetrical Go/Nogo (SGN) tasks. In the PGN task, the cab was always placed in one of the four corners of the field. One three-dimensional object, selected from a total of four was presented, at one time or another, in each place. Each of the objects, which were rectangular parallelepipeds made of differently colored blocks, predicted a reward only when it was presented in one corresponding place (Go trial), and the same object predicted nonreward in all other places (Nogo trials). For example, object one meant a reward in place I, and nonreward in places II, III or IV, and object two meant a reward only in place II, etc. Consequently, each object presented two variable response/reinforcement contingencies depending on the location in which the monkey encountered it; there was a Go/reward contingency and a Nogo/nonreward contingency.

Each trial consisted of four phases (Fig. 1C). The control phase was a period of 4 s before each trial. A trial was initiated by turning off the inside cab lamps with a click sound. While the cab lamps were turned off for 4 s (place phase), the monkey could see outside the cab through the half-mirrored cab walls. In the following object phase, a front lamp was lit for 1 s to illuminate an object behind the half-mirrored shutter. If a given object meant a reward in that place (Go trial), the monkey was required to initiate a response within 7 s after the end of the object phase and hold the bar down for 3 s. After the bar had been pressed for 3 s, approximately 0.3 ml juice reward was presented for 4 s. If a given object meant a nonreward in that place (Nogo trial) and the monkey pressed the bar during the 7-s period after the object phase, a buzzer tone was presented for 1 s to indicate a nonreward (an error Nogo trial). If the bar had not been pressed for the 7-s period in a Go or Nogo trial, the trial finished without any event (an error Go trial and correct Nogo trial). The cab was moved to the next location after the end of each trial during an intertrial interval (ITI) of 30–40 s and arrived before the control phase of the next trial. In training, the same Nogo trial was repeated when the monkey pressed the bar in a Nogo trial so that the monkey could learn to withhold an improper Go response. Moreover, errors were punished by prolonging the ITI to a 1-min time-out both in the Nogo trial (error of commission) and in the Go-trial (error of omission).

In the AGN task, the cab was placed in any one of the four locations utilized in the PGN task, and two objects other than those used in the PGN task were presented at random. Regardless of the monkey's location, a white object is always associated with a reward, and a yellow object indicated no reward. Except for this point, the other task sequences, such as presentation of the outside views (place phase), presentation of the objects (object phase), bar pressing in a Go trial, and delivery of juice reward only in a Go trial (response/reward phase), were the same as those of the PGN task. Since the object-response/reward contingency is constant in this task, this task was regarded as an unconditional place-independent version of the PGN task.

The SGN task was similar to the AGN task except that correct response in a Nogo trial was associated with juice reward as well as in a Go trial. The white object which was used in the AGN task was also presented in a Go trial and a black object used only in this task was presented in a Nogo trial. The white and black objects differed in response contingency, but were the same in reward contingency. That is, the black object was constantly associated with a Nogo/reward contingency.

Results

Monkey HF neuronal activity

Of 329 neurons recorded from the HF during the PCR task, 88 (26.8%) responded during the object phase. Of these 88 neurons, 49 (14.9%) also responded during the place phase, and 39 (11.9%) responded exclusively during the object phase. These neurons were classified into four subgroups, based on their responsiveness in the object phase: specific combination, object differential, behavior differential, and nondifferential. The responses of specific-combination neurons (n = 17) depended differentially on specific combinations of object, place and behavior (i.e., Go or Nogo response) in the PCR task. The responses of object-differential neurons (n = 12) differentiated between kinds of objects in the PCR task. On the other hand, behavior-differential neurons (n = 16) responded to behavior (Go or Nogo) during the object phase, regardless of the kind of object or the monkey's location, in both the PCR and SCR tasks. Nondifferential neurons (n = 43) indiscriminately responded during the object phase, in both PCR and SCR tasks.

Place-dependent monkey HF neuronal activity

The specific-combination neuron shown in Fig. 2A specifically responded to place II during the place phase (Fig. 2Ac,d). This neuron also responded to the red block, which indicated a Nogo response, presented at place II (Ac), but did not respond to other combinations of place-behavioral response object during the object phase (Fig. 2Aa,b,d). In the SCR task, the neuron did not respond to the red block which indicated a Go response (not shown). In the reversed SCR task, the neuron did not respond to the combination of Red/Nogo (not shown), although the object-behavioral response contingency was the same as in the PCR task. These results indicate that the selective neuronal response to the combination of a red object and a Nogo response in the PCR task was further constrained by the monkey's location (place II).

The object-differential neuron, shown in Fig. 2B, demonstrated significant differential responses to a red block in the object phase of the PCR task, regardless of the behavioral response contingency or the monkey's location (Fig. 2Ba,c), although it displayed no significant responses in the place phase of the PCR task (Fig. 2Ba–d). It should be noted that this neuron ceased to respond to the red object in the SCR task as well as in the reversal version of the SCR task (not shown). Thus, these neuronal responses were specific to the PCR task. These PCR task-specific

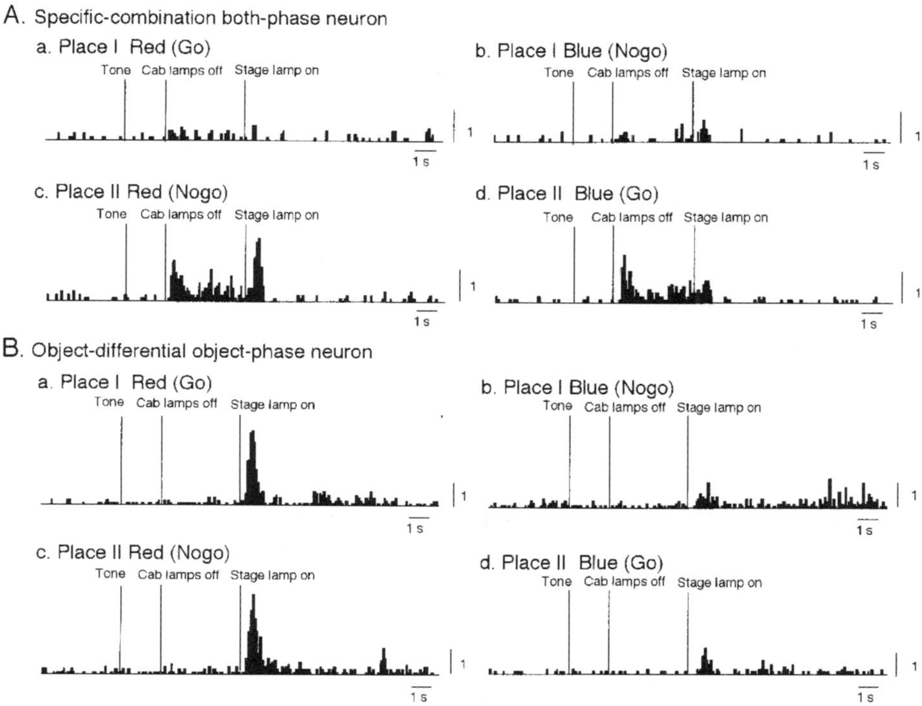

Fig. 2. Two examples of monkey HF neurons that responded specifically during the PCR task. **A**: Responses of a typical specific-combination both-phase neuron in the PCR task. This neuron responded differentially in the place phase to outside views of place II, and responded differentially in the object phase to one specific combination: the place II-Red block-Nogo response. **B**: Responses of a typical object-differential object-phase neuron. The neuron did not respond during the place phase, but it responded differentially to the red block in the object phase regardless of the required behavior or location. a–d: Cumulative 100-ms bin, 10-trial histogram of neuronal activity during trials of each place-object-behavioral response combination.

object-differential responses were not constrained by the monkey's location since neuronal responses to the same object occurred in both locations. This result indicates that the conditional structure of the PCR task, rather than the monkey's location, is what constrained the neuron's responses.

Responses of behavior-differential neurons were correlated in the object phase to the behavior required, regardless of the kind of object or the monkey's location (Fig. 3A). This neuron had significantly stronger responses to the Nogo stimulus in the object phase of the PCR task regardless of the object or the monkey's location. On the other hand, responses of this neuron in the place phase of the PCR task were not significantly different. This neuron also responded more strongly to the Nogo stimulus in the object phase of the SCR task, regardless of the object (not shown). Thus, neuronal activity of this neuron is differentially associated with the behavioral response contingency of the object regardless of the object or the task.

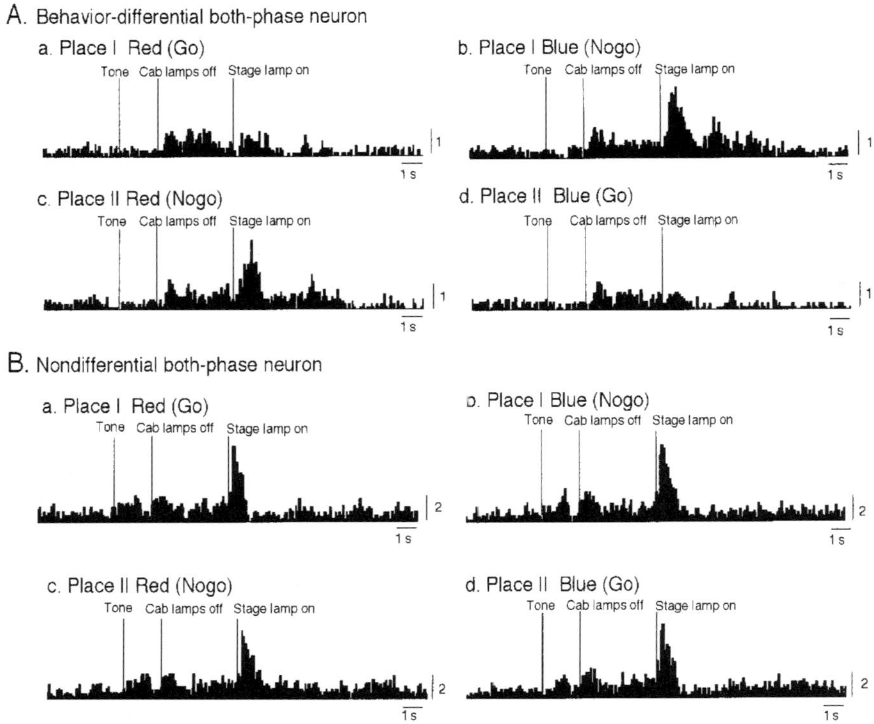

Fig. 3. Two examples of monkey HF neurons that responded similarly during both the PCR and SCR tasks. **A**: Responses of a typical behavior-differential both-phase neuron in the PCR task. This neuron responded nondifferentially in the place phase, but responded differentially in the object phase to one specific behavior: Nogo response. **B**: Responses of a typical nondifferential both-phase neuron. The neuron responded nondifferentially during both the place and object phases. Other descriptions as for Fig. 2.

Responses of nondifferential neurons were not significantly different among the four combinations in both phases (Fig. 3B). This neuron responded nondifferentially in the place phase to outside views from places I and II; and nondifferentially in the object phase to both objects regardless of behavioral response contingency or the monkey's location in the PCR task. This neuron also responded nondifferentially to four objects in the SCR task (not shown). Thus, this neuron was not specific to the PCR task.

It has been reported that animals with HF lesions were specifically deficient in performing various conditional tasks, but had no problems in performing simple discriminative tasks [37–40]. Preferential involvement of the HF in the conditional task suggests that the HF is involved in retrieval processes. Previous recording studies reported that a large number of HF neurons responded to a particular association in a simple conditional stimulus-behavioral response association task when unfamiliar objects were used [41], while neuronal responses to stimulus-behavioral response association were fairly rare when familiar learned objects were used [42]. In the present study, we used the same completely familiar pairs of objects as stimuli in both the PCR and the SCR tasks. Existence of two types of PCR-selective neurons (i.e., specific-combination and object-differential neurons) suggest that the HF is not generally involved in the retrieval process, but could be involved in an ongoing retrieval process only in a situation such as in the PCR task when familiar learned objects were used. The PCR task differed from the SCR tasks in the two points: 1) the PCR task is a spatial task while the SCR task is nonspatial task; and 2) association structure of the task is more complicated in the PCR task than in the SCR task. Characteristics of these two neuron types suggest that HF neurons might link diverse information; specific-combination neurons might connect information for a stimulus itself, the place where the stimulus is presented, and behavioral response while object-differential neurons connect information for a stimulus itself and the conditional relation imposed in the task.

Functional subpopulations of the monkey HF neurons

To quantitatively compare the HF neuronal responses in the object phase to the PCR and the SCR tasks, as well as the reversal of the SCR task, the Task Specificity Index (TSI) was introduced. The TSI was defined by $(|P| - |S|)/(|P| + |S|)$, where P and S were maximal responses during the object phase in PCR and SCR tasks, respectively. The value around TSI = 0.0 means that the neuron responds similarly during the object phase of both the PCR and SCR tasks, while the value around TSI = 1.0 means that the neuron responds during the object phase specifically in the PCR task. The results indicated two HF neuron types (Fig. 4A). There was one group with relatively low values around TSI = 0.0, which was composed mainly of nondifferential and behavior-differential neurons. The other group with TSI > 0.6, was composed mainly of specific-combination and object-differential neurons. Among the four neuronal types, there was a significant difference in TSI values (one-way ANOVA, $p < 0.01$). The post-hoc test indicated that the mean TSI values of the specific-combination and object-differential neurons were significantly larger than those of the behavior-differential and nondifferential neurons (Newman-Keuls test, $p < 0.01$). This indicates two HF neuron types (HF neurons with high and low-TSI values: specific-combination and object-differential vs. behavior-differential and nondifferential). Other neurophysiological indices also support this dichotomy. The spontaneous firing rates of the specific-combination (3.3 ± 0.7 spikes/s; mean ± SEM) and object-differential (2.4 ± 1.0 spikes/s) neurons, as well as behavior-differential neurons (2.9 ± 0.5 spikes/s), were significantly larger than that of nondifferential neurons (10.2 ± 1.2 spikes/s) (Fig. 4B). The mean spike widths of the specific-combination (267.9 ± 14.8 ms) and object-differential (275.0 ± 11.2 ms) neurons, as well as of the behavior-differential neurons (273.8 ± 19.3 ms), were significantly longer than that of nondifferential neurons (217.0 ± 6.0 ms) (Fig. 4C). The mean response latencies of the specific-combination (300.7 ± 10.1 ms) and object-differential (293.3 ± 12.0 ms) neurons were longer than those of behavior-differential (228.8 ± 8.9 ms) and

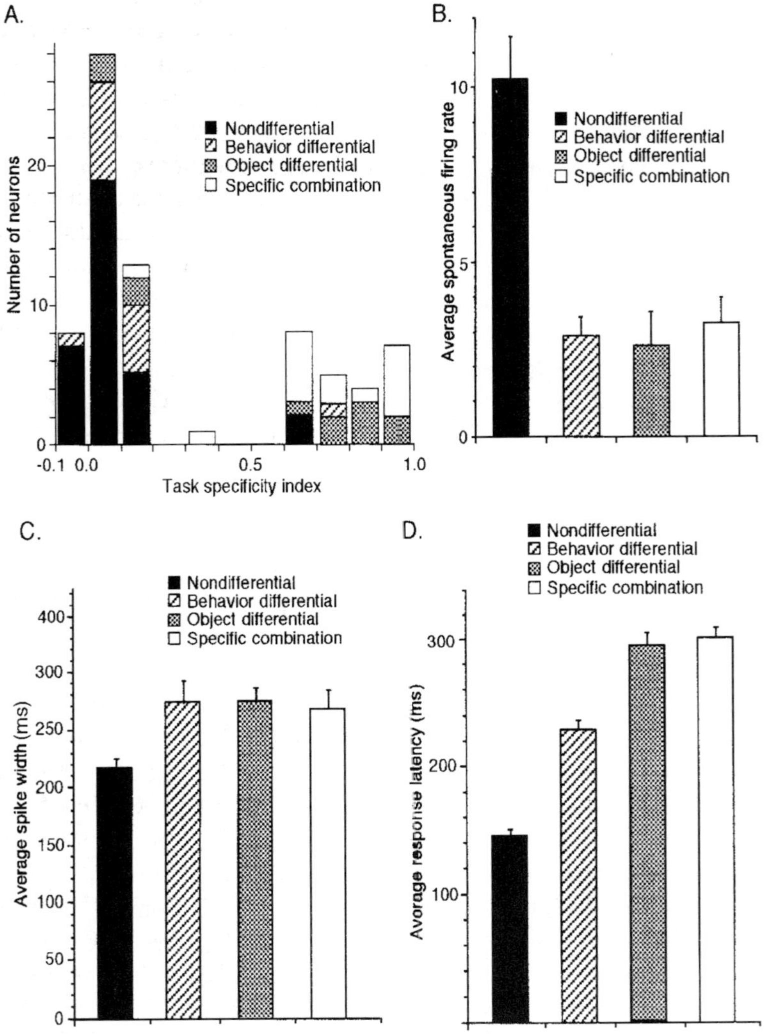

Fig. 4. Functional classification of the monkey HF neurons. **A**: Distributions of the task specificity index (TSI) of the monkey HF neurons. Two groups appear, one around TSI = 0.0, is composed mainly of nondifferential neurons and behavior-differential neurons; the other is a group with TSI > 0.33, composed mainly of specific-combination and object-differential neurons. **B**: Mean spontaneous activity (mean ± SEM) of each neuron type in the HF. Spontaneous activity of nondifferential neurons is significantly higher ($p < 0.01$) than the spontaneous activity of specific-combination, object-differential, and behavior-differential neurons. **C**: Mean spike width (mean ± SEM) of each neuron type in the HF. The spike width of specific-combination, object-differential, and behavior-differential neurons is significantly longer ($p < 0.01$) than the spike width of nondifferential neurons. **D**: Mean neuronal response latency (mean ± SEM) of each neuron type during the object phase in the HF. Response latency of specific-combination and object-differential neurons is significantly longer ($p < 0.01$) than that of behavior-differential neurons. At the same time, the response latency of behavior-differential neurons is significantly longer ($p < 0.01$) than that of nondifferential neurons.

nondifferential (145.7 ± 4.6 ms) neurons (Fig. 4D). A comparison of the mean response latencies, spontaneous-firing rates, and spike widths of the four types of HF neurons indicated the existence of two functional neuronal types: pyramidal and θ-cells.

It has been proposed that at least two types of neurons can be distinguished functionally in the rodent HF [43]. One type includes complex spike cells that are regarded as pyramidal cells. This type has a typically low spontaneous firing rate. The other group includes the so-called "θ" cells that have relatively higher spontaneous firing rates, and have been considered to correspond to inhibitory interneurons [32]. In rats, most θ cells are broadly tuned or not spatially selective to space, while complex cells were selective to space [44,45]. Similarity of selectivity to external stimuli and neurophysiological data suggest the dichotomy of the monkey HF neurons (specific-combination and object-differential neurons vs. behavior differential and nondifferential neurons) might be functionally comparable to that of the rat-complex spike cell vs. θ cell dichotomy. It is suggested that these two types of HF neurons are important to the conjunctive encoding of incoming multiple information. These findings might well correspond to the neurophysiological bases of human episodic memory suggested by the theoretical computation studies [32,46].

Monkey SP neuronal activity

The activity of 430 neurons was recorded from the SP in two monkeys while they performed the PGN task. Of these neurons, 14 responded in the place phase but not in the object phase (place-phase neurons), 47 responded in the object phase but not in the place phase (object-phase neurons), and 129 responded in both place and object phases (both-phase neurons). In the present paper, we analyzed the responses of these 190 visually responsive neurons during the place and/or object phases.

Place-differential responses in the place phase

Responses during the place phase were significantly place differential in 58 (13 place-phase and 45 both-phase) neurons. Responses of the remaining 85 neurons were nondifferential. Response of a typical place-differential neuron is shown in Fig. 5A. This neuron responded during the place phase only when the cab was located in place III in the PGN task (Fig. 5Aa). The mean magnitudes of the responses during the place phase are presented in Fig. 5Ab. This place-differential response was independent on a specific distal cue, since the responses were not affected by the environmental manipulation when the monkey learned the new environment in which a table in front of the place II was moved to the front of place I (not shown).

PGN-specific responses in the object phase

For each of the object- and both-phase neurons with object-phase responses (n = 176), responses to the four objects in the PGN task were individually analyzed and compared with respect to the response/reinforcement contingency. Of these 176 neurons, 91 (34 of 47 object-phase; 57 of 129 both-phase) exhibited a contingency-differential (i.e., Go/reward- and Nogo/nonreward-selective) responses to one or more objects. Of 91 contingency-differential neurons, 53 Go/reward-selective neurons responded more in Go trials, and 38 Nogo/nonreward-selective neurons responded

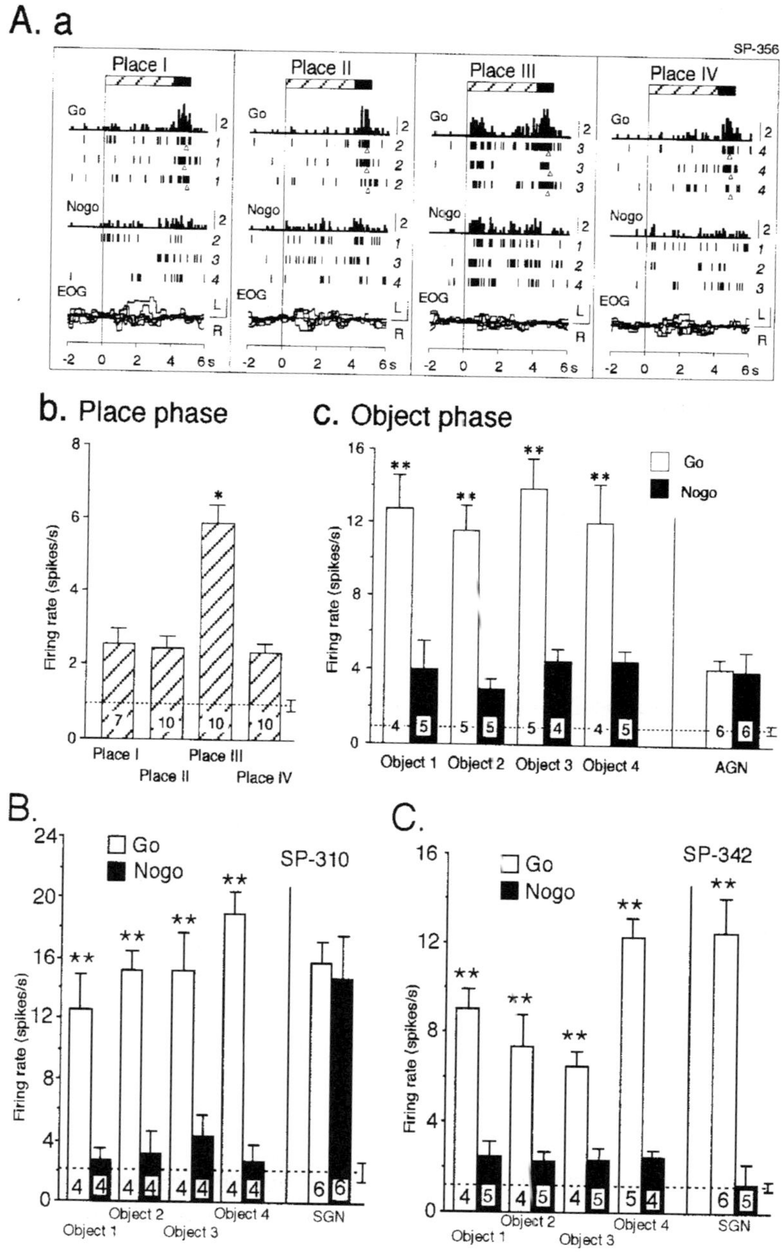

Fig. 5. Three types of the SP neurons identified during the PGN task. **A**: Activity of a place-differential response and PGN-specific Go/reward-selective response. a: Differential responses to place III and to objects indicating Go/reward contingency. Horizontal hatched rectangle, place phase; filled rectangle, object phase; open triangle below a raster display, onset of bar press; number at the right of each raster line, number of corresponding object; calibration bar for EOG, 1 mV. b: Place-phase responses (mean ± SEM) for four places. *Place phase activity in place III significantly greater than spontaneous and other place activity. The number near the bottom of each column shows the number of trials. c: Object-phase responses (mean ± SEM) to sight of each object for Go and Nogo trials regardless of monkey's location in the PGN (object 1–4) and place-independent asymmetrical Go/Nogo (AGN) tasks. The neuron differentially responded to objects indicating Go/reward contingency in the PGN task but not in the AGN task. **B and C**: Activities of reward-related (B) and Go-related (C) responses in the object phase. Histograms of mean firing rate to sight of each object for Go and Nogo trials regardless of monkey's location in the PGN (object 1–4) and place-independent symmetrical Go/Nogo (SGN) tasks. **Significantly different responses to objects indicating go/reward and Nogo/nonreward contingencies (Mann-Whitney U test, $p < 0.01$).

more in Nogo trials. The neuron shown in Fig. 5A also showed typical Go/reward-selective responses in the object phase. Responses of this neuron were stronger to each object in Go trials than in Nogo trials (Fig. 5Aa,c).

To test whether or not contingency-differential response during the object phase in the PGN task depended on conditional place-object relations, 72 neurons (37 of 53 Go/reward-selective; 35 of 38 Nogo/nonreward-selective) were tested in the unconditional place-independent AGN task. Of these 72 neurons, 22 Go/reward-selective and 21 Nogo/nonreward-selective neurons responded differentially only in the PGN task in which discrimination of conditional place-object relations were required (PGN-specific). Typical examples are shown in Fig. 5Ac. The Go/reward-selective neuron did not respond differentially to a Go/reward object (white object) in the AGN task. The remaining 29 neurons (15 Go/reward-selective; 14 Nogo/nonreward-selective) were contingency-differential responsive in both the AGN and the PGN tasks (PGN-nonspecific).

Reward- or behavior-related responses in the object phase

Of the 15 PGN-nonspecific Go/reward-selective neurons, nine were further tested in the SGN task, in which Go and Nogo responses were both reinforced to test whether this contingency-differential response reflected behavior (Go response) or reward availability. Figures 5B and C show the results for two different Go/reward-selective neurons. Both responded strongly during the object phase when the objects were rewarding, depending on their place-object relations in the PGN task (Fig. 5B and C). The neuron in Fig. 5B responded to the object indicating Nogo/reward contingency as well as the object indicating Go/reward contingency in the SGN task (reward-related neuron). On the contrary, the neuron in Fig. 5C did not respond to the object indicating Nogo/reward contingency, but responded to the object indicating Go/reward contingency in the SGN task (Go-related neuron). Of the nine Go/reward-selective neurons that could be tested on the SGN task, seven were reward-related, and two were Go-related.

Of the 14 PGN-nonspecific Nogo/nonreward-selective neurons, 12 were similarly tested with the SGN task. All of these 12 neurons were concluded to be nonreward-related neurons, since they did not respond to the object indicating Nogo/reward contingency in the SGN task. Thus, most of the PGN-nonspecific neurons (19/21, 90.5%) which could be tested in the SGN task were related to the reward/nonreward contingency, rather than to the response contingency.

Location of each neuron type

Many of the contingency-differential neurons (PGN-specific, reward-related, Go-related, and nonreward-related) were densely packed in the dorsal and intermediate parts of lateral SP and some PGN-specific neurons were in the medial SP. Most neurons with place-differential responses were located in the area dorsal to the anterior commissure in the middle planes.

Functional roles of the SP and the HF neurons

In the place phase, both the SP and the HF neurons responded differentially to the monkey's location. Since the lateral SP receives direct inputs from the pyramidal neurons in the hippocampal CA1 and CA3 subfields [29,30], place-differential responses in the SP might be attributed to the hippocampal afferents. In the object phase, there were some differences in neuronal responses to objects between the HF and SP. The HF neurons responded differentially to the kinds of the objects (object-differential neurons). On the other hand, the SP neurons respond nondifferentially to the kind of the object, but differentially to reward contingency of the objects. This evidence is consistent with the idea that individual information is relationally encoded in the HF [8–10], while the SP might be involved in the integration and transformation of conditional relations between place and object into a more simple form of information such as reward contingency. This information might be transferred, to manifest appropriate behavioral responses in response to environmental stimuli, from the SP to the lateral hypothalamic area which receives strong anatomical connections with the SP, and is important to recognize reward contingency of the stimulus [47–49].

Conclusion

Results in the present neurophysiological studies using monkeys suggested that the SP-HF neurons encoded multiple factors such as reward, place, stimulus (object), behavior, and their conditional relations. The spatial factors (place) were commonly encoded conjunctively with other factors. The results are consistent with the idea that the SP-HF system encodes incoming information into memory indexes based on spatial-temporal context. It has been reported that lesions of the HF or electrical shock induced temporally graded retrograde amnesia in animals and humans [50–53].

This evidence suggests that memory indexes encoding a pattern of neocortical activation are at least temporally stored in the HF. By these memory indexes, the HF contributes to retrieval of stored engram [16,54]. That is, neuronal activation in one neocortical area can evoke neuronal activation in other neocortical areas through the HF. This idea is consistent with the contextual retrieval theory of target information [55], in which target information is retrieved based on preceding contextual information which refers to the target information. Furthermore, the selectivity to external stimuli and the neurophysiological characteristics of the HF neurons suggest the dichotomy of the monkey HF neurons which might be functionally comparable to that of the rat-complex spike cell (pyramidal neuron) vs. θ cell. The HF neural circuit composed of pyramidal and inhibitory interneurons might enable the HF to encode an episode consisting of these multiple factors by conjunctive association [32,46].

Acknowledgements

This work was supported partly by Uehara Memorial Foundation, and partly by the Japanese Ministry of Education, Science and Culture Grants-in-Aid for Scientific Research (08408036, 08279105, 08279215, 08680884 and 08234209).

References

1. Scoville WB, Milner B. Loss of recent memory after bilateral hippocampal lesions. J Neurol Neurosurg Psychiat 1957;20:11–21.
2. Squire, LR. Memory and Brain. New York: Oxford University Press, 1987.
3. Squire LR, Zola-Morgan S. The medial temporal lobe memory system. Science 1991;253:1380–1385.
4. Cohen NJ, Eichenbaum H. Memory, Amnesia and the Hippocampal System. Cambridge: MIT Press, 1993.
5. Eichenbaum H, Otto T, Cohen NJ. The hippocampus – what does it do? Behav Neural Biol 1992; 57:2–36.
6. Olton DS, Becker JT, Handelmann GE. Hippocampus, space, and memory. Behav Brain Sci 1979; 2:313–365.
7. O'Keefe J, Nadel L. The Hippocampus as a Cognitive Map. Oxford: Clarendon Press, 1978.
8. Eichenbaum H, Stewart C, Morris RGM. Hippocampal representation in spatial learning. J Neurosci 1990;10:331–339.
9. Eichenbaum H. Thinking about brain cell assemblies. Science 1993;261:993–994.
10. Bunsey, M, Eichenbaum H. Conservation of hippocampal memory function in rats and humans. Nature 1996;379:255–257.
11. Sutherland RJ, Rudy JW. Configural association theory: the role of the hippocampal formation in learning, memory, and amnesia. Psychobiology 1989;17:129–144.
12. Amaral DG. Memory: anatomical organization of candidate brain regions. In: Plum F, Mountcastle VB (eds) Handbook of Physiology 5. Bethesda: American Physiological Society, 1987;211–294.
13. Amaral DG, Insausti R. Hippocampal formation. In: Paxinos G (ed) The Human Nervous System. San Diego: Academic Press, 1990;711–755.
14. Felleman DJ, Van Essen DC. Distributed hierarchical processing in the primate cerebral cortex. Cerebral Cortex 1991;1:1–47.
15. Teyler TJ, Discenna P. The hippocampal memory indexing theory. Behav Neurosci 1986;100: 147–154.

16. Alvarez P, Squire LR. Memory consolidation and the medial temporal lobe: a simple network model. Proc Natl Acad Sci USA 1994;91:7041–7045.
17. Wible CG, Findling RL, Shapiro M, Lang EJ, Crane S, Olton DS. Mnemonic correlates of unit activity in the hippocampus. Brain Res 1986;399:97–110.
18. Wiener SI, Paul CA, Eichenbaum H. Spatial and behavioral correlates of hippocampal neuronal activity. J Neurosci 1989;9:2737–2763.
19. Sharp PE, Kubie JL, Muller RU. Firing properties of hippocampal neurons in a visually symmetrical environment: contributions of multiple sensory cues and mnemonic processes. J Neurosci 1990; 10:3093–3105.
20. Young BJ, Fox GD, Eichenbaum H. Correlates of hippocampal complex-spike cell activity in rats performing a non-spatial radial maze task. J Neurosci 1994;14:6553–6563.
21. Watanabe T, Niki H. Hippocampal unit activity and delayed response in the monkey. Brain Res 1985;325:241–254.
22. Rolls ET, Miyashita Y, Cahusac PMB, Kesner RP. Niki H, Feigenbaum J, Bach L. Hippocampal neurons in the monkey with activity related to the place in which a stimulus is shown. J Neurosci 1989;9:1835–1845.
23. Tamura R, Ono T, Fukuda M, Nakamura K. Spatial responsiveness of monkey hippocampal neurons to various visual and auditory stimuli. Hippocampus 1992;2:307–322.
24. Ono T, Nakamura K, Fukuda M, Tamura R. Place recognition responses of neurons in monkey hippocampus. Neurosci Lett 1991;121:194–198.
25. Ono T, Nakamura K, Nishijo H, Eifuku S. Monkey hippocampal neuron related to spatial and non spatial functions. J Neurophysiol 1993;70:1516–1529.
26. Nishijo H, Ono T, Tamura R, Nakamura K. Amygdalar and hippocampal neuron responses related to recognition and memory in monkey. Prog Brain Res 1993;95:339–357.
27. Gray JA, McNaughton N. Comparison between the behavioural effects of septal and hippocampal lesions: a review. Neurosci Behav Rev 1983;7:119–188.
28. Gaffan D, Harrison S. Place memory and scene memory: effects of fornix transection in the monkey. Exp Brain Res 1989;74:202–212.
29. Swanson LW, Köhler C, Björklund A. The limbic region. I: The septohippocampal system In: Björklund A, Hökfelt T, Swanson LW (eds) Handbook of Chemical Neuroanatomy, vol 5. New York: Elsevier, 1987;125–277.
30. Raisman G. The connexions of the septum. Brain 1966;89:317–348.
31. Lynch G, Baudry M. The biochemistry of memory: a new and specific hypothesis. Science 1984; 224:1057–1063.
32. McNaughton BL, Morris RGM. Hippocampal synaptic enhancement and information storage within a distributed memory system. Trends Neurosci 1987;10:408–415.
33. Stevens DR, Cotman CW. Long-term potentiation of excitatory and slow inhibitory synaptic potentials in the hippocampal-septal projection of the rat. Brain Res 1991;558:120–122.
34. Ono T, Eifuku S, Nakamura K, Nishijo H. Monkey hippocampal neuron responses related to spatial and non-spatial influence. Neurosci Lett 1993;159:75–78.
35. Eifuku S, Nishijo H, Kita T, Ono T. Neuronal activity in the primate hippocampal formation during a conditional association task based on the subject's location. J Neurosci 1995;15:4952–4969.
36. Kita T, Nishijo H, Eifuku S, Terasawa K, Ono T. Place and contingency differential responses of monkey septal neurons during conditional place-object discrimination. J Neurosci 1995;15: 1683–1703.
37. Hirsh R. The hippocampus, conditional operations, and cognition. Physiol Psychol 1980;8:175–182.
38. Ross RT, Orr WB, Holland PC, Berger TW. Hippocampectomy disrupts acquisition and retention of learned conditional responding. Behav Neurosci 1984;98:211–225.
39. Good M, Honey RC. Conditioning and contextual retrieval in hippocampal rats. Behav Neurosci 1991;105:499–509.
40. Ridley RM, Timothy CJ, Maclean CJ, Baker HF. Conditional learning and memory impairments following neurotoxic lesion of the CA1 field of the hippocampus. Neuroscience 1995;67:263–275.

41. Miyashita Y, Rolls ET, Cahusac PMB, Niki H, Feigenbaum JD. Activity of hippocampal formation neurons in the monkey related to a conditional spatial response task. J Neurophysiol 1989;61: 669–678.
42. Tamura R, Ono T, Fukuda M, Nakamura K. Responsiveness of monkey hippocampal and entorhinal neurons in learning behavior. Jpn J Physiol Suppl 1989;39:S120.
43. Fox SE, Ranck JB Jr. Electrophysiological characteristics of hippocampal complex-spike cells and theta cells. Exp Brain Res 1981;41:399–410.
44. Kubie JL, Muller RU, Bostock E. Spatial firing properties of hippocampal theta cells. J Neurosci 1990;10:1110–1123.
45. Jung MW, McNaughton BL. Spatial selectivity of unit activity in the hippocampal granular layer. Hippocampus 1993;3:165–182.
46. McClelland JL, McNaughton BL, O'Reilly RC. Why there are complementary learning systems in the hippocampus and neocortex: insights from the successes and failures of connectionist models of learning and memory. Psychol Rev 1995;102:419–457.
47. Rolls ET, Rolls BJ, Rowe EA. Sensory-specific and motivation-specific satiety for the sight and taste of food and water in man. Physiol Behav 1983;30:185–192.
48. Ono T, Nishino H, Sasaki K, Fukuda M, Muramoto K. Monkey lateral hypothalamic neuron response to sight of food, and during bar press and ingestion. Neurosci Lett 1981;21:99–104.
49. Ono T, Nishino H, Fukuda M, Sasaki K. Monkey amygdala, lateral hypothalamus and prefrontal cortex roles in food discrimination, motivation to bar press, and ingestion reward. In: Bandler R (ed) Modulation of Sensorimotor Activity During Alternations in Behavioral States. New York: Alan R. Liss, 1984;251–268.
50. Squire LR, Cohen NJ, Zouzounis JA. Preserved memory in retrograde amnesia: sparing of a recently acquired skill. Neuropsychologia 1984;22:145–152.
51. Zola-Morgan S, Squire LR, Amaral DG. Human amnesia and the medial temporal region: enduring memory impairment following a bilateral lesion limited to field CA1 of the hippocampus. J Neurosci 1986;6:2950–2967.
52. Zola-Morgan SM, Squire LR. The primate hippocampal formation: evidence for a time-limited role in memory storage. Science 1990;250:288–291.
53. Kim JJ, Fanselow MS. Modality-specific retrograde amnesia of fear. Science 1992;256:675–677.
54. Teyler TJ, Discenna P. The hippocampal memory indexing theory. Behav Neurosci 1986;100: 147–154.
55. Hirsh R. The hippocampus and contextual retrieval of information from memory: a theory. Behav Biol 1979;12:421–444.

1996 Elsevier Science B.V.
Integrative and Molecular Approach to Brain Function.
M. Ito and Y. Miyashita, editors.

Functions of the inferior temporal cortex and hippocampus in memory

Edmund T. Rolls
University of Oxford, Department of Experimental Psychology, Oxford, UK

Abstract. By combining neurophysiological and computational approaches it is possible to produce a theory for how neural networks in the temporal cortical visual areas could function to stored representations of objects in the world which can later be recognised when the object is presented in a different view, different size, and in a different position on the retina. A similar approach is then described which leads to a theory of how the hippocampus could rapidly store memories, especially episodic memories including spatial context for where and when an event happened, and how that information could later be recalled to the neocortex.

Introduction

Advances are being made in understanding how the brain could perform some of the processing involved in perception and memory. These advances come in part from neurophysiological experiments in which the processing involved in vision and memory is analysed by recording the activity of single neurons in primates during these types of processing, and incorporating this information into computational models at the network level which provide an account of the ways in which many neurons in a network could perform the required computations. Examples of this approach are described here. Many of the experimental investigations described here were performed on nonhuman primates, macaque monkeys, because they have highly developed visual and cortical areas, which make them appropriate when developing an understanding how the corresponding areas of the human brain work.

Brain mechanisms involved in visual object recognition

I focus on the nature of the representation of objects in the world that is found by the time information has reached the higher parts of the visual system, the temporal cortical visual areas; and on how that representation is built by passing through multiple connected cortical processing stages. Particular attention is paid to neural systems involved in processing information about faces, because with the large number of neurons devoted to this class of stimuli, this system has proved amenable

Address for correspondence: Edmund T. Rolls, University of Oxford, Department of Experimental Psychology, South Parks Road, Oxford OX1 3UD, UK. Tel.: +44-1865-271348. Fax: +44-1865-310447. E-mail: Edmund.Rolls@psy.ox.ac.uk to experimental analysis; and because of the importance of face recognition and expression identification in primate behaviour. However, the ideas described here are also thought to be involved in the recognition of objects [1]. Further elaboration of these ideas is provided by Rolls [2–4].

It is important to know how information is represented about objects and faces in the brain, for the representation is the key to understanding how the information is stored and later accessed. For this reason, in some of the next sections, we consider whether information about each object is represented by a single neuron in the brain, or is instead distributed across many neurons; and the exact way in which the representations found in the brain are invariant with respect to size, position on the retina, angle of view, etc.

Visual cortical areas in the temporal lobes

The visual pathways project by a number of cortico-cortical stages from the primary visual cortex until they reach the temporal lobe visual cortical areas [5–7]. The inferior temporal visual cortex, area TE, is divided into a set of subareas, and in addition there is a set of different areas in the cortex in the superior temporal sulcus [5,8] (Fig. 1). Of these latter areas, TPO receives inputs from temporal, parietal and occipital cortex; PGa and IPa from parietal and temporal cortex; and TS and TAa primarily from auditory areas [5]. There is considerable specialization of function in these areas [8]. For example, areas TPO, PGa and IPa are multimodal, with neurons

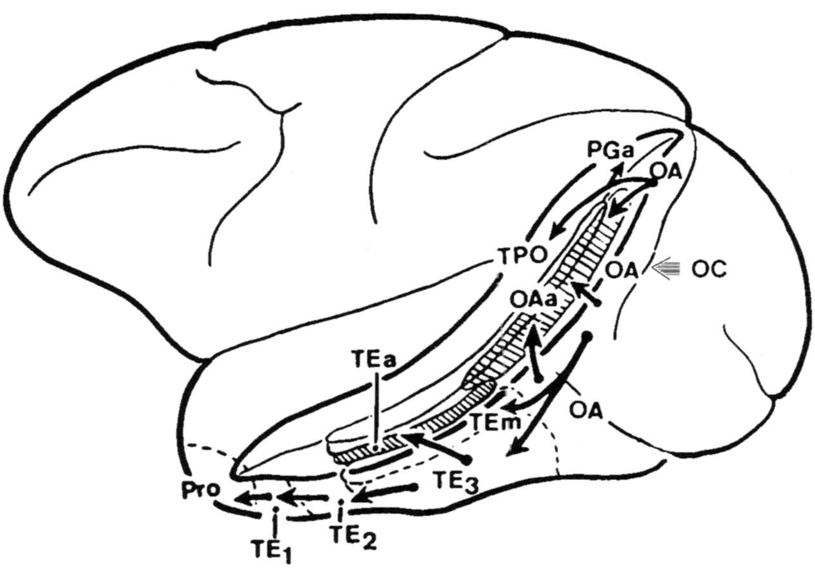

Fig. 1. Forward projections to and within the temporal lobe from earlier visual areas in the macaque. OC is the primary (striate) visual cortex. The TE areas form the inferior temporal gyrus, and the superior temporal sulcus (the margins of which are shown by a thick line) has been opened to reveal some of the areas within it. (From [5]).

which respond to visual, auditory and/or somatosensory inputs; the inferior temporal gyrus and adjacent areas (TE3, TE2, TE1, TEa and TEm) are primarily unimodal visual areas; areas in the cortex in the anterior and dorsal part of the superior temporal sulcus (e.g., TPO, IPa and IPg) have neurons specialised for the analysis of moving visual stimuli; and neurons responsive primarily to faces are found more frequently in areas TPO, TEa and TEm [8], where they comprise approximately 20% of the visual neurons responsive to stationary stimuli, in contrast to the other temporal cortical areas in which they comprise 4–10%. The neurons which respond to nonface stimuli often require two or more simple features to be present in the correct spatial relationship in order to respond [9].

Distributed encoding of face and object identity

Some neurons described in the temporal cortical visual areas have responses selective for faces in that they respond 2–20 times more to faces than to a wide range of simple geometrical stimuli, or complex three-dimensional objects [8,10,11]. These neurons reflect information not just that a face has been seen, but about which face has been seen: they respond differently to different faces. An important question for understanding brain computation is whether a particular object (or face) is represented in the brain by the firing of one or a few gnostic (or “grandmother” or “cardinal”) cells [12] (a local representation), or whether instead the firing of a group or ensemble of cells each with somewhat different responsiveness provides the representation (a distributed representation). For faces [10] and objects [1] it has been shown that even at the highest part of the visual system, the representation is distributed, in that each neuron responds to many different stimuli in a large set of stimuli, with a characteristic response rate to each stimulus in the set [13]. Such a distributed representation would be ideal for discrimination, for the maximum information suitable for comparing fine differences between different stimuli would be made available across the population (if 50% were active to each stimulus).

The approach so far has focused on how visual information about what is being looked at in the world is represented by the activity of individual neurons. How does the process scale when we take large numbers of such visual neurons? We have recently obtained evidence that the information available about which visual stimulus (which of 20 equiprobable faces) had been shown increases linearly with the number of neurons in the sample [14,15]. Because information is a logarithmic measure (see the papers referred to), this indicates that the number of stimuli encoded rises approximately exponentially, as the number of cells in the sample increases. The consequence of this is that large numbers of stimuli, and fine discriminations between them, can be represented without having to measure the activity of an enormous number of neurons. For example, the results of the experiments of Rolls, Tovee and Treves [16] indicate that the activity of 15 neurons would be able to encode 192 face stimuli (at 50% accuracy), of 20 neurons 768 stimuli, of 25 neurons 3,072 stimuli, of 30 neurons 12,288 stimuli, and of 35 neurons 49,152 stimuli [15]. This means that it is now neurophysiologically possible to read the code from the end of this part at least of the visual system. By measuring the firing rates of relatively small numbers (tens) of neurons, we know which (of potentially hundreds or thousands) of visual stimuli are being looked at by a monkey.

A neuronal representation of visual stimuli showing invariance

One of the major problems which must be solved by a visual system used for object recognition is the building of a representation of visual information which allows recognition to occur relatively independently of size, contrast, position on the retina, and angle of view, etc. We have shown that many of the neurons whose responses reflect face identity have responses that are relatively invariant with respect to size and contrast [17]; spatial frequency [18,19]; and retinal translation, i.e., position in the visual field [20]. Some of these neurons even have relatively view invariant responses, responding to different views of the same face but not of other faces [21]. The same has recently been shown for objects [1]. It is clearly important that invariance in the visual system is made explicit in the neuronal responses, for this simplifies greatly the output of the visual system to memory systems such as the hippocampus and amygdala, which can then remember or form associations about objects. The function of these memory systems would be almost impossible if there were no consistent output from the visual system about objects (including faces), for then the memory systems would need to learn about all possible sizes, positions, etc. of each object, and there would be no easy generalisation from one size or position of an object to that object when seen with another retinal size or position.

Although the neurons just described have view-invariant responses, there is another population of face-selective neurons, found particularly in the cortex in the superior temporal sulcus, which tends to have view-dependent responses [21]. Some of these neurons have responses which reflect the facial expression but not the facial identity of the stimulus [21]. These neurons could be useful in providing information of potential use in social interactions [22–25]. Damage to this population and to brain areas to which these neurons project may contribute to the deficits in social and emotional behaviour produced by temporal or ventral frontal lobe damage [22–24, 26–30].

Learning of new representations in the temporal cortical visual areas

Given the fundamental importance of a computation which results in neurons which can be relatively finely tuned but which still use distributed encoding, investigations have been performed to determine whether such neurons can be built rapidly by visual experience. To test this, Rolls et al. [31] investigated whether the responses of temporal cortex face-selective neurons were at all altered by the presentation of new faces which the monkey had never seen before. It might be, for example, that the population would make small adjustments in the responsiveness of its individual neurons, so that neurons would become tuned to stimuli in a way which would enable the population as a whole to discriminate between the faces actually seen. Rolls et al. [31] thus investigated whether when a set of totally novel faces was introduced, the responses of these neurons were fixed and stable from the first presentation, or instead whether there was some adjustment of responsiveness over repeated presentations of the new faces. First, it was shown for each neuron tested that its responses were stable over five to 15 repetitions of a set of familiar faces. Then a set of new faces was shown in random order (with 1 s for each presentation), and the set was repeated with a new random order over many iterations. Some of the neurons studied in this way altered the relative degree to which they responded to the different members of the set of novel faces over the first few (1–2) presentations of the set [31]. If in a different experiment a single novel face was introduced when the responses of a neuron to a set of familiar faces was being recorded, it was found that the responses to the set of familiar faces were not disrupted, while the responses to the novel face became stable within a few presentations. Thus, there is now some evidence from these experiments that the response properties of neurons in the temporal lobe visual cortex are modified by experience, and that the modification is such that when novel faces are shown, the relative responses of individual neurons to the new faces alter [31]. It is suggested that alteration of the tuning of individual neurons in this way results in a good discrimination over the population as a whole of the faces known to the monkey. This evidence is consistent with the categorisation being performed by self-organising competitive neuronal networks, as described by Rolls [32–35].

Further evidence that these neurons can learn new representations very rapidly comes from an experiment in which binarised black and white images of faces which blended with the background were used. These did not activate face-selective neurons. Full grey-scale images of the same photographs were then shown for 10 0.5-s presentations. It was found in a number of cases, if the neuron happened to be responsive to the face, that when the binarised version of the same face was shown next, the neurons responded to it [36]. This is a direct parallel to the same phenomenon which is observed psychophysically, and provides dramatic evidence that these neurons are influenced by only a very few seconds (in this case 5 s) of experience with a visual stimulus.

Such rapid learning of representations of new objects appears to be a major type of learning in which the temporal cortical areas are involved. Ways in which this learning could occur are considered below. It is also the case that there is a much shorter term form of memory in which some of these neurons are involved, for whether a particular visual stimulus (such as a face) has been seen recently, in that some of these neurons respond differently to recently seen stimuli in short-term visual memory tasks [37,38]. A tendency of some temporal cortical neurons to associate together visual stimuli when they have been shown over very many repetitions separated by several seconds has also been described by Miyashita [39]. When a visual stimulus is shown that has been repeatedly paired with another visual stimulus, inferior temporal cortex neurons may produce a response consistent with the associated not the seen visual stimulus [40].

The speed of processing in the temporal cortical visual areas

Given that there is a whole sequence of visual cortical processing stages including V1, V2, V4, and the posterior inferior temporal cortex to reach the anterior temporal cortical areas, and that the response latencies of neurons in V1 are about 40–50 ms, and in the anterior inferior temporal cortical areas approximately 80–100 ms, each stage may need to perform processing for only 15–30 ms before it has performed sufficient processing to start influencing the next stage. Consistent with this, response latencies between V1 and the inferior temporal cortex increase from stage to stage [41]. This seems to imply very fast computation by each cortical area, and therefore to place constraints on the type of processing performed in each area that is necessary for final object identification from the stored information. Rapid identification of visual stimuli is important in social and many other situations, and that there must be strong selective pressure for rapid identification. For these reasons, the speed of processing by the cortical visual system has been investigated quantitatively, as described next.

In a first approach, Tovee et al. [42] measured the information available in short temporal epochs of the responses of temporal cortical face-selective neurons about which face had been seen. They found that if a period of the firing rate of 50 ms was taken, then this contained 84.4% of the information available in a much longer period of 400 ms about which of four faces had been seen. If the epoch was as little as 20 ms, the information was 65% of that available from the firing rate in the 400-ms period [42]. These high-information yields were obtained with the short epochs taken near the start of the neuronal response, e.g., in the poststimulus period 100–120 ms. Moreover, they were able to show that the firing rate in short periods taken near the start of the neuronal response was highly correlated with the firing rate taken over the whole response period, so that the information available was stable over the whole response period of the neurons [42]. They were able to extend this finding to the case when a much larger stimulus set, of 20 faces, was used. Again, they found that the information available in short (e.g., 50 ms) epochs was a considerable proportion (e.g., 65%) of that available in a 400-ms long firing-rate analysis period [43]. These investigations thus showed that there was considerable information about which stimulus had been seen in short time epochs near the start of the response of temporal cortex neurons.

The next approach was to address the issue of for how long a cortical area must be active to mediate object recognition. This approach used a visual backward masking paradigm. In this paradigm there is a brief presentation of a test stimulus which is rapidly followed (within 1–100 ms) by the presentation of a second stimulus (the mask), which impairs or masks the perception of the test stimulus. This paradigm used psychophysically leaves unanswered for how long visual neurons actually fire under the masking condition at which the subject can just identify an object. Although there has been a great deal of psychophysical investigation with the visual masking paradigm [44–46], there is very little direct evidence on the effects of visual masking on neuronal activity. For example, it is possible that if a neuron is well tuned to one class of stimulus, such as faces, a pattern mask which does not activate the neuron, will leave the cell firing for some time after the onset of the pattern mask. In order to obtain direct neurophysiological evidence on the effects of backward masking on neuronal activity, Rolls and Tovee [47] analysed the effects of backward masking with a pattern mask on the responses of single neurons to faces. This was performed to clarify both what happens with visual backward masking, and to show how long neurons may respond in a cortical area when perception and identification of a stimulus, which includes accessing its stored representation in memory, are just possible. When there was no mask the cell responded to a 16-ms presentation of the test stimulus for 200–300 ms, far longer than the presentation time. It is suggested that this reflects the operation of a short-term memory system implemented in cortical circuitry, the importance of which is considered below. If the mask was a stimulus which did not stimulate the cell (either a nonface pattern mask consisting of black and white letters N and O, or a face which was a noneffective stimulus for that cell), then as the interval between the onset of the test stimulus and the onset of the mask stimulus (the Stimulus Onset Asynchrony, SOA) was reduced, the length of time for which the cell fired in response to the test stimulus was reduced. This reflected an abrupt interruption of neuronal activity produced by the effective face stimulus. When the SOA was 20 ms, face-selective neurons in the inferior temporal cortex of macaques responded for a period of 20–30 ms before their firing was interrupted by the mask [47]. They went on to show that under these conditions (a test-mask stimulus onset asynchrony of 20 ms), human observers looking at the same displays could just identify which of six faces was shown [48].

These results provide evidence that a cortical area can perform the computation necessary for the recognition of a visual stimulus in 20–30 ms, and provide a fundamental constraint which must be accounted for in any theory of cortical computation. The results emphasise just how rapidly cortical circuitry can operate. This rapidity of operation has obvious adaptive value, and allows the rapid behavioural responses to the faces and face expressions of different individuals which are a feature of primate social and emotional behaviour. Moreover, although this speed of operation does seem fast for a network with recurrent connections (mediated by, e.g., recurrent collateral or inhibitory interneurons), recent analyses of networks with analog membranes which integrate inputs, and with spontaneously active neurons, shows that such networks can operate very rapidly [49,50].

These experiments also have implications for visual processing in relation to top-down processing (see [51]). The evidence just described indicates that visual recognition can occur (the subjects say which face they saw) with largely feedforward processing. There is not time in the experiments described for visual information to pass from V1 to V2 to V4 and thus to posterior and then anterior inferior temporal cortex, and back again all the way to V1, before V1 has started to process the second visual input, that is to have its processing of the first visual stimulus cut off by the mask.

Possible computational mechanisms in the visual cortex for learning invariant representations

The neurophysiological findings described above on the existence of ensemble-encoded invariant representations in the inferior temporal cortical areas (see also [2,26,52]), and wider considerations on the possible computational properties of the cerebral cortex [4,11,32,33], lead to the following outline working hypotheses on face and object recognition by visual cortical mechanisms.

Cortical visual processing for object recognition is considered to be organised as a set of hierarchically connected cortical regions consisting at least of V1, V2, V4, posterior inferior temporal cortex (TEO), inferior temporal cortex (e.g., TE3, TEa and TEm), and anterior temporal cortical areas (e.g., TE2 and TE1). (This stream of processing has many connections with a set of cortical areas in the anterior part of the superior temporal sulcus, including area TPO.) There is convergence from each small part of a region to the succeeding region (or layer in the hierarchy) in such a way that the receptive field sizes of neurons (e.g., 1° near the fovea in V1) become larger by a factor of approximately 2.5 with each succeeding stage (and the typical parafoveal receptive field sizes found would not be inconsistent with the calculated approximations of, e.g., 8° in V4, 20° in TEO, and 50° in inferior temporal cortex [53]) (Fig. 2). Such zones of convergence would overlap continuously with each other (Fig. 2). This connectivity would be part of the architecture by which translation invariant representations are computed. Each layer is considered to act partly as a set of local self-organising competitive neuronal networks with overlapping inputs. (The region within which competition would be implemented would depend on the spatial

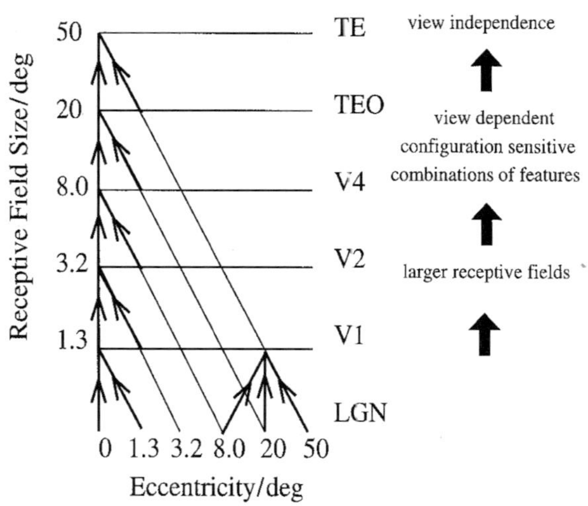

Fig. 2. Schematic diagram showing convergence achieved by the forward projections in the visual system, and the types of representation that may be built by competitive networks operating at each stage of the system from the primary visual cortex (V1) to the inferior temporal visual cortex (area TE) (see text). LGN – lateral geniculate nucleus. Area TEO forms the posterior inferior temporal cortex. The receptive fields in the inferior temporal visual cortex (e.g., in the TE areas) cross the vertical midline (not shown).

properties of inhibitory interneurons, and might operate over distances of 1–2 mm in the cortex.) These competitive nets operate by a single set of forward inputs leading to (typically nonlinear, e.g., sigmoid) activation of output neurons; of competition between the output neurons mediated by a set of feedback inhibitory interneurons which receive from many of the principal (in the cortex, pyramidal) cells in the net and project back to many of the principal cells which serves to decrease the firing rates of the less-active neurons relative to the rates of the more active neurons (i.e., soft competition); and then of synaptic modification by a modified Hebb rule, such that synapses to strongly activated output neurons from active input axons strengthen, and from inactive input axons weaken (see [34,35]). (A biologically plausible form of this learning rule that operates well in such networks is:

$$\delta w_{ij} = k \cdot m \, (r'_j - w_{ij})$$

where k is a constant, δw_{ij} is the change of synaptic weight, r'_j is the firing rate of the j'th axon, and m is a nonlinear function of the output activation of neuron i which mimics the operation of the NMDA receptors in learning; see Rolls, 1989c,b,a; Rolls and Treves, 1997). Such competitive networks operate to detect correlations between the activity of the input neurons, and to allocate output neurons to respond to each cluster of such correlated inputs. These networks thus act as categorisers. In relation to visual information processing, they would remove redundancy from the input representation, and would develop low-entropy representations of the information (*cf.* [54,55]). Such competitive nets are biologically plausible, in that they utilise Hebb-modifiable forward excitatory connections, with competitive inhibition mediated by cortical inhibitory neurons. The competitive scheme suggested would not result in the formation of "winner-take-all" or "grandmother" cells, but would instead result in a small ensemble of active neurons representing each input [32–35].

Translation invariance would be computed in such a system by utilising competitive learning to detect regularities in inputs when real objects are translated in the physical world. The hypothesis is that because objects have continuous properties in space and time in the world, an object at one place on the retina might activate feature analysers at the next stage of cortical processing, and when the object was translated to a nearby position, because this would occur in a short period (e.g., 0.5 s), the membrane of the postsynaptic neuron would still be in its "Hebb-modifiable" state (caused, for example, by calcium entry as a result of the voltage-dependent activation of NMDA receptors), and the presynaptic afferents activated with the object in its new position would thus become strengthened on the still-activated postsynaptic neuron. It is suggested [11] that the short-temporal window (e.g., 0.5 s) of Hebb-modifiability helps neurons to learn the statistics of objects moving in the physical world, and at the same time to form different representations of different feature combinations or objects, as these are physically discontinuous and present less-regular correlations to the visual system. Foldiak [56] has proposed computing an average activation of the postsynaptic neuron to assist with the same problem. One idea here is that the temporal properties of the biologically implemented learning mechanism are such that it is well suited to detecting the relevant continuities in the world of real objects. Another suggestion is that a memory trace for what has been seen in the last 300 ms appears to be implemented by a mechanism as simple as continued firing of inferior temporal neurons after the stimulus has disappeared, as we have shown in masking experiments (see also [47,48]). I also suggest that other invariances, e.g., size, spatial frequency, and rotation invariance, could be learned by a comparable process. (Early processing in V1 which enables different neurons to represent inputs at different spatial scales would allow combinations of the outputs of such neurons to be formed at later stages. Scale invariance would then result from detecting at a later stage which neurons are almost conjunctively active as the size of an object alters.) It is suggested that this process takes place at each stage of the multiple-layer cortical processing hierarchy, so that invariances are learned first over small regions of space, and then over successively larger regions. This limits the size of the connection space within which correlations must be sought.

Increasing complexity of representations could also be built in such a multiple-layer hierarchy by similar mechanisms. At each stage or layer the self-organising competitive nets would result in combinations of inputs becoming the effective stimuli for neurons. In order to avoid the combinatorial explosion, it is proposed that low-order combinations of inputs would be what is learned by each neuron. (Each stimulus would not be represented by activity in a single input axon, but instead by activity in a set of active input axons.) Evidence consistent with this suggestion that neurons are responding to combinations of a few variables represented at the preceding stage of cortical processing is that some neurons in V2 and V4 respond to end-stopped lines, to tongues flanked by inhibitory subregions, or to combinations of colours (see references cited by Rolls [26]); in posterior inferior temporal cortex to stimuli which may require two or more simple features to be present [9]; and in the temporal cortical face processing areas to images that require the presence of several features in a face (such as eyes, hair, and mouth) in order to respond (see [57,58]). (Precursor cells to face-responsive neurons might, it is suggested, respond to combinations of the outputs of the neurons in V1 that are activated by faces, and might be found in areas such as V4.) It is an important part of this suggestion that some local spatial information would be inherent in the features which are being combined. For example, cells might not respond to the combination of an edge and a small circle unless they were in the correct spatial relation to each other. (This is in fact consistent with the data of Tanaka et al. [9], and with the data on face neurons, in that some face neurons require the face features to be in the correct spatial configuration, and not jumbled – see Perrett et al. [57], Rolls, Tovee, Purcell et al. [48].) The local spatial information in the features being combined would ensure that the representation at the next level would contain some information about the (local) arrangement of features. Further, low-order combinations of such neurons at the next stage would include sufficient local spatial information so that an arbitrary spatial arrangement of the same features would not activate the same neuron, and this is the proposed, and limited, solution which this mechanism would provide for the feature binding problem (*cf.* [59]). By this stage of processing a view-dependent representation of objects suitable for view-dependent processes such as behavioural responses to face expression and gesture would be available.

It is suggested that view-independent representations could be formed by the same type of computation, operating to combine a limited set of views of objects. The plausibility of providing view-independent recognition of objects by combining a set of different views of objects has been proposed by a number of investigators [60]. Consistent with the suggestion that the view-independent representations are formed by combining view-dependent representations in the primate visual system, is the fact that in the temporal cortical visual areas, neurons with view-independent representations of faces are present in the same cortical areas as neurons with view-dependent representations (from which the view-independent neurons could receive inputs) [61,62]. This solution to "object-based" representations is very different from that traditionally proposed for artificial vision system, in which the coordinates in three-dimensional space of objects are stored in a database, and general-purpose algorithms operate on these to perform transforms such as translation, rotation, and scale change in three-dimensional space. In the present, much more limited but more biologically plausible scheme, the representation would be suitable for recognition of an object, and for linking associative memories to objects, but would be less effective for making actions in three-dimensional space to particular parts of, or inside, objects, as the three-dimensional coordinates of each part of the object would not be explicitly available. It is therefore proposed that visual fixation is used to locate in foveal vision part of an object to which movements must be made, and that local disparity and other measurements of depth then provide sufficient information for the motor system to make actions relative to the small part of space in which a local, view-dependent, representation of depth would be provided (*cf.* [63]).

The computational processes proposed above operate by an unsupervised learning mechanism, which utilises regularities in the physical environment to enable representations with low entropy to be built. In some cases it may be advantageous to utilise some form of mild teaching input to the visual system, to enable it to learn, for example, that rather similar visual inputs have very different consequences in the world, so that different representations of them should be built. In other cases, it might be helpful to bring representations together, if they have identical consequences, in order to use storage capacity efficiently. It is proposed elsewhere [32,33] that the backprojections from each adjacent cortical region in the hierarchy (and from the amygdala and hippocampus to higher regions of the visual system) play such a role by providing guidance to the competitive networks suggested above to be important in each cortical area. This guidance, and also the capability for recall, are it is suggested implemented by Hebb-modifiable connections from the backprojecting neurons to the principal (pyramidal) neurons of the competitive networks in the preceding stages [32,33].

The computational processes outlined above use coarse coding with relatively finely tuned neurons with a graded response region centred about an optimal response achieved when the input stimulus matches the synaptic weight vector on a neuron.

The coarse coding and fine tuning would help to limit the combinatorial explosion, to keep the number of neurons within the biological range. The graded response region would be crucial in enabling the system to generalise correctly to solve, for example, the invariances. However, such a system would need many neurons, each with considerable learning capacity, to solve visual perception in this way. This is fully consistent with the large number of neurons in the visual system, and with the large number of, probably modifiable, synapses on each neuron (in the order of 5,000). Further, the fact that many neurons are tuned in different ways to faces is consistent with the fact that in such a computational system, many neurons would need to be sensitive (in different ways) to faces, in order to allow recognition of many individual faces when all share a number of common properties.

A computational model of invariant visual object recognition

To test and clarify the hypotheses just described about how the visual system may operate to learn invariant object recognition, Rolls and colleagues have performed a simulation which implements many of the ideas just described, and is consistent and based on much of the neurophysiology summarised above. The network simulated can perform object, including face, recognition in a biologically plausible way, and after training shows, for example, translation and view invariance [2,3,64].

In the four-layer network, the successive layers correspond approximately to V2, V4, the posterior temporal cortex, and the anterior temporal cortex. The forward connections to a cell in one layer are derived from a topologically corresponding region of the preceding layer, using a Gaussian distribution of connection probabilities to determine the exact neurons in the preceding layer to which connections are made. This schema is constrained to preclude the repeated connection of any cells. Each cell receives 50 connections from the 32×32 cells of the preceding layer, with a 67% probability that a connection comes from within four cells of the distribution centre. Figure 3 shows the general convergent network architecture used, and may be compared with Fig. 2. Within each layer, lateral inhibition between neurons has a radius of effect just greater than the radius of feedforward convergence just defined. The lateral inhibition is simulated via a linear local contrast enhancing filter active on each neuron. (Note that this differs from the global “winner-take-all” paradigm implemented by Foldiak [56]). The cell activation is then passed through a nonlinear cell output activation function, which also produces contrast enhancement of the firing rates.

In order that the results of the simulation might be made particularly relevant to understanding processing in higher cortical visual areas, the inputs to layer 1 come from a separate input layer which provides an approximation to the encoding found in visual area 1 (V1) of the primate visual system. These response characteristics of neurons in the input layer are provided by a series of spatially tuned filters with image contrast sensitivities chosen to accord with the general tuning profiles observed in the simple cells of V1 [64].

The synaptic learning rule used can be summarized as follows:

$$\delta w_{ij} = k.\ m.\ r'_j$$

and

$$m_i^{t} = (1 - \eta)\ r_i^{(t)} + \eta m_i^{(t-1)}$$

where r'_j is the jth input to the neuron, r_i is the output of the ith neuron, w_{ij} is the jth weight on the ith neuron, η governs the relative influence of the trace and the new input (typically 0.4–0.6), and $m^{(t)}$ represents the value of the ith cell's memory trace at time t. In the simulation the neuronal learning was bounded by normalisation of each cell's dendritic weight vector. An alternative, more biologically relevant implementation, using a local weight-bounding operation, has in part been explored using a version of the Oja update rule [65,66]. To train the network to produce a translation invariant representation, one stimulus was placed successively in a sequence of seven positions across the input, then the next stimulus was placed successively in the same sequence of seven positions across the input, etc., through the set of stimuli. The idea was to enable the network to learn whatever was common at each stage of the network about a stimulus shown in different positions. To train on view invariance, different views of the same object were shown in succession, then different views of the next object were shown in succession, etc.

One test of the network used a set of three nonorthogonal stimuli, based upon probable three-dimensional edge cues (such as "T, L and +" shapes). During training these stimuli were chosen in random sequence to be swept across the "retina" of the network, a total of 1,000 times. In order to assess the characteristics of the cells

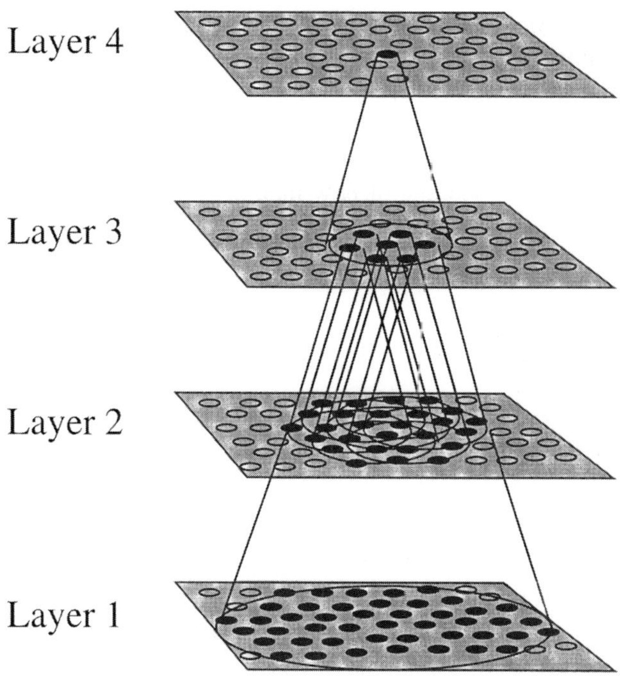

Fig. 3. Hierarchical network structure used in the model of invariant visual object recognition.

within the net, a two-way analysis of variance was performed on the set of responses of each cell, with one factor being the stimulus type and the other the position of the stimulus on the "retina". A high F ratio for stimulus type (F_s), and low F ratio for stimulus position (F_p) would imply that a cell had learned a position invariant representation of the stimuli. The discrimination factor of a particular cell was then simply the ratio F_s:F_p (a factor useful for ranking at least the most invariant cells). To assess the utility of the trace learning rule, nets trained with the trace rule were compared with nets trained without the trace rule and with untrained nets (with the initial random weights). The result of the simulations, illustrated in Fig. 4, show that networks trained with the trace learning rule do have neurons in layer 4 with much higher values of the discrimination factor. An example of the responses of one such cell in layer 4 are illustrated in Fig. 5. Similar position invariant encoding has been demonstrated for a stimulus set consisting of eight faces. View invariant coding has also been demonstrated for a set of five faces each shown in four views (see further [67]).

These results show that the proposed learning mechanism and neural architecture can produce cells with responses selective for stimulus type with considerable position or view invariance. The ability of the network to be trained with natural scenes is currently helping to advance our understanding of how representations of objects are built and encoded in the primate visual system.

This combined neurophysiological and computational approach is, thus, leading to biologically plausible theories about how the brain operates when it performs face or object recognition. In addition, we now have quite considerable evidence about what happens in our higher cortical visual areas when we recognise faces, and how information about at least some classes of object in the world is represented in the visual system. We now contrast processing in the visual system, in which a major transformation from a retinal representation to one in which an invariant representation of objects is provided, with the hippocampal memory system, which appears to be involved in the rapid storage of new memories using already highly processed cortical inputs.

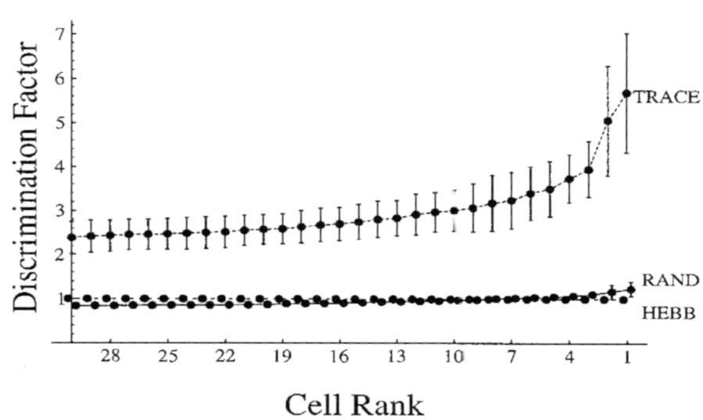

Fig. 4. Comparison of network discrimination when trained with the trace learning rule, with a Hebb rule (No trace), and when not trained (Random). The values of discrimination factor (see text) for the 50 most invariant cells in layer 4 are shown.

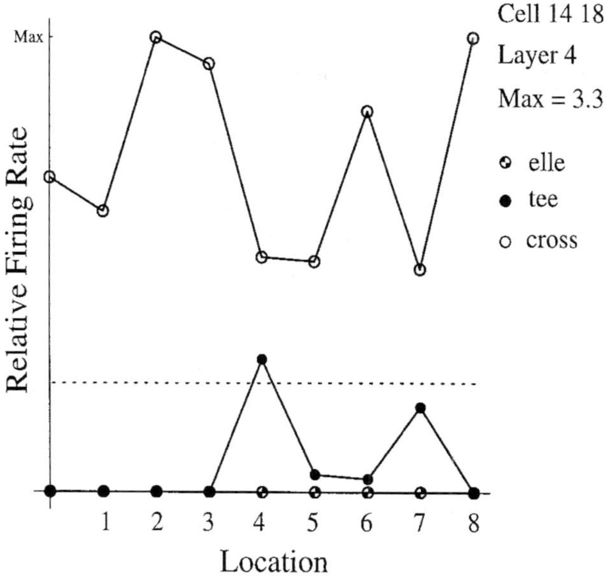

Fig. 5. The responses of a layer 4 cell in the simulation. The cell had a translation invariant response to stimulus 1.

The hippocampus and memory

What functions are performed by the hippocampus? Evidence from the effects of damage to the brain

In humans with bilateral damage to the hippocampus and nearby parts of the temporal lobe, amnesia is produced. The amnesia is anterograde, i.e., for events which happen after the brain damage. Although there is usually some retrograde amnesia, that is for events before the brain damage, this may not be severe, and indeed its severity appears to depend on how much damage there is to the cortical areas adjacent and connected to the hippocampus such as the entorhinal, perirhinal and parahippocampal cortex. These effects of damage to the hippocampus indicate that the very long-term storage of information is not in the hippocampus, at least in humans. On the other hand, the hippocampus does appear to be necessary for the storage of certain types of information (characterised by the description declarative, or knowing that, as contrasted with procedural, or knowing how, which is spared in amnesia). Declarative memory includes what can be declared or brought to mind as a proposition or an image. Declarative memory includes episodic memory, that is memory for particular episodes, and semantic memory, that is memory for facts [68–70].

Because of the clear importance of the hippocampus and nearby connected structures in memory, there have been many studies to analyse its role in memory.

In monkeys, damage to the hippocampus or to some of its connections such as the fornix produces deficits in learning about where objects are and where responses must be made (see [4,23,26,71]. For example, macaques and humans with damage to the hippocampus or fornix are impaired in object-place memory tasks in which not only the objects seen, but where they were seen, must be remembered [72–74]. Such object-place tasks require a whole-scene or snapshot-like memory [75]. Also, fornix lesions impair conditional left-right discrimination learning, in which the visual appearance of an object specifies whether a response is to be made to the left or the right [76]. A comparable deficit is found in humans [77]. Fornix sectioned monkeys are also impaired in learning on the basis of a spatial cue which object to choose (e.g., if two objects are on the left, choose object A, but if the two objects are on the right, choose object B) [78]. Further, monkeys with fornix damage are also impaired in using information about their place in an environment. For example, Gaffan and Harrison [79] found learning impairments when which of two or more objects the monkey had to choose depended on the position of the monkey in the room. Rats with hippocampal lesions are impaired in using environmental spatial cues to remember particular places (see [80,81]).

It appears that the deficits in "recognition" memory (e.g., for visual stimuli seen recently) produced by damage to this brain region are related to damage to the perirhinal cortex, which receives from high-order association cortex and has connections to the hippocampus (Fig. 6) [82,83].

Relation between spatial and nonspatial aspects of hippocampal function, episodic memory, and long-term semantic memory

One way of relating the impairment of spatial processing to other aspects of hippocampal function (including the memory of recent events or episodes in humans) is to note that this spatial processing involves a snapshot type of memory, in which one whole scene with its often unique set of parts or elements must be remembered. This memory may then be a special case of episodic memory, which involves an arbitrary association of a set of spatial and/or nonspatial events which describe a past episode. Indeed, the nonspatial tasks impaired by damage to the hippocampal system may be impaired because they are tasks in which a memory of a particular episode or context rather than of a general rule is involved [84]. Further, the deficit in paired associate learning in humans may be especially evident when this involves arbitrary associations between words, for example window – lake.

Rolls [3,32,33,85,86] has suggested that the reason why the hippocampus is used for the spatial and nonspatial types of memory described above, and the reason that makes these two types of memory so analogous, is that the hippocampus contains one stage, the CA3 stage, which acts as an autoassociation memory. It is suggested that an autoassociation memory implemented by the CA3 neurons equally enables whole (spatial) scenes or episodic memories to be formed, with a snapshot quality which depends on the arbitrary associations which can be made and the short-temporal window which characterises the synaptic modifiability in this system (see below and

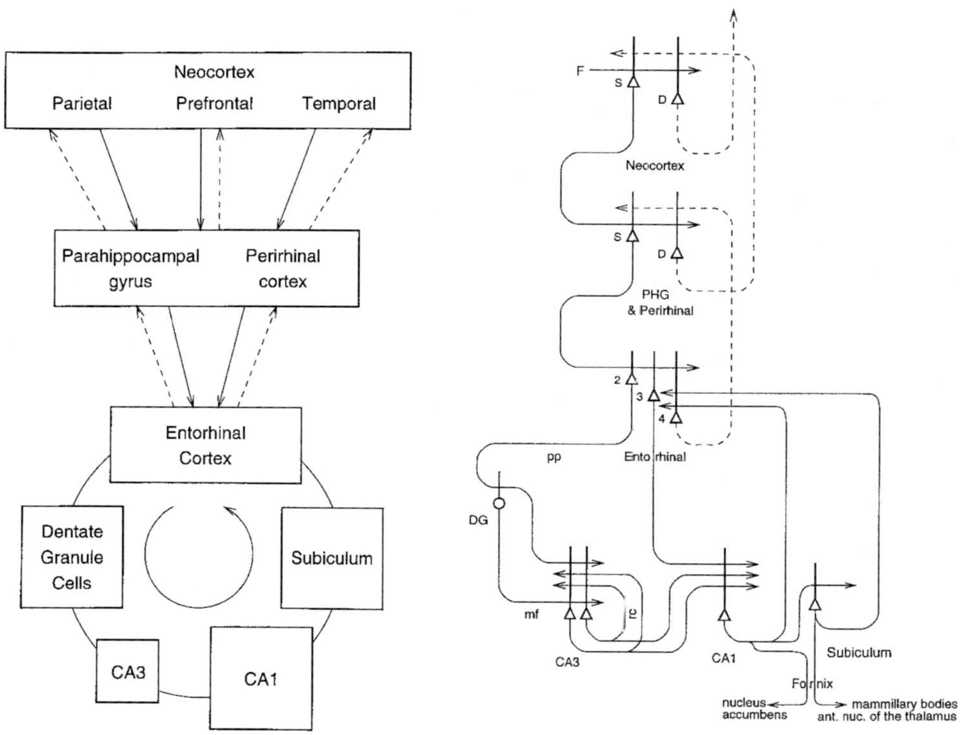

Fig. 6. Forward connections (solid lines) from areas of cerebral association neocortex via the parahippocampal gyrus and perirhinal cortex, and entorhinal cortex, to the hippocampus; and backprojections (dashed lines) via the hippocampal CA1 pyramidal cells, subiculum, and parahippocampal gyrus to the neocortex. There is great convergence in the forward connections down to the single network implemented in the CA3 pyramidal cells; and great divergence again in the backprojections. Left: block diagram. Right: more detailed representation of some of the principal excitatory neurons in the pathways. Abbreviations: D = deep pyramidal cells; DG = dentate granule cells; F = forward inputs to areas of the association cortex from preceding cortical areas in the hierarchy; mf = mossy fibres; PHG = parahippocampal gyrus and perirhinal cortex; pp = perforant path; rc = recurrent collateral of the CA3 hippocampal pyramidal cells; S = superficial pyramidal cells; 2 = pyramidal cells in layer 2 of the entorhinal cortex; 3 = pyramidal cells in layer 3 of the entorhinal cortex. The thick lines above the cell bodies represent the dendrites.

Rolls [32,33,85]). The hypothesis is that the autoassociation memory enables arbitrary sets of concurrent activity, involving for example the spatial context where an episode occurred, the people present during the episode, and what was seen during the episode, to be associated together and stored as one event. Later recall of that episode from the hippocampus in response to a partial cue can then lead to reinstatement of the activity in the neocortex that was originally present during the episode. The readout of information from the hippocampus could be useful in a number of ways. One is in organising action based on the stored information, as described later.

Another is in providing information useful to build long-term memories, as outlined next.

The information about episodic events recalled from the hippocampus could be used to help form semantic memories [23,32,33,87]. For example, remembering many particular journeys could help to build a geographic cognitive map in the neocortex. The important constraint here is to add information to an existing (neocortical, semantic) network without overwriting or disrupting its existing contents. This is probably best performed by making a succession of small adjustments to the existing semantic network [88]. In any case, the process is very different from that involved in forming an episodic memory "on the fly" (in real time), in which the whole collection of subevents present in the episode must be stored together strongly so that what happened on a single occasion can later be recalled correctly, and must be kept separate from other episodes, even if the episodes are somewhat similar. The hippocampus and neocortex would, thus, be complementary memory systems, with the hippocampus being used for rapid, real-time, unstructured storage of information involving activity potentially arriving from many areas of the neocortex; while the neocortex would gradually build and adjust on the basis of much accumulating information the semantic representation [33,87,88]. The suggested role of the hippocampus in this process is that it acts as a temporary buffer store, the contents of which can be recalled, and used to help long-term storage elsewhere, in the neocortex, if the information is useful. The idea that the hippocampus may act as a fast but temporary buffer store to complement long-term memory storage in the neocortex is an idea which is consistent with the fact that the hippocampus is necessary for the new learning but not (at least in humans) for the long-term storage of memories, and has been developed by many authors (e.g., [23,32–34,52,88–90]).

This raises the issue of the gradient of retrograde amnesia following hippocampal/temporal lobe damage. Memories stored some time before hippocampal damage may be less impaired than more recent memories [68]. This suggests that recently stored information may be retrieved from the hippocampus if it is needed, allowing the possibility of incorporating the retrieved information into neocortical memory stores. Whether or not there is a strong gradient of retrograde amnesia, there is in any case the necessity of a mechanism to retrieve information stored in the hippocampus, and to use this retrieved information to affect neural circuits elsewhere. (If this were not the case, information stored in the hippocampus could never be used for anything). A theory described below shows how information could be retrieved within the hippocampus, and how this retrieved information could enable the activity in neocortical areas that was present during the original storage of the episodic event to be reinstated, thus implementing recall. The backprojections from the hippocampus to the neocortex are one of the two major outputs of the hippocampus (Fig. 6). The backprojections are most likely to be involved in what is described by humans as recall. As a result of such neocortical recall, information may be incorporated into long-term, neocortical stores, but also action may be initiated. Other major sets of outputs from the hippocampus project via the fimbria/fornix system to the anterior nucleus of the thalamus (both directly and via the mammillary bodies), which in turn projects to the cingulate cortex; and via the subiculum to the basal ganglia. These may provide an output for very action-directed use of information stored in the hippocampus, e.g., in the initiation of conditional spatial responses in a visual conditional spatial response task [76,91]. In such a task, a rapid mapping must be learned between a visual stimulus and a spatial response, and a new mapping must be learned each day. The hippocampus is involved in this rapid visual to spatial response mapping [76], and the way in which hippocampal circuitry may be appropriate for this is that the CA3 region enables signals originating from very different parts of the cerebral cortex to be associated rapidly together (see below).

Systems-level connections of the hippocampus

Inputs converge into the hippocampus, via the parahippocampal gyrus and perirhinal cortex, which in turn feed into the entorhinal cortex and, thus, into the hippocampus, from virtually all association areas in the neocortex, including areas in the parietal cortex concerned with spatial function, temporal lobe areas concerned with vision and hearing, and the frontal lobes (Fig. 6A). An extensively divergent system of output projections enables the hippocampus to feed back into most of the cortical areas from which it receives inputs.

Given that some topographic segregation is maintained in the afferents to the hippocampus through the perirhinal and parahippocampal cortices [92–94], it may be that these areas are able to subserve memory within one of these topographically separated areas, whereas the final convergence afforded by the hippocampus into a single network in CA3 (Fig. 6 and below) is especially for an event or episodic memory typically involving arbitrary associations between any of the inputs to the hippocampus, e.g., spatial, visual object, and auditory (see below).

Information represented in the hippocampus

In order to understand the role of the hippocampus in memory, it is necessary to know what information is represented in the hippocampus. Whole series of findings have shown that hippocampal pyramidal cells (e.g., CA3 and CA1 neurons) in rats respond when the rat is in a particular place in a spatial environment. In monkeys, it has been shown that there is a rich representation of space outside the monkey, in that (in the order of 10% of) hippocampal neurons in monkeys respond to particular positions in space [23,26,28,52,86,95]. In many cases, this spatial encoding has been shown to be allocentric, that is coded in world coordinates, not in egocentric coordinates relative to the body axis [96]. Moreover, 2.4% of hippocampal neurons combine information about the object shown and the place where it is shown, in that they respond in an object-place memory task differentially to the novel and familiar presentations of an object shown in a particular place [31]. Consistent with this, in a cue-controlled environment, some hippocampal neurons in primates respond to views of particular parts of the environment, irrespective of where the monkey is [95]. Further, it has recently been possible to show that during active locomotion, "spatial view" neurons respond to the location in space at which the monkey is looking (experiments of Rolls, Robertson and Georges-Francois [97]; see Rolls [4,71]). These findings together show that there is a representation of space outside the organism in primates. This may be important, in that primates can explore space by moving their eyes to look at different places in the environment, and can remember the location of what they see there. In contrast, the hippocampal spatial cells found so far in rats respond when the rat is in a particular place, perhaps related to the fact that rats normally need to visit places in order to explore and store information about them, consistent with their use of olfactory cues, and their relatively poor vision compared to primates.

Another type of spatial information that has been found in primates is information about whole-body motion [98]. Such cells, e.g., respond to clockwise but not counter-clockwise whole-body motion, or to forwards but not backwards linear translation. In some cases, these neurons appear to be driven by vestibular information (in that they respond when the monkey is moved on a robot without performing the movement in the dark), while others are driven by the optic flow that is induced by whole-body motion [98]. These cells may be important in short-range navigation, for which a memory of recent movements is needed. In addition, in monkeys, when conditional spatial response tasks are being performed, some hippocampal neurons respond to combinations of the visual image and the spatial response which must be linked by memory for the task to be performed correctly, and during the learning of such visual to spatial response associations, the responses of some hippocampal neurons become modified [91,99].

These findings about neuronal activity in the hippocampus, and the effects of damage to the hippocampus, suggest that the importance of the hippocampus in spatial and other memories is that it can rapidly form "episodic" representations of information originating from many areas of the cerebral cortex. For this learning, synaptic modification is needed, and is present in the hippocampus [100–104]. In order to understand how the hippocampus could store the information it receives from the cerebral neocortex, and form, e.g., episodic memories which link spatial information with information about objects which originate from different regions of the cerebral cortex, we turn now to consider the internal circuitry of the hippocampus, and how it could operate (see Rolls [26,28,32,33,71,86], Treves and Rolls [87], *cf.* Marr [89]).

Hippocampal CA3 circuitry (Fig. 6)

Projections from the entorhinal cortex reach the granule cells (of which there are 10^6 in the rat) in the dentate gyrus (DG) via the perforant path (pp). The granule cells project to CA3 cells via the mossy fibres (mf), which provide a sparse but possibly powerful connection to the $3 \cdot 10^5$ CA3 pyramidal cells in the rat. Each CA3 cell receives approximately 50 mossy fibre inputs, so that the sparseness of this connectivity is thus 0.005%. By contrast, there are many more – possibly weaker – direct perforant path inputs onto each CA3 cell, in the rat of the order of $4 \cdot 10^3$.

The largest number of synapses (about $1.2 \cdot 10^4$ in the rat) on the dendrites of CA3 pyramidal cells is, however, provided by the (recurrent) axon collaterals of CA3 cells themselves (rc). It is remarkable that the recurrent collaterals are distributed to other CA3 cells throughout the hippocampus [92,105,106], so that effectively the CA3 system provides a single network, with a connectivity of approximately 4% between the different CA3 neurons.

CA3 as an autoassociation memory

Many of the synapses in the hippocampus show associative modification as shown by long-term potentiation, and this synaptic modification appears to be involved in learning [100]. On the basis of the evidence summarised above, Rolls [26,32–34,85] has suggested that the CA3 stage acts as an autoassociation memory which enables episodic memories to be formed and stored for an intermediate term in the CA3 network, and that subsequently, the extensive recurrent collateral connectivity allows for the retrieval of a whole representation to be initiated by the activation of some small part of the same representation (the cue).

The crucial synaptic modification for this autoassociation memory is in the recurrent collateral synapses (rc in Fig. 6). The idea is that the recurrent collateral connections allow, via associative synaptic modification, the different neurons firing to represent the different parts of an episodic memory to develop during learning of the episode strong connections between each other. Then later, when only a part of the original memory is presented, the whole memory can be recalled because of the strengthened connections between the neurons that were coactive during the original formation of the episodic memory. For example, three different sets of neurons might be active during an episode being learned, e.g., where one is for lunch, with whom, and what was being served for lunch. The synapses between these neurons would become strengthened during the original learning. Later, if we see the person again, we may be able to recall from the strengthened connections where we had lunch with them, and what was for lunch. The operation of these autoassociative memories (also called attractor networks, because during recall they are attracted towards a stable recall state), is described in more detail by Hertz, Krogh and Palmer [107] and Rolls and Treves [35].

An important part of the theory of how the hippocampus operates is that because the CA3 operates effectively as a single network, it can allow arbitrary associations between inputs originating from very different parts of the cerebral cortex, to be formed. These might involve associations between information originating in the temporal visual cortex about the presence of an object, and information originating in the parietal cortex about where it is.

To define better and quantitatively the operation of such a memory system, Treves and Rolls [108,109] performed quantitative analyses of the storage and retrieval processes in the CA3 network. They extended previous formal models of autoassociative memory (see Amit [110]) by analysing a network with graded response units, so as to represent more realistically the continuously variable rates at which neurons fire, and with incomplete connectivity [108,111]. They found that in general the maximum number p_{max} of firing patterns that can be (individually) retrieved is proportional to the number C^{RC} of (associatively) modifiable RC synapses per cell, by a factor that increases roughly with the inverse of the sparseness a of the neuronal representation. (The sparseness of the representation provided by a neuron might be measured by the proportion of stimuli in a set to which it responds.) Approximately,

$$p_{\max} \simeq \frac{C^{RC}}{a \ln(1/a)} k$$

where k is a factor that depends weakly on the detailed structure of the rate distribution, on the connectivity pattern, etc., but is roughly in the order of 0.2–0.3 [108].

The main factors that determine the maximum number of memories that can be stored in an autoassociative network are, thus, the number of connections on each neuron devoted to the recurrent collaterals, and the sparseness of the representation. For example, for C^{RC} = 12,000 and a = 0.02 (realistic estimates for the rat), p_{max} is calculated to be approximately 36,000.

The operation of the other parts of the hippocampal circuitry are considered in detail elsewhere. The mossy inputs to CA3 are thought, e.g., to be important during learning, by forcing a new pattern of firing onto the CA3 neurons to represent the new episodic memory [32,33,71,109]. The dentate granule cells are thought to set up the sparse representation needed for the mossy fibres to operate correctly. The direct perforant path system to CA3 (Fig. 6) is thought to be necessary to trigger recall. It is suggested that the CA1 cells help to form the efficient representation needed as the first stage of the multistage backprojection pathways to the neocortex involved in recall (see further Rolls [71]).

Backprojections to the neocortex

It is suggested that the hippocampus is able to recall the whole of a previously stored episode for a period of days, weeks or months after the episode, when even a fragment of the episode is available to start the recall. This recall from a fragment of the original episode would take place particularly as a result of completion produced by the autoassociation implemented in the CA3 network. It would then be the role of the hippocampus to reinstate in the cerebral neocortex the whole of the episodic memory. The cerebral cortex would then, with the whole of the information in the episode now producing firing in the correct sets of neocortical neurons, be in a position to incorporate the information in the episode into its long-term store in the neocortex, or use the recalled information for planning or action (see Rolls [71]).

It has been suggested that during recall, the connections from CA3 via CA1 and the subiculum would allow activation of at least the pyramidal cells in the deep layers of the entorhinal cortex (Fig. 6). These neurons would then, by virtue of their backprojections to the parts of cerebral cortex that originally provided the inputs to the hippocampus, terminate in the superficial layers of those neocortical areas, where synapses would be made onto the distal parts of the dendrites of the cortical pyramidal cells (see Rolls [32–34]).

The architecture with which this would be achieved is shown in Fig. 6. The feedforward connections from association areas of the cerebral neocortex (solid lines in Fig. 6), show major convergence as information is passed to CA3, with the CA3 autoassociation network having the smallest number of neurons at any stage of the processing. The backprojections allow for divergence back to neocortical areas. The way in which Rolls [32,33] has suggested that the backprojection synapses are set up to have the appropriate strengths for recall is as follows (see also Treves and Rolls [87]). During the setting up of a new episodic memory, there would be strong feedforward activity progressing towards the hippocampus. During the episode, the CA3 synapses would be modified, and via the CA1 neurons and the subiculum, a pattern of activity would be produced on the backprojecting synapses to the entorhinal cortex. Here, the backprojecting synapses from active backprojection axons onto pyramidal cells being activated by the forward inputs to entorhinal cortex would be associatively modified. A similar process would be implemented at preceding stages of neocortex, that is in the parahippocampal gyrus/perirhinal cortex stage, and in association cortical areas.

Quantitative constraints on the connectivity of backprojections

How many backprojecting fibres does one need to synapse on any given neocortical pyramidal cell, in order to implement the mechanism outlined above? Consider a polysynaptic sequence of backprojecting stages, from hippocampus to neocortex, as a string of simple (hetero-)associative memories (see Rolls and Treves [35]) in which, at each stage, the input lines are those coming from the previous stage (closer to the hippocampus). Implicit in this framework is the assumption that the synapses at each stage are modifiable and have been modified at the time of first experiencing each episode, according to some Hebbian associative plasticity rule. A plausible requirement for a successful hippocampo-directed recall operation, is that the signal generated from the hippocampally retrieved pattern of activity, and carried backwards towards neocortex, remain undegraded when compared to the noise due, at each stage, to the interference effects caused by the concurrent storage of other patterns of activity on the same backprojecting synaptic systems. That requirement is equivalent to that used in deriving the storage capacity of such a series of hetero-associative memories, and it was shown in Treves and Rolls [108] that the maximum number of independently generated activity patterns that can be retrieved is given, essentially, by the same formula as above:

$$p \simeq \frac{C}{a \ln(1/a)} k'$$

where, however, a is now the sparseness of the representation at any given stage, and C is the average number of (back-)projections each cell of that stage receives from cells of the previous one. (k' is a similar slowly varying factor to that introduced above.) If p is equal to the number of memories held in the hippocampal buffer, it is limited by the retrieval capacity of the CA3 network, p_{max}. Putting together the formula for the latter with that shown here, one concludes that, roughly, the requirement implies that the number of afferents of (indirect) hippocampal origin to a given neocortical stage(C^{HBP}), must be $C^{HBP} = C^{RC} a_{\mathrm{nc}} / a_{\mathrm{CA3}}$, where C^{RC} is the number of recurrent collaterals to any given cell in CA3, the average sparseness of a representation is a_{nc}, and a_{CA3} is the sparseness of memory representations there in CA3 [87].

One is led to a definite conclusion: a mechanism of the type envisaged here could not possibly rely on a set of monosynaptic CA3-to-neocortex backprojections. This would imply that, to make a sufficient number of synapses on each of the vast number of neocortical cells, each cell in CA3 has to generate a disproportionate number of synapses (i.e., C^{HBP} × the ratio between the number of neocortical cells and the number of CA3 cells). The required divergence can be kept within reasonable limits only by assuming that the backprojecting system is polysynaptic, provided that the number of cells involved grows gradually at each stage, from CA3 back to neocortical association areas (*cf.* Fig. 6).

This theory of the recall of recent memories in the neocortex from the hippocampus provides a clear view about why backprojections should be as numerous as forward projections in the cerebral cortex. The reason suggested for this is that as many representations may need to be accessed by backprojections for recall and related functions (see Rolls [32–34]) in the population of cortical pyramidal cells as can be accessed by the forward projections, and this limit is given by the number of inputs onto a pyramidal cell (and the sparseness of the representation), irrespective of whether the input is from a forward or a backprojection system.

Conclusions

In the steps being taken to understand memory systems in the brain are described. The work on the hippocampus illustrates how information may be stored, and later recalled. The research on the inferior temporal visual cortex illustrates that difficult problems are solved by some other memory systems in the brain, for in this case the input representation at the retina must be transformed into a quite different representation at the object level in order for objects and faces to be recognised independently of where they are on the retina, their size, and their viewing angle.

Acknowledgements

The author has worked on some of the experiments described here with G.C. Baylis, M.J. Burton, M.E. Hasselmo, C.M. Leonard, F. Mora, D.I. Perrett, M.K. Sanghera, T.R. Scott, S.J. Thorpe, and F.A.W. Wilson, and their collaboration is sincerely acknowledged. Some of the research described was supported by the Medical Research Council (PG8513579), and by The Human Frontier Science Program.

References

1. Rolls ET, Booth MCA, Treves A. View-invariant representations of objects in the inferior temporal visual cortex. Soc Neurosci Abstr 1996;22:(In press).
2. Rolls ET. Brain mechanisms for invariant visual recognition and learning. Behav Process 1994;33:113–138.
3. Rolls ET. Learning mechanisms in the temporal lobe visual cortex. Behav Brain Res 1995a;66: 177–185.
4. Rolls ET. A neurophysiological and computational approach to the functions of the temporal lobe cortical visual areas in invariant object recognition. In: Harris L, Jenkin M (eds) Computational and Biological Mechanisms of Visual Coding. Cambridge: Cambridge University Press, 1996b;(In press).
5. Seltzer B, Pandya DN. Afferent cortical connections and architectonics of the superior temporal sulcus and surrounding cortex in the rhesus monkey. Brain Res 1978;149:1–24.
6. Maunsell JHR, Newsome WT. Visual processing in monkey extra-striate cortex. Ann Rev Neurosci 1987;10:363–401.
7. Baizer JS, Ungerleider LG, Desimone R. Organization of visual inputs to the inferior temporal and posterior parietal cortex in macaques. J Neurosci 1991;11:168–190.
8. Baylis GC, Rolls ET, Leonard CM. Functional subdivisions of temporal lobe neocortex. J Neurosci 1987;7:330–342.
9. Tanaka K, Saito C, Fukada Y, Moriya M. Integration of form, texture, and color information in the inferotemporal cortex of the macaque. In: Iwai E, Mishkin M (eds) Vision, Memory and the Temporal Lobe. New York: Elsevier, 1990;101–109.
10. Baylis GC, Rolls ET, Leonard CM. Selectivity between faces in the responses of a population of neurons in the cortex in the superior temporal sulcus of the monkey. Brain Res 1985;342:91–102.
11. Rolls ET. Neurophysiological mechanisms underlying face processing within and beyond the temporal cortical visual areas. Phil Trans R Soc 1992b;335:11–21.
12. Barlow HB. Single units and sensation: a neuron doctrine for perceptual psychology? Perception 1972;1:371–394.
13. Rolls ET, Tovee MJ. Sparseness of the neuronal representation of stimuli in the primate temporal visual cortex. J Neuro Physiol 1995;73:713–726.
14. Rolls ET, Treves A, Tovee MJ. The representational capacity of the distributed encoding of information provided by populations of neurons in the primate temporal visual cortex. 1996;(In press).
15. Abbott LA, Rolls ET, Tovee MJ. Representational capacity of face coding in monkeys. Cereb Cortex 1996;6:(In press).
16. Rolls ET, Tovee M, Treves A. Information in the neuronal representation of individual stimuli in the primate temporal visual cortex. 1996.
17. Rolls ET, Baylis GC. Size and contrast have only small effects on the responses to faces of neurons in the cortex of the superior temporal sulcus of the monkey. Exp Brain Res 1986;65:38–48.
18. Rolls ET, Baylis GC, Leonard CM. Role of low and high spatial frequencies in the face-selective responses of neurons in the cortex in the superior temporal sulcus. Vis Res 1985;25:1021–1035.
19. Rolls ET, Baylis GC, Hasselmo ME. The responses of neurons in the cortex in the superior temporal sulcus of the monkey to band-pass spatial frequency filtered faces. Vis Res 1987;27: 311–326.
20. Tovee MJ, Rolls ET, Azzopardi P. Translation invariance and the responses of neurons in the temporal visual cortical areas of primates. J Neuro Physiol 1994;72:1049–1060.
21. Hasselmo ME, Rolls ET, Baylis GC, Nalwa V. Object-centered encoding by face-selective neurons in the cortex in the superior temporal sulcus of the monkey. Exp Brain Res 1989;75:417–429.
22. Rolls ET. Neurons in the cortex of the temporal lobe and in the amygdala of the monkey with responses selective for faces. Hum Neurobiol 1984;3:209–222.
23. Rolls ET. A theory of emotion, and its application to understanding the neural basis of emotion.

Cognition and Emotion 1990b;4:161–190.

24. Rolls ET. Neurophysiology and functions of the primate amygdala. In: Aggleton JP (ed) The Amygdala. New York: Wiley–Liss, 1992a;143–165.
25. Perrett DI, Smith PAJ, Mistlin AJ, Chitty AJ, Head AS, Potter DD, Broennimann R, Milner AD, Jeeves MA. Visual analysis of body movements by neurons in the temporal cortex of the macaque monkey: a preliminary report. Behav Brain Res 1985;16:153–170.
26. Rolls ET. Neural organisation of higher visual functions. Curr Opin Neurobiol 1991;1:274–278.
27. Rolls ET. A theory of emotion and consciousness, and its application to understanding the neural basis of emotion. In: Gazzaniga MS (ed) The Cognitive Neurosciences. Cambridge, MA: MIT Press, 1995b;1091–1106.
28. Rolls ET. The orbitofrontal cortex. Phil Trans R Soc B 1996d;351:(In press).
29. Leonard CM, Rolls ET, Wilson FAW, Baylis GC. Neurons in the amygdala of the monkey with responses selective for faces. Behav Brain Res 1985;15:159–176.
30. Hornak J, Rolls ET, Wade D. Face and voice expression identification and their association with emotional and behavioural changes in patients with frontal lobe damage. Neuropsychologia 1996;34:247–261.
31. Rolls ET, Baylis GC, Hasselmo ME, Nalwa V. The effect of learning on the face-selective responses of neurons in the cortex in the superior temporal sulcus of the monkey. Exp Brain Res 1989;76:153–164.
32. Rolls ET. Functions of neuronal networks in the hippocampus and neocortex in memory. In: Byrne JH, Berry WO (eds) Neural Models of Plasticity: Experimental and Theoretical Approaches. San Diego: Academic Press, 1989a;240–265.
33. Rolls ET. The representation and storage of information in neuronal networks in the primate cerebral cortex and hippocampus. In: Durbin R, Miall C, Mitchison G (eds) The Computing Neuron. Wokingham, England: Addison-Wesley, 1989b;125–159.
34. Rolls ET. Functions of neuronal networks in the hippocampus and cerebral cortex in memory. In: Cotterill RMJ (ed) Models of Brain Function. Cambridge: Cambridge University Press, 1989c;15–33.
35. Rolls ET, Treves A. Neural networks and the brain. 1997.
36. Rolls ET, Tovee MJ, Ramachandran VS. Visual learning reflected in the responses of neurons in the temporal visual cortex of the macaque. Soc Neurosci Abstr 1993;19:27.
37. Baylis GC, Rolls ET. Responses of neurons in the inferior temporal cortex in short term and serial recognition memory tasks. Exp Brain Res 1987;65:614–622.
38. Miller EK, Desimone R. Parallel neuronal mechanisms for short-term memory. Science 1994;263: 520–522.
39. Miyashita Y. Neuronal correlates of visual associative long-term memory in the primate temporal cortex. Nature 1988;335:817–820.
40. Miyashita Y. Neuronal origins of visual imagery. In: Ito M, Miyashita Y, Rolls ET (eds) Cognition, Computation, and Consciousness. Oxford: Oxford University Press, 1996.
41. Thorpe SJ, Imbert M. Biological constraints on connectionist models. In: Pfeifer R, Schreter Z, Fogelman-Soulie F (eds) Connectionism in Perspective. Amsterdam: Elsevier, 1989;63–92.
42. Tovee MJ, Rolls ET, Treves A, Bellis RP. Information encoding and the responses of single neurons in the primate temporal visual cortex. J Neuro Physiol 1993;70:640–654.
43. Tovee MJ, Rolls ET. Information encoding in short firing rate epochs by single neurons in the primate temporal visual cortex. Visual Cognition 1995;2:35–58.
44. Turvey MT. On the peripheral and central processes in vision: inferences from an information processing analysis of masking with patterned stimuli. Psychol Rev 1973;80:1–52.
45. Breitmeyer BG. Unmasking visual masking: a look at the "why" behind the veil of the "how". Psychol Rev 1980;87:52–69.
46. Humphreys GW, Bruce V. Visual cognition. Hove: Erlbaum, 1989.
47. Rolls ET, Tovee MJ. Processing speed in the cerebral cortex, and the neurophysiology of visual masking. Proc R Soc B 1994;257:9–15.

48. Rolls ET, Tovee MJ, Purcell DG, Stewart AL, Azzopardi P. The responses of neurons in the temporal cortex of primates, and face identification and detection. Exp Brain Res 1994;101: 474–484.
49. Treves A. Mean-field analysis of neuronal spike dynamics. Network 1993;4:259–284.
50. Simmen MW, Rolls ET, Treves A. On the dynamics of a network of spiking neurons. In Eekman FH, Bower JM (eds) Computations and Neuronal Systems: Proceedings of CNS95. Boston: Kluwer, 1996;(In press).
51. Rolls ET. Brain mechanisms involved in perception and memory, and their relation to consciousness. In: Ito M, Miyashita Y, Rolls ET (eds) Cognition, Computation, and Consciousness. Oxford: Oxford University Press, 1996e;(In press).
52. Rolls ET. Functions of the primate hippocampus in spatial processing and memory. In: Olton DS, Kesner RP (eds) Neurobiology of Comparative Cognition. Hillsdale, NJ: Lawrence Erlbaum, 1990a;339–362.
53. Boussaoud D, Desimone R, Ungerleider LG. Visual topography of area TEO in the macaque. J Comput Neurol 1991;306:554–575.
54. Barlow HB. Cerebral cortex as model builder. In: Rose D, Dobson VG (eds) Models of the Visual Cortex. Chichester: Wiley, 1985;37–46.
55. Barlow HB, Kaushal TP, Mitchison GJ. Finding minimum entropy codes. Neural Computat 1989;1:412–423.
56. Foldiak P. Learning invariance from transformation sequences. Neural Computat 1991;3:193–199.
57. Perrett DI, Rolls ET, Caan W. Visual neurons responsive to faces in the monkey temporal cortex. Exp Brain Res 1982;47:329–342.
58. Yamane S, Kaji S, Kawano K. What facial features activate face neurons in the inferotemporal cortex of the monkey? Exp Brain Res 1988;73:209–214.
59. Malsburg C von der. A neural architecture for the representation of scenes. In: McGaugh JL, Weinberger NM, Lynch G (eds) Brain Organization and Memory: Cells, Systems and Circuits. New York: Oxford University Press, 1990;356–372.
60. Poggio T, Edelman S. A network that learns to recognize three-dimensional objects. Nature 1990;343:263–266.
61. Hasselmo ME, Rolls ET, Baylis GC. The role of expression and identity in the face-selective responses of neurons in the temporal visual cortex of the monkey. Behav Brain Res 1989;32:203–218.
62. Perrett DI, Mistlin AJ, Chitty AJ. Visual neurons responsive to faces. Trends Neurosci 1987;10: 358–364.
63. Ballard DH. Animate vision uses object-centred reference frames. In: Eckmiller, R (ed) Advanced Neural Computers. Amsterdam: North-Holland, 1990;229–236.
64. Wallis G, Rolls ET, Foldiak P. Learning invariant responses to the natural transformations of objects. Int Joint Conf on Neural Networks 1993;2:1087–1090.
65. Oja E. A simplified neuron model as a principal component analyser. J Math Biol 1982;15: 267–273.
66. Kohonen T. Self-organization and associative memory, 2nd edn. New York: Springer-Verlag, 1988.
67. Wallis G, Rolls ET. A model of invariant object recognition in the visual system. J Comput Neurosci 1996.
68. Squire LR. Memory and the hippocampus: A synthesis from findings with rats, monkeys and humans. Psychol Rev 1992;99:195–231.
69. Squire LR, Knowlton BJ. Memory, hippocampus, and brain systems. In: Gazzaniga M (ed) The Cognitive Neurosciences. Cambridge, MA: MIT Press, 1994;825–837.
70. Squire LR, Shimamura AP, Amaral DG. Memory and the hippocampus. In: Byrne J, Berry WO (eds) Neural Models of Plasticity: Theoretical and Empirical Approaches. New York: Academic Press, 1989;208–239.
71. Rolls ET. A theory of hippocampal function in memory. Hippo Campus 1996a;(In press).
72. Gaffan D, Saunders RC. Running recognition of configural stimuli by fornix transected monkeys.

Q J Exp Psychol 1985;37B:61–71.
73. Parkinson JK, Murray EA, Mishkin M. A selective mnemonic role for the hippocampus in monkeys: memory for the location of objects. J Neurosci 1988;8:4059–4167.
74. Smith ML, Milner B. The role of the right hippocampus in the recall of spatial location. Neuropsychologia 1981;19:781–793.
75. Gaffan D. Scene-specific memory for objects: a model of episodic memory impairment in monkeys with fornix transection. J Cogn Neurosci 1994;6:305–320.
76. Rupniak NMJ, Gaffan D. Monkey hippocampus and learning about spatially directed movements. J Neurosci 1987;7:2331–2337.
77. Petrides M. Deficits on conditional associative-learning tasks after frontal- and temporal-lobe lesions in man. Neuropsychologia 1985;23:601–614.
78. Gaffan D, Harrison S. A comparison of the effects of fornix section and sulcus principalis ablation upon spatial learning by monkeys. Behav Brain Res 1989a;31:207–220.
79. Gaffan D, Harrison S. Place memory and scene memory: effects of fornix transection in the monkey. Exp Brain Res 1989b;74:202–212.
80. O'Keefe J, Nadel L. The Hippocampus as a Cognitive Map. Oxford: Clarendon Press, 1978.
81. Jarrard EL. On the role of the hippocampus in learning and memory in the rat. Behav Neural Biol 1993;60:9–26.
82. Zola-Morgan S, Squire LR, Amaral DG, Suzuki WA. Lesions of perirhinal and parahippocampal cortex that spare the amygdala and hippocampal formation produce severe memory impairment. J Neurosci 1989;9:4355–4370.
83. Zola-Morgan S, Squire LR, Ramus SJ. Severity of memory impairment in monkeys as a function of locus and extent of damage within the medial temporal lobe memory system. Hippocampus 1994;4:483–494.
84. Gaffan D, Saunders RC, Gaffan EA, Harrison S, Shields C, Owen MJ. Effects of fornix transection upon associative memory in monkeys: role of the hippocampus in learned action. Q J Exp Psychol 1984;26B:173–221.
85. Rolls ET. Information representation, processing and storage in the brain: analysis at the single neuron level. In: Changeux J-P, Konishi M (eds) The Neural and Molecular Bases of Learning. Chichester: Wiley, 1987;503–540.
86. Rolls ET. The representation of space in the primate hippocampus, and episodic memory. In: Ono T, McNaughton BL, Molotchnikoff S, Rolls ET, Nishijo H (eds) Perception, Memory and Emotion: Frontier in Neuroscience. Amsterdam: Elsevier, 1996c;(In press).
87. Treves A, Rolls ET. A computational analysis of the role of the hippocampus in memory. Hippocampus 1994;4:374–391.
88. McClelland JL, McNaughton BL, O'Reilly RC. Why there are complementary learning systems in the hippocampus and neocortex: insights from the successes and failures of connectionist models of learning and memory. Psychol Rev 1995;102:419–457.
89. Marr D. Simple memory: a theory for archicortex. Phil Trans R Soc Lond B 1971;262:24–81.
90. Rolls ET. Theoretical and neurophysiological analysis of the functions of the primate hippocampus in memory. Cold Spring Harbor Symp Quant Biol 1990c;55:995–1006.
91. Miyashita Y, Rolls ET, Cahusac PMB, Niki H, Feigenbaum JD. Activity of hippocampal neurons in the monkey related to a conditional spatial response task. J Neuro Physiol 1989;61:669–678.
92. Amaral DG, Witter MP. The three-dimensional organization of the hippocampal formation: a review of anatomical data. Neurosci 1989;31:571–591.
93. Suzuki W, Amaral DG. Topographic organisation of the reciprocal connections between the monkey entorhinal cortex and the perirhinal and parahippocampal cortices. J Neurosci 1994a;14: 1856–1877.
94. Suzuki W, Amaral DG. Perirhinal and parahippocampal cortices of the macaque monkey: cortical afferents. J Comput Neurol 1994b;350:497–533.
95. Rolls ET, O'Mara S. Neurophysiological and theoretical analysis of how the hippocampus functions in memory. In: Ono T, Squire LR, Raichle ME, Perrett DI, Fukuda M (eds) Brain Mechanisms of

Perception and Memory: From Neuron to Behaviour. New York: Oxford University Press, 1993;276–300.

96. Feigenbaum JD, Rolls ET. Allocentric and egocentric spatial information processing in the hippocampal formation of the behaving primate. Psychobiol 1991;19:21–40.
97. Rolls ET, Robertson R, Georges-François P. The representation of space in the primate hippocampus. Soc Neurosci Abstr 1995;21:586.10
98. O'Mara SM, Rolls ET, Berthoz A, Kesner RP. Neurons responding to whole-body motion in the primate hippocampus. J Neurosci 1994;14:6511–6523.
99. Cahusac PMB, Rolls ET, Miyashita Y, Niki H. Modification of the responses of hippocampal neurons in the monkey during the learning of a conditional spatial response task. Hippocampus 1993;3:29–42.
100. Morris RGM. Does synaptic plasticity play a role in information storage in the vertebrate brain? In: Morris RGM (ed) Parallel Distributed Processing: Implications for Psychology and Neurobiology. Oxford: Oxford University Press, 1989;248–285.
101. Brown TH, Kairiss EW, Keenan CL. Hebbian synapses: biophysical mechanisms and algorithms. Ann Rev Neurosci 1990;13:475–511.
102. Miles R. Plasticity of recurrent excitatory synapses between CA3 hippocampal pyramidal cells. Soc Neurosci Abstr 1988;14:19.
103. Levy WB, Desmond NL. The rules of elemental synaptic plasticity. In: Levy WB, Anderson JA, Lehmkuhle S (eds) Synaptic Modification, Neuron Selectivity, and Nervous System Organization. Hillsdale, New Jersey: Erlbaum, 1985;105–121.
104. Levy WB, Colbert CM, Desmond NL. Elemental adaptive processes of neurons and synapses: a statistical/computational perspective. In: Gluck M, Rumelhart D (eds) Neuroscience and Connectionist Theory. Hillsdale, NJ: Erlbaum, 1990;187–235.
105. Ishizuka N, Weber J, Amaral DG. Organization of intrahippocampal projections originating from CA3 pyramidal cells in the rat. J Comput Neurol 1990;295:580–623.
106. Amaral DG, Ishizuka N, Claiborne B. Neurons, numbers and the hippocampal network. Prog Brain Res 1990;83:1–11.
107. Hertz J, Krogh A, Palmer RG. Introduction to the Theory of Neural Computation. Wokingham, UK: Addison-Wesley, 1991.
108. Treves M, Rolls ET. What determines the capacity of autoassociative memories in the brain? Network 1991;2:371–397.
109. Treves A, Rolls ET. Computational constraints suggest the need for two distinct input systems to the hippocampal CA3 network. Hippo Campus 1992;2:189–199.
110. Amit DJ. Modelling brain function. New York: Cambridge University Press, 1989.
111. Treves A. Graded-response neurons and information encodings in autoassociative memories. Physical Rev A 1990;42:2418–2430.

1996 Elsevier Science B.V.
Integrative and Molecular Approach to Brain Function.
M. Ito and Y. Miyashita, editors.

New concepts regarding the organization of basal ganglia and cerebellar output

Frank A. Middleton[2] and Peter L. Strick[1,2,3]

Departments of [1]Neurosurgery and [2]Physiology, and [3]Research Service, Veterans Administration Medical Center, State University of New York Health Science Center at Syracuse, New York, New York, USA

Abstract. *Background.* The "classical" view of basal ganglia and cerebellar loops with the cerebral cortex is that they gather information from widespread regions of the cerebral cortex, perform some sensorimotor transformation on it, and then use the result to direct motor output through thalamocortical projections to the primary motor cortex. Recent observations have led us to question this view. The neurons in the output nuclei of the basal ganglia and cerebellum project to a diverse set of thalamic nuclei. We have examined whether these projections provide the basal ganglia and cerebellum with access to widespread regions of the cerebral cortex.

Methods. We examined the organization of basal ganglia-thalamocortical and cerebello-thalamocortical pathways using retrograde transneuronal transport of herpes simplex virus type I. In separate animals, localized injections of virus were made into different regions of the frontal lobe (regions of primary motor cortex, premotor cortex, prefrontal cortex, and the frontal eye field) and into an area of inferotemporal cortex.

Results. We found that each of these cortical areas receives a projection, via the thalamus, from a distinct set of neurons in one or more of the output nuclei of the basal ganglia or cerebellum.

Conclusions. These observations indicate that basal ganglia and cerebellar loops direct their output to multiple cortical areas. Consequently, these circuits have the potential to influence not only the generation and control of movement, but also higher order aspects of cognitive behavior.

Key words: *Cebus apella*, cerebral cortex, dentate nucleus, globus pallidus, substantia nigra, virus transport.

Introduction

The cerebral cortex of mammals is massively interconnected with two large groups of subcortical structures, the basal ganglia and cerebellum. The basal ganglia and cerebellum both have an "input layer", which receives direct input from the cerebral cortex, and an "output layer," which projects back to the cerebral cortex via the thalamus. Thus, a major structural feature of basal ganglia and cerebellar anatomy is their participation in multiple "loops" with the cerebral cortex.

An important issue concerning the organization of these loops is the degree of anatomical convergence or segregation that takes place within them. The "classical" view of these loops is that they gathered information from widespread cortical areas. This information was then "funneled" through basal ganglia or cerebellar circuits

Address for correspondence: Dr Peter L. Strick, Research Service (151S), VA Medical Center, 800 Irving Ave., Syracuse, NY 13210, USA.

where it ultimately converged on output nuclei that innervate the ventrolateral thalamus [1—4]. The ventrolateral thalamus was thought to project to a single cortical area, the primary motor cortex (M1). Thus, basal ganglia and cerebellar loops were thought to function largely in the domain of motor control.

It is now clear, however, that basal ganglia and cerebellar projections are not limited to a single region of the ventrolateral thalamus [5—15]. Some of these thalamic regions have been shown to innervate cortical areas outside the primary motor cortex (e.g., [14—23]). As a result, it is possible that basal ganglia and cerebellar output influences more widespread regions of the cerebral cortex than previously recognized.

In the past, attempts to determine the cortical targets of basal ganglia and cerebellar output were faced with a number of technical and conceptual limitations, including difficulties with thalamic nomenclature and boundaries (see [14,15]). To overcome many of these problems, we used transneuronal transport of herpes simplex virus type I (HSV1) to map basal ganglia- and cerebello-thalamocortical pathways in primates [24—30]. In this chapter, we will present some of our findings on the organization of these pathways. Our results indicate that basal ganglia and cerebellar outputs are not limited to the primary motor cortex. Instead, these subcortical nuclei innervate multiple regions of premotor and prefrontal cortex. We also have found evidence for basal ganglia projections to a region of inferotemporal cortex. Importantly, the projections to a given cortical area appear to originate from neurons that are clustered in distinct portions of each output nucleus. We have proposed that these clusters form distinct "output channels" [27,30—33]. In a final section of this chapter, we will present some of our physiological observations that indicate individual output channels are concerned with different aspects of motor and/or cognitive behavior [26,34,35].

Anatomical studies

We have found that distinct strains of HSV1 are transported transneuronally in different directions within the primate motor system [24]. The H129 strain is transported transneuronally in the anterograde direction, whereas the McIntyre-B strain of HSV1 is transported transneuronally in the retrograde direction [24]. With the careful adjustment of survival time, it is possible to identify circuits at least three neurons in length [36]. We have worked extensively with the McIntyre-B strain to map basal ganglia and cerebellar projections, via the thalamus, to different cortical areas (Fig. 1) [24—32].

Primary motor cortex

Our first experiments examined inputs to M1 [24,26,27]. Three days after injections of McIntyre-B into the arm area of M1 (Fig. 1), densely labeled neurons were found in subdivisions of the ventrolateral thalamus that are known to innervate M1, such as ventralis posterior lateralis pars oralis (VPLo) and ventralis lateralis pars oralis (VLo) (e.g., [37,38]). Following survival times of 5 days, virus was transported

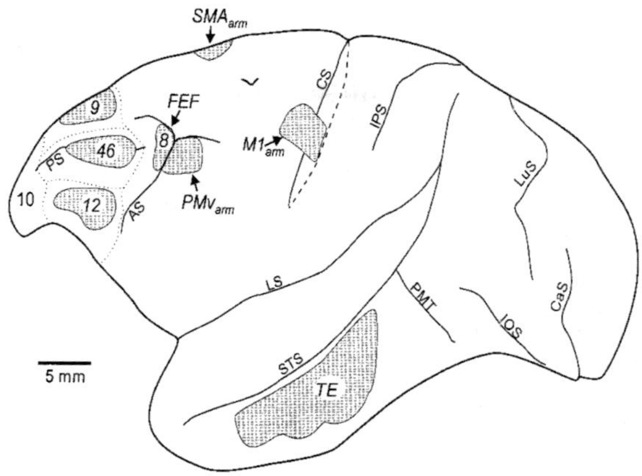

Fig. 1. Location of virus injection sites in the cerebral cortex. Lateral view of a *cebus* monkey brain. The shaded areas indicate the spread of HSV1 from injections into each cortical area. The numbers 8, 9, 10, 12 and 46 refer to cytoarchitectonic areas of the frontal lobe according to Walker [47]. The dotted lines define the borders between areas. Abbreviations: AS, arcuate sulcus; CaS, calcarine sulcus; CS, central sulcus; FEF, frontal eye field; IOS, inferior occipital sulcus; IPS, intraparietal sulcus; LS, lateral sulcus; LuS, lunate sulcus; $M1_{arm}$, arm area of the primary motor cortex; PMT, posterior middle temporal sulcus; PMv_{arm}, arm area of the ventral premotor area; PS, principal sulcus; SMA_{arm}, arm area of the supplementary motor area; STS, superior temporal sulcus; TE, area of inferotemporal cortex.

transneuronally in the retrograde direction from first-order neurons in the ventrolateral thalamus to second-order neurons in the deep cerebellar nuclei (dentate and interpositus) and to second-order neurons in output nuclei of the basal ganglia (the internal segment of the globus pallidus (GPi)).

Labeled neurons in interpositus were largely confined to caudal portions of the anterior division of the nucleus, and those in the dentate were restricted to dorsal portions of the nucleus at mid-rostrocaudal levels (Fig. 2, "$M1_{arm}$"). These regions of

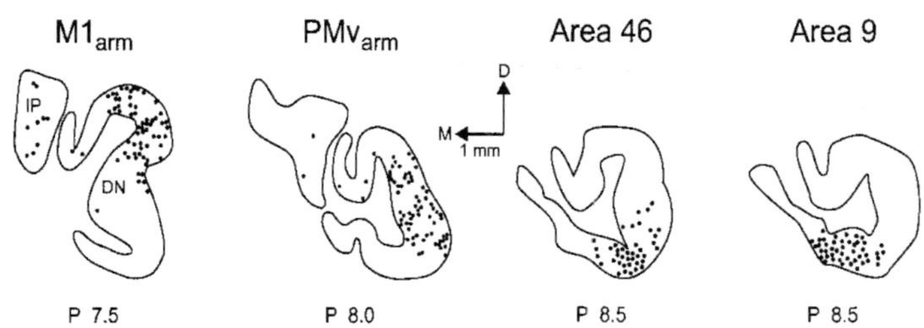

Fig. 2. Origin of cerebellar projections to M1, PMv, area 46 and area 9. Representative coronal sections through the dentate and/or interpositus nuclei of animals that received injections of the McIntyre-B strain of HSV1 into different cortical areas (see Fig. 1). The solid dots indicate the positions of neurons labeled by retrograde transneuronal transport of virus. The maps display labeled neurons found on 2–3 adjacent sections whose approximate anterior-posterior location is indicated at the bottom of each section outline. Abbreviations: D, dorsal; DN, dentate nucleus; IP, interpositus nucleus; M, medial. Adapted from [24,26,27,29,32].

the dentate and interpositus correspond to the sites in these nuclei where neurons related to arm movements have been recorded in electrophysiological studies (e.g., [39–41]). These observations suggest that cerebellar control of activity in the arm area of M1 originates from highly localized regions in dentate and interpositus.

In GPi, neurons labeled by the M1 injections were confined almost exclusively to the middle of the nucleus rostrocaudally, and formed two distinct clusters in the outer and inner portions of GPi (Fig. 3, "$M1_{arm}$"). The location of these labeled neurons was comparable to the region where electrophysiological studies have found GPi neurons that were inhibited by stimulation of the primary motor cortex [42], or that displayed changes in activity related to arm movements (e.g., [42–46]). These findings confirm that the arm area of M1 is the target of output from the basal ganglia (for references and discussion see [38]). Furthermore, these observations indicate that the projection from GPi through the ventrolateral thalamus to the arm area of M1 originates from a specific region within GPi.

Premotor cortex

Our next experiments examined inputs to the arm representations of two premotor areas in the frontal lobe, the ventral premotor area (PMv) and the supplementary motor area (SMA) (Fig. 1) [26,27]. Following injections into the PMv, labeled neurons were found in a 1–2 mm region in the middle of the dentate rostrocaudally (Fig. 2, "PMv_{arm}"). Moreover, these labeled neurons were located ventral to the region of the dentate that contained labeled neurons after virus injections into the arm area of M1 (compare Fig. 2, "$M1_{arm}$" and "PMv_{arm}"). Thus, the arm areas in M1 and PMv receive input from different portions of the dentate.

In GPi, injections into either the PMv or SMA consistently labeled neurons in a 1–2 mm region in the middle of the nucleus rostrocaudally. Within this region, the dorsoventral location of labeled neurons varied depending on the location of the cortical injection site. SMA injections labeled neurons in a middorsal region of GPi (Fig. 3, "SMA_{arm}"). In contrast, injections into PMv labeled neurons mainly in

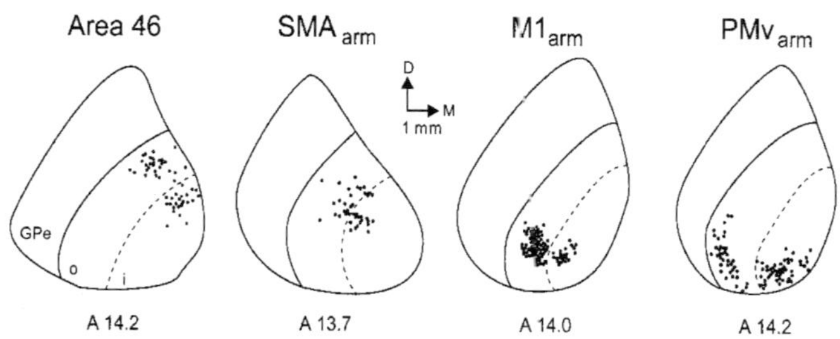

Fig. 3. Pallidal projections to M1, PMv, SMA, and area 46. Representative coronal sections through the globus pallidus of animals that received injections of the McIntyre-B strain of HSV1 into different cortical areas (Fig. 1). See Fig. 2 for conventions. GPe, external segment of globus pallidus; o, outer portion of the internal segment of globus pallidus; i, inner portion of the internal segment of globus pallidus. Adapted from [24,26].

ventrolateral portions of GPi (Fig. 3, "PMv_{arm}"). Neurons labeled after injections into the arm area of M1 were located between those labeled by the SMA and PMv injections (Fig. 3). These observations lead to two important conclusions. First, pallido-thalamocortical projections target premotor, as well as primary motor areas of the cerebral cortex. Secondly, the arm representation of each of these motor areas receives input from a topographically distinct set of GPi neurons. We have proposed that this anatomical arrangement creates distinct "output channels" in the sensorimotor portion of GPi [27].

The frontal eye field

We have also examined subcortical inputs to the frontal eye field (FEF) located in Walker's area 8 [28,47]. This cortical region has long been recognized as an important component of the cortical system that controls voluntary eye movements in primates (for references and review, see [48,49]). As in our studies of skeletomotor areas, injections of virus were made into the FEF after it had been physiologically mapped (Fig. 1). The FEF injections labeled neurons in the dentate nucleus of the cerebellum, the pars reticulata of the substantia nigra (SNpr), and the superior colliculus.

Within the dentate, labeled neurons were found only in the most caudal third of the nucleus (see [28]). Prior studies have shown that this region contains some neurons that display changes in activity correlated with saccadic eye movements (e.g., [41]). In addition, the region of the dentate labeled after FEF injections was strikingly different from the regions of the dentate labeled after injections into M1 or PMv. Thus, separate regions of the dentate project to skeletomotor and oculomotor areas of frontal cortex.

Within the basal ganglia, FEF injections labeled neurons in a lateral region of SNpr (Fig. 4, "FEF") and not in GPi [28]. These labeled neurons were located mainly in the posterior two-thirds of the SNpr. The lateral region of SNpr that contained labeled neurons has previously been designated *pars lateralis* by some investigators [50,51]. Neurons in the pars lateralis have been observed to display changes in activity related to saccadic eye movements [52]. This region of the nigra is also the

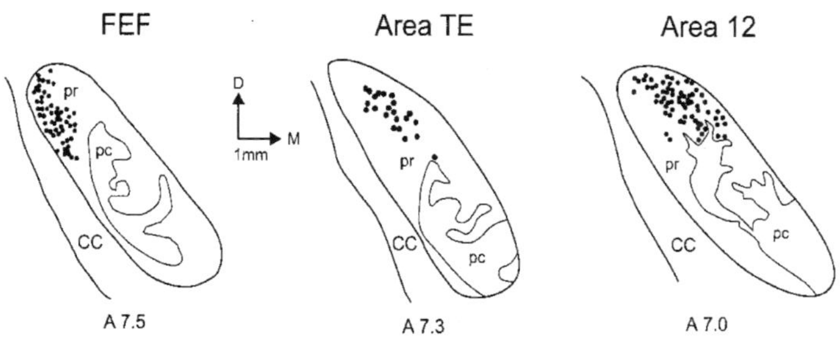

Fig. 4. Nigral projections to the FEF, area TE, and area 12. Coronal sections indicating the location of labeled neurons in the SNpr following HSV1 injections into the FEF, area TE, or area 12. CC, crus cerebri; pc, pars compacta; pr, pars reticulata. Adapted from [28,30,31].

origin of projections to the oculomotor layers of the superior colliculus in monkeys [53–55].

Prefrontal cortex

To test some of the hypotheses about basal ganglia and cerebellar involvement in higher cognitive function (e.g., [56–58]), we have begun to examine subcortical projections to regions of prefrontal cortex [29,31,32]. Our initial experiments focused on cerebellar and basal ganglia projections to Walker's areas 9, 12, and 46 (Fig. 1) [47]. Each of these areas appears to be involved in aspects of "working memory" and is thought to guide behavior based on transiently stored information rather than immediate external cues (e.g., [59–64]).

Injections of HSV1 into areas 9 and 46 (but not area 12) labeled many neurons in the dentate nucleus. These labeled neurons were confined to the most ventral portions of the dentate and were concentrated rostrocaudally in the middle third of the nucleus (Fig. 2, "Area 46" and "Area 9"). The neurons labeled after area 9 injections were found largely medial to the neurons labeled by area 46 injections. The ventral regions of the dentate that project to areas 9 and 46 clearly differ from the more dorsal regions of this nucleus that innervate M1 or the PMv (Fig. 2) and the more caudal region of the dentate that innervates the FEF [28].

In the basal ganglia, many neurons were labeled after injections of HSV1 into any of the prefrontal areas. Virus injections into area 12 labeled neurons in the caudal and dorsomedial portion of SNpr, but not GPi (Fig. 4, "Area 12"). In contrast, virus injections into area 46 led to labeled neurons mainly in dorsomedial portions of GPi (Fig. 3, "Area 46") and only a few labeled neurons in SNpr (not illustrated). Injections into area 9 labeled many neurons in both GPi and SNpr. Each cortical area appeared to receive input from topographically separate regions of GPi or SNpr. For example, in GPi, the neurons that influence area 9 were located dorsal to those that innervate area 46. Similarly, in SNpr, the neurons that influence area 9 were located rostral and lateral to those that innervate area 12.

Our observations on inputs to prefrontal cortex suggest three important conclusions. First, multiple areas of prefrontal cortex are the target of output from the basal ganglia and/or cerebellum. Second, the portions of the basal ganglia and cerebellum that influence prefrontal cortex are separate from those that influence the cortical motor areas. Third, distinct "output channels" exist in the nonmotor portions of GPi, SNpr and dentate that are comparable to the output channels in the motor portions of these nuclei.

Inferotemporal cortex

Each of the cortical regions that we have shown is a target of basal ganglia and/or cerebellar output also is known to send an input to these subcortical nuclei (e.g., [1,65–70]). This anatomical arrangement suggests that many areas in the frontal lobe participate in a "closed" loop with the basal ganglia and cerebellum. We tested whether this arrangement extends to areas outside the frontal lobe by examining subcortical inputs to a region of inferotemporal cortex, area TE [71]. TE has long been recognized as a source of input to the basal ganglia (see [72–74]). This cortical area also is of interest because of evidence that it plays a critical role in the visual recognition and discrimination of objects, and in visual memory (e.g., [75–79]).

We thought it possible that TE was the target of basal ganglia output because of recent reports that it receives thalamic input from the nucleus ventralis anterior pars magnocellularis (VAmc) [72]. This thalamic nucleus is known to be a major site of termination of efferents from the SNpr (e.g., [7,14]). To test this hypothesis, we made multiple injections of HSV1 throughout area TE (Fig. 1). This led to a distinct cluster of labeled neurons in SNpr (Fig. 4 "Area TE"). Most of these neurons were located dorsally in the caudal third of the nucleus. This portion of the SNpr appears to be separate from the regions of this nucleus that influence either the FEF or regions of prefrontal cortex (Fig. 4). Thus, TE is both a source of input to, and target of output from, a distinct portion of the basal ganglia.

The region of SNpr that influences TE has been shown to contain some neurons that display changes in activity related to the presentation of visual stimuli [52]. This observation, together with our anatomy, suggests that the basal ganglia may be involved in higher order aspects of visual processing, in addition to their involvement in motor and cognitive function (see [30] for references and discussion). Many cortical areas other than the ones we have examined project to the basal ganglia and/or to the cerebellum. Whether the basal ganglia and cerebellum participate in closed loops with these cortical areas remains to be determined.

Physiological studies

The anatomical findings described above raise an important question. Do individual output channels send the same or different information to the cortical areas they innervate? For example, do neurons in the dentate output channel that projects to the PMv have the same or different relations to movement as the neurons in the output channel to M1? Similarly, do the neurons in the pallidal output channel that innervates the SMA have the same or different relations to movement as those in the output channel to M1? To begin to address this question, we have recorded the activity of single neurons in the dentate nucleus and globus pallidus of awake primates trained on complex motor tasks [34,35,80].

Single neuron recording in awake trained primates

Monkeys were trained to perform sequential pointing movements under two task conditions (for a more detailed description of the task and training, see [34]). In both conditions, the monkey faced a panel with five touch pads which were numbered one to five (left to right). A small red light emitting diode (LED) was located over each touch pad. The monkey began a trial by placing his right hand on a hold key in front of him for a variable "Hold" period. In the Remembered Sequence Task (REM task,

LEDs over three touch pads were illuminated in a sequence as an instruction to the monkey. At the end of a variable "Instruction" period, an auditory "Go" signal told the monkey to release the hold key and press the three touch pads according to the instructed sequence (i.e., in the same order that the LEDs were illuminated). Thus, the specific sequence of movements that the monkey performed during each trial of the REM task was initially stored in "working memory" and then, internally cued.

In the "Tracking Task" (TRACK task), an LED over a single touch pad was illuminated after the "Hold" period. The auditory "Go" signal was turned on at the same time. Following the onset of this signal, the monkey was required to release the hold key and press the indicated touch pad. As soon as the monkey contacted the first touch pad, a second LED over another touch pad was illuminated. The monkey was required to move quickly from the first to the second touch pad. Then, when the monkey contacted the second touch pad, a third LED over another touch pad was illuminated and the monkey was required to move to the third touch pad. Thus, the specific sequence of movements that the monkey performed during each trial of the TRACK task was externally cued.

Movement-related activity in the dentate nucleus

We recorded from 172 neurons in the dentate that were task-related during the RT period [34]. Most task-related neurons were found in approximately the middle third of the dentate nucleus, rostrocaudally. Approximately 60% of the task-related neurons (102/72) were classified as task-independent. These neurons displayed comparable patterns of movement-related activity during the RT period of both the REM and TRACK tasks. The majority of the task-independent neurons were located in a dorsal region of the dentate nucleus. This region is likely to be within the output channel that projects to M1. Based on their firing patterns, the neurons in this output channel appear to participate in the generation and control of sequential movements that are internally cued, as well as those that are externally instructed.

On the other hand, the other 40% of the task-related neurons in the dentate (70/172) were considered task-dependent. More than 75% of these neurons (54/70) were termed TRACK neurons because they displayed activity changes during the RT period only during the TRACK task or their changes in activity were more pronounced (≥50%) during the TRACK task than during the REM task. An example of a TRACK neuron is shown in Fig. 5. The rasters and averages are aligned on the hold key release. They illustrate the response of this neuron during trials that began with a movement to touch pad 4 (4-5-3, 4-3-5, 4-3-1, 4-2-1). The individual trials in the rasters have been sorted according to the length of the RT period. This neuron displayed little or no modulation in its activity during the RT period of the REM task (Fig. 5, left). In contrast, the same neuron showed a clear increase in activity during the RT period of the TRACK task (Fig. 5, right). Many of the TRACK neurons were located ventral and lateral to dentate neurons which were task-related, but task-independent. This localization suggests that TRACK neurons are within the output channel that innervates the PMv (see Fig. 2). Thus, the neurons in this output channel

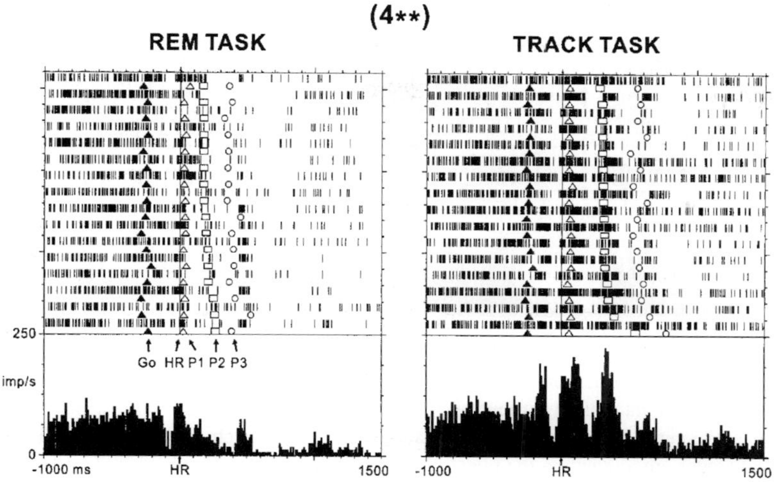

Fig. 5. Responses of a TRACK neuron in the dentate. The rasters and averages are aligned on the hold key release (HR). They illustrate the response of this neuron during trials that began with a movement to touch pad 4 (4**, i.e., 4-5-3, 4-3-5, 4-3-1, 4-2-1). The individual trials in the rasters have been sorted according to the length of the interval between the onset of the "Go" signal and HR (i.e., the RT period). Symbols indicate the onset of different behavioral events: filled triangle, "Go" signal; open triangle, press of the first touch pad in the sequence (P1); open square, press of the second touch pad (P2); open circle, press of the third touch pad (P3). Note that this neuron displayed a phasic increase in activity in the RT period only during the TRACK task. Adapted from [34].

appear to be preferentially involved in the generation and control of sequential movements that are visually guided.

Movement-related activity in globus pallidus

In the basal ganglia, we recorded more than 230 pallidal neurons that displayed a significant change in activity during the performance of the REM and/or TRACK task [35]. Most of these neurons were found in approximately the middle third of the globus pallidus, rostrocaudally. Two-thirds of the task-related neurons were considered task-dependent because they displayed an exclusive or enhanced (>50%) change in activity for one of the two tasks. Over 65% of the task-dependent neurons displayed exclusive or enhanced changes in activity for the REM task (REM neurons) (Fig. 6, left). For example, the neuron illustrated in Fig. 6 displayed little or no modulation in its activity during any period of the TRACK task (Fig. 6, right). In contrast, the same neuron showed a clear decrease in activity during the reaction time period ("Go" signal to Hold key release) and execution period (Hold key release to reward) of the REM task (Fig. 6, left). Interestingly, many of the REM neurons were located dorsal and medial to GP neurons that were task-related, but not task-dependent (Fig. 7) [81]. The GP region containing REM neurons is potentially within

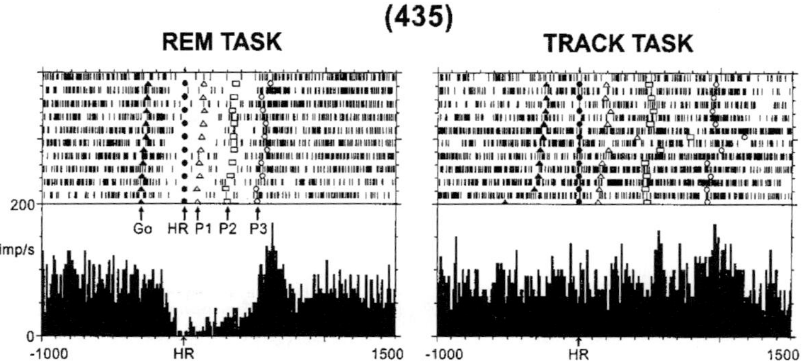

Fig. 6. Responses of REM neuron in GP during the REM and TRACK tasks. These averages illustrate the response of a pallidal neuron during the REM and TRACK tasks. The animal performed the same sequence of movements for both sets of trials and first pressed touch pad 4, then 3 and finally 5. The rasters and averages are aligned on the release of the hold key (HR). Activity 1,000 ms before and 1,500 ms after HR is shown. Individual trials in the rasters have been sorted according to the length of the interval between the onset of the go signal and HR (i.e., the reaction time period). Symbols used are the same as in Fig. 5. The maximum scale of the averages (impulses/s) is 200 and the bin width is 10 ms. Note that this neuron displayed a phasic decrease in activity only during the REM task. Adapted from [35].

the output channel that innervates the SMA (compare Figs. 3 and 7). Thus, this output channel may be specifically involved in the guidance of sequential movements based on internal cues.

Another group of pallidal neurons displayed preferential changes in activity during

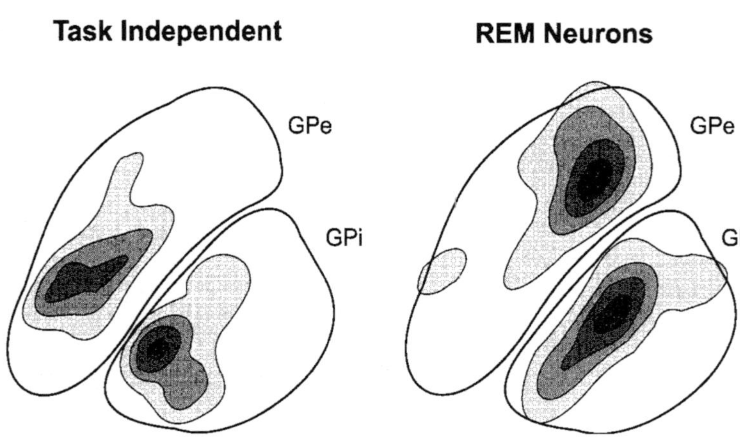

Fig. 7. Distribution of movement-related neurons in the globus pallidus. The relative density of task-related neurons in the two pallidal segments is indicated by shading (darker shading indicates greater density). Task-independent neurons displayed comparable changes in activity during the REM and TRACK tasks. REM neurons displayed exclusive or enhanced (>50%) changes in activity during the REM task (see text for details). Note the tendency for REM neurons to be in pallidal regions that were dorsal to task-independent neurons. Adapted from [81].

the TRACK task (TRACK neurons, not illustrated). These neurons were found more ventrally in GPi, potentially in the output channel that innervates the PMv (see Fig. 3). Consequently, this output channel may be particularly concerned with guiding movement based on external cues.

These physiological results provide some important insights into the motor functions of the basal ganglia. These subcortical nuclei have been thought to be critical for movements under internal control as opposed to movements guided by stimuli in extrapersonal space (e.g., [52,82,83]). In addition, there have been numerous suggestions that the basal ganglia are involved in the generation and control of sequential limb movements (e.g., [84,85]). The REM neurons we have recorded display patterns of activity that are precisely what one might expect to find in an area involved in the central control of movement sequences. On the other hand, it is important to note that a substantial number of GP neurons in our sample was either task independent or preferentially active during the TRACK task. This observation implies that pallidal output is not solely concerned with movements that are internally generated, but is also involved in other aspects of motor behavior, such as the generation of specific movement parameters and the guidance of movement based on external cues. This view is consistent with our anatomical results that indicate pallidal output innervates multiple motor areas including M1, PMv and SMA.

Instruction-related activity in the dentate and globus pallidus

Approximately 15% of the task-related neurons in dentate and globus pallidus were "instruction-related" (I-related), that is, they displayed changes in activity during the instructed delay period (e.g., Fig. 8) [80]. Some of these I-related neurons displayed transient changes in activity immediately after the presentation of visual cues (Fig. 8, "Cue" neuron). Other I-related neurons displayed changes in activity only during the delay period following the illumination of the three instruction LEDs (Fig. 8, "Delay" neuron). Some of the I-related activity was specific for the particular remembered sequence of movements that the animal was preparing to perform. Still other I-related neurons displayed changes in activity during both task periods (Fig. 8, "Cue + Delay" neuron).

I-related neurons tended to be located in ventral regions of dentate and dorsomedial regions of the globus pallidus. These sites appear to be within the output channels that innervate motor areas concerned with motor preparation (e.g., the presupplementary motor area, see [86]), and channels that innervate prefrontal areas involved in working memory (areas 46 and 9, see Fig. 3). Thus, pallidal and dentate activity during instructed delay periods could participate in higher order motor and/or cognitive functions.

Discussion

The anatomical and physiological results we have briefly described reveal important features of the basic structural framework that connects basal ganglia and cerebellar

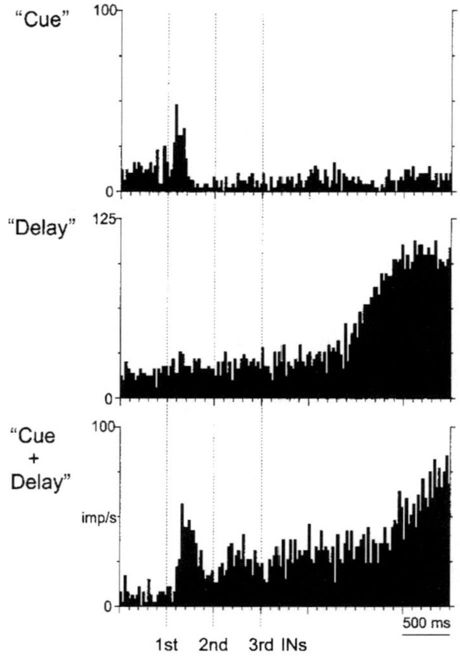

Fig. 8. Responses of I-related neurons in the dentate. The rasters and averages illustrate the activity of three different types of I-related neurons during the instructed delay period (first, second and third INs). The trials are aligned on the presentation of the third instruction. The bin width for the averages is 20 ms. The trials illustrated all began with the illumination of LED 4. Adapted from [80].

output with the cerebral cortex. We propose that the clustering of output neurons that project to a single cortical area creates distinct output channels in the basal ganglia and cerebellum [27,30–33]. Output channels related to the cortical motor areas appear to be topographically separate from each other and from those associated with the prefrontal or inferotemporal cortex (Figs. 2–4). In addition, our physiological results suggest that individual output channels are concerned with distinct aspects of motor and/or cognitive behavior [26,34,35,80]. For example, the output channels in dentate and globus pallidus that project to M1 may be concerned with the control of specific movement parameters. Other output channels, such as those innervating the PMv, may be concerned with other levels of motor behavior like the guidance of sequential movements based on external cues. Output channels that influence the prefrontal cortex may be involved in cognitive aspects of behavior such as working memory. Finally, the output channel in the nigra which targets area TE may be involved in higher order aspects of visual processing.

These results lead to the prediction that localized dysfunction of different regions within the basal ganglia or cerebellum would lead to distinct changes in behavior. For example, there is abundant evidence that lesions of sensorimotor regions of the basal ganglia or cerebellum produce alterations in motor behavior. In contrast, there are reports that lesions at other sites within the basal ganglia or cerebellum result in cognitive dysfunction (e.g., [57,58,87–92]). Lesions in portions of the striatum that are part of the basal ganglia loop with TE have been shown to produce profound impairments in the learning and retention of visual discriminations of objects [87,93]. Thus, the behavioral deficits associated with lesions of the basal ganglia or cerebellum are likely to depend on the extent of damage to a given subcortical loop and dysfunction of its output channel.

Acknowledgements

This work was supported by funds from an Established Investigator Award from the National Alliance for Research on Schizophrenia and Depression (PLS), the Veterans Administration Medical Research Service (PLS), US Public Health Service Grant NS24328 (PLS), and Predoctoral Fellowship MH11262 from the National Institute of Mental Health (FAM).

References

1. Kemp JM, Powell TPS. The connexions of the striatum and globus pallidus: synthesis and speculation. Phil Trans R Soc London Ser B 1971;262:441–457.
2. Allen GI, Tsukahara N. Cerebrocerebellar communication systems. Physiol Rev 1974;54:957–1006.
3. Brooks VB, Thach WT. Cerebellar control of posture and movement. In: Brooks VB (ed) Handbook of Physiology. Section I: The Nervous System, vol II: Motor Control, Part I. Bethesda: American Physiological Society, 1981;877–946.
4. Asanuma C, Thach WT, Jones EG. Distribution of cerebellar terminations in the ventral lateral thalamic region of the monkey. Brain Res Rev 1983;5:237–265.
5. Nauta WJH, Mehler WR. Projections of the lentiform nucleus in the monkey. Brain Res 1966;1:3–42.
6. Kusama T, Mabuchi M, Sumino T. Cerebellar projections to the thalamic nucleus in the monkey. Proc Jpn Acad 1971;47:505–510.
7. Carpenter MB, Nakano K, Kim R. Nigrothalamic projections in the monkey demonstrated by autoradiographic technics. J Comp Neurol 1976;165:401–416.
8. Kuo JS, Carpenter MB. Organization of pallido-thalamic projections in the rhesus monkey. J Comp Neurol 1973;151:201–236.
9. Kim R, Nakano K, Jayarman A, Carpenter MB. Projections of the globus pallidus adjacent structures: an autoradiographic study in the monkey. J Comp Neurol 1976;169:263–290.
10. Percheron G. The thalamic territory of cerebellar afferents and the lateral region of the thalamus of the macaque in stereotaxic ventricular coordinates. J Hirnforsch 1977;18:375–400.
11. Stanton G. Topographical organization of ascending cerebellar projections from the dentate and interposed nuclei in *Macaca mulatta*: an anterograde degeneration study. J Comp Neurol 1980; 190:699–731.
12. Kalil K. Projections of the cerebellar and dorsal column nuclei upon the thalamus of the rhesus monkey. J Comp Neurol 1981;195:25–50.
13. DeVito JL, Anderson ME. An autoradiographic study of efferent connections of the globus pallidus in *Macaca mulatta*. Exp Brain Res 1982;46:107–117.
14. Ilinsky IA, Jouandet M, Goldman-Rakic PS. Organization of the nigrothalamocortical system in the rhesus monkey. J Comp Neurol 1985;236:315–330.
15. Yamamoto T, Yoshida K, Yoshikawa Y, Kishimoto Y, Oka H. The medial dorsal nucleus is one of the thalamic relays of the cerebellocerebral response to the frontal association cortex in the monkey: horseradish peroxidase and fluorescent dye double staining study. Brain Res 1992;579:315–320.

16. Kievit J, Kuypers HGJM. Organization of thalamo-cortical connexions to the frontal lobe in the rhesus monkey. Exp Br Res 1977;29:299–322.
17. Miyata M, Sasaki K. HRP studies on thalamocortical neurons related to the cerebellocerebral projection in the monkey. Brain Res 1983;274:213–224.
18. Schell GR, Strick PL. The origin of thalamic inputs to the arcuate premotor and supplementary motor areas. J Neurosci 1984;4:539–560.
19. Goldman-Rakic PS, Porrino LJ. The primate mediodorsal (MD) nucleus and its projection to the frontal lobe. J Comp Neurol 1985;242:535–560.
20. Wiesendanger R, Wiesendanger M. The thalamic connections with medial area 6 (supplementary motor cortex) in the monkey (Macaca fascicularis). Exp Brain Res 1985;59:91–104.
21. Orioli PJ, Strick PL. Cerebellar connections with the motor cortex and the arcuate premotor area: an analysis employing retrograde transneuronal transport of WGA-HRP. J Comp Neurol 1989;288: 612–626.
22. Matelli M, Luppino G, Fogassi L, Rizzolatti G. Thalamic input to inferior area 6 and area 4 in the macaque monkey. J Comp Neurol 1989;280:468–488.
23. Barbas H, Haswell Henion TH, Dermon CR. Diverse thalamic projections to the prefrontal cortex in the rhesus monkey. J Comp Neurol 1991;313:65–94.
24. Zemanick MC, Strick PL, Dix RD. Transneuronal transport of herpes simplex virus type 1 in the primate motor system: transport direction is strain dependent. Proc Natl Acad Sci USA 1991;88: 8048–8051.
25. Strick PL, Card JP. Transneuronal mapping of neural circuits with alpha herpesviruses. In: Bolam JP (ed) Experimental Neuroanatomy: a Practical Approach. Oxford: Oxford University Press, 1992; 81–101.
26. Strick PL, Hoover JE, Mushiake H. Evidence for "output channels" in the basal ganglia and cerebellum. In: Mano N, Hamada I, DeLong MR (eds) Role of the Cerebellum and Basal Ganglia in Voluntary Movement. Amsterdam: Elsevier, 1993;171–180.
27. Hoover JE, Strick PL. Multiple output channels in the basal ganglia. Science 1993;259:819–821.
28. Lynch JC, Hoover JE, Strick PL. Input to the primate frontal eye field from the substantia nigra, superior colliculus, and dentate nucleus demonstrated by transneuronal transport. Exp Brain Res 1994;100:181–186.
29. Middleton FA, Strick PL. Anatomical evidence for cerebellar and basal ganglia involvement in higher cognitive function. Science 1994;266:458–461.
30. Middleton FA, Strick PL. The temporal lobe is a target of output from the basal ganglia. Proc Natl Acad Sci USA 1996;93:8683–8687.
31. Middleton FA, Strick PL. New concepts about the organization of basal ganglia output. Adv Neurol (In press).
32. Middleton FA, Strick PL. Dentate output channels: motor and cognitive components. Prog Brain Res (In press).
33. Strick PL, Dum RP, Picard N. Macro-organization of the circuits connecting the basal ganglia with the cortical motor areas. In: Houk JC, Davis JL, Beiser DG (eds) Models of Information Processing in the Basal Ganglia. Cambridge: MIT Press, 1995;117–130.
34. Mushiake H, Strick PL. Preferential activity of dentate neurons during limb movements guided by vision. J Neurophysiol 1993;70:2660–2664.
35. Mushiake H, Strick PL. Pallidal neuron activity during sequential arm movements. J Neurophysiol 1995;74:2754–2758.
36. Hoover JE, Strick PL. Retrograde transneuronal transport of HSV-1 from primary motor cortex to cerebellar deep nuclei and purkinje cells. Soc Neurosci Abstr 1993;19:1590.
37. Olszewski J. The thalamus of *Macaca mulatta*. An atlas for use with the stereotaxic instrument. Basel: Karger, 1952.
38. Holsapple JW, Preston JB, Strick PL. The origin of thalamic inputs to the "hand" representation in the primary motor cortex. J Neurosci 1991;11:2644–2654.
39. Thach WT. Correlation of neural discharge with pattern and force of muscular activity, joint position,

and direction of intended next movement in motor cortex and cerebellum. J Neurophysiol 1978; 41:654–676.
40. Wetts R, Kalaska JF, Smith AM. Cerebellar nuclear activity during contraction and reciprocal inhibition of forearm muscles. J Neurophysiol 1985;54:231–244.
41. van Kan PLE, Houk JC, Gibson AR. Output organization of intermediate cerebellum of the monkey. J Neurophysiol 1993;69:57–73.
42. Yoshida S, Nambu A, Jinnai K. The distribution of the globus pallidus neurons with input from various cortical areas in the monkeys. Brain Res 1993;611:170–174.
43. DeLong MR, Crutcher MD, Georgopoulos AP. Primate globus pallidus and subthalamic nucleus: functional organization. J Neurophysiol 1985;53:530–543.
44. Anderson ME, Horak FB. Influence of the globus pallidus on arm movements in monkeys. III. Timing of movement-related information. J Neurophysiol 1985;54:433–448.
45. Hamada I, DeLong MR, Mano N. Activity of identified wrist-related pallidal neurons during step and ramp wrist movements in the monkey. J Neurophysiol 1990;64:1892–1906.
46. Mink JW, Thach WT. Basal ganglia motor control: I. Nonexclusive relation of pallidal discharge to five movement modes. J Neurophysiol 1991;65:273–300.
47. Walker AE. A cytoarchitectural study of the prefrontal area of the macaque monkey. J Comp Neurol 1940;73:59–86.
48. Bruce CJ, Goldberg ME. Primate frontal eye fields. I. Single neurons discharging before saccades. J Neurophysiol 1985;53:603–635.
49. Bruce CJ, Goldberg ME, Bushnell MC, Stanton GB. Primate frontal eye fields. II. Physiological and anatomical correlates of electrically evoked eye movements. J Neurophysiol 1985;54:714–734.
50. Ma TP. Identification of the substantia nigra pars lateralis in the macaque using cytochrome oxidase and fiber stains. Brain Res 1989;48:305–311.
51. Yelnik J, Francois C, Percheron G, Heyner S. Golgi study of the primate substantia nigra I. Quantitative morphology and typology of nigral neurons. J Comp Neurol 1987;265:455–472.
52. Hikosaka O, Wurtz RH. Visual and oculomotor functions of monkey substantia nigra pars reticulata. I. Relation of visual and auditory responses to saccades. J Neurophysiol 1983;49:1230–1253.
53. Beckstead RM, Frankfurter A. The distribution and some morphological features of substantia nigra neurons that project to the thalamus, superior colliculus and pedunculopontine nucleus in the monkey. Neuroscience 1982;7:2377–2388.
54. Francois C, Percheron G, Yelnik J. Localization of nigrostriatal, nigrothalamic and nigrotectal neurons in ventricular coordinates in macaques. Neuroscience 1984;13:61–76.
55. May PJ, Hall WC, Porter JD, Sakai ST. The comparative anatomy of nigral and cerebellar control over tectally initiated orienting movements. In: Mano N, Hamada I, DeLong MR (eds) Role of the Cerebellum and Basal Ganglia in Voluntary Movement. Amsterdam: Elsevier, 1993;221–231.
56. Alexander GE, DeLong MR, Strick PL. Parallel organization of functionally segregated circuits linking basal ganglia and cortex. Annu Rev Neurosci 1986;9:357–381.
57. Leiner HC, Leiner AL, Dow RS. The human cerebro-cerebellar system: its computing, cognitive, and language skills. Behav Brain Res 1991;44:113–128.
58. Leiner HC, Leiner AL, Dow RS. Cognitive and language functions of the human cerebellum. Trends Neurosci 1993;16:444–447.
59. Mishkin M, Manning FJ. Non-spatial memory after selective prefrontal lesions in monkeys. Brain Res 1978;143:313–323.
60. Fuster JM. The Prefrontal Cortex, 2nd edn. New York: Raven Press, 1989.
61. Funahashi S, Bruce CJ, Goldman-Rakic PS. Mnemonic coding of visual space in the monkeys dorsolateral prefrontal cortex. J Neurophysiol 1989;61:331–349.
62. Funahashi S, Bruce CJ, Goldman-Rakic PS. Dorsolateral prefrontal lesions and oculomotor delayed-response: evidence for mnemonic "scotomas". J Neurosci 1993;13:1479–1497.
63. Wilson FA, O'Scalaidhe SP, Goldman-Rakic PS. Dissociation of object and spatial processing domains in primate prefrontal cortex. Science 1994;260:1995–1997.
64. Petrides M. Impairments on nonspatial self-ordered and externally-ordered working memory tasks

after lesions of the mid-dorsal part of the lateral frontal cortex in the monkey. J Neurosci 1995; 15:359–375.
65. Goldman PS, Nauta WJH. An intricately patterned prefronto-caudate projection in the rhesus monkey. J Comp Neurol 1977;171:369–386.
66. Brodal P. The corticopontine projection in the rhesus monkey. Origin and principles of organization. Brain 1978;101:251–283.
67. Brodal P. The pontocerebellar projection in the rhesus monkey: an experimental study with retrograde axonal transport of horseradish peroxidase. Neurosci 1979;4:193–208.
68. Glickstein M, May JG, Mercier BE. Corticopontine projection in the macaque: the distribution of labelled cortical cells after large injections of horseradish peroxidase in the pontine nuclei. J Comp Neurol 1985;235:343–359.
69. Selemon LD, Goldman-Rakic PS. Longitudinal topography and interdigitation of corticostriatal projections in the rhesus monkey. J Neurosci 1985;5:776–794.
70. Schmahmann JD, Pandya DN. Prefrontal cortex projections to the basilar pons in rhesus monkey: implications for the cerebellar contributions to higher function. Neurosci Lett 1995;199:175–178.
71. von Bonin G, Bailey P. The neocortex of *Macaca mulatta*. Urbana, IL: Univ. Illinois Press, 1947.
72. Webster MJ, Bachevalier J, Ungerleider LG. Subcortical connections of inferior temporal areas TE and TEO in macaque monkeys. J Comp Neurol 1993;335:73–91.
73. Saint-Cyr JA, Ungerleider LG, Desimone R. Organization of visual cortical inputs to the striatum and subsequent outputs to the pallido-nigral complex in the monkey. J Comp Neurol 1990;298:129–156.
74. Yeterian EH, Pandya DN. Corticostriatal connections of extrastriate visual areas in rhesus monkeys. J Comp Neurol 1995;352:436–457.
75. Gross CG. Visual functions of inferotemporal cortex. In: Jung R (ed) Handbook of Sensory Physiology. Berlin: Springer-Verlag, 1972;451–482.
76. Ungerleider LG, Mishkin M. Two cortical visual systems. In: Ingle DJ, Goodale MA, Mansfield RJW (eds) Analysis of Visual Behavior. Cambridge: MIT 1982;549–586.
77. Mishkin M. A memory system in the monkey. Phil Trans Royal Soc London B 1982;298:85–95.
78. Tanaka K, Saito H-A, Fukuda Y, Moriya M. Coding visual images of objects in the inferotemporal cortex of the macaque monkey. J Neurophysiol 1991;66:170–189.
79. Miyashita Y. Inferior temporal cortex: where visual perception meets memory. Annu Rev Neurosci 1993;16:245–263.
80. Mushiake H, Strick PL. Cerebellar and pallidal activity during instructed delay periods. Soc Neurosci Abstr 1995;21:411.
81. Strick PL, Dum RP, Mushiake H. Basal ganglia 'loops' with the cerebral cortex. In: Kimura M, Graybiel AM (eds) Functions of the Cortico-Basal Ganglia Loop. Tokyo: Springer-Verlag, 1995; 106–124.
82. Flowers KA. Visual "closed-loop" and "open-loop" characteristics of voluntary movement in patients with parkinsonism and intention tremor. Brain 1976;99:269–310.
83. Flowers KA. Lack of prediction in the motor behaviour of parkinsonism. Brain 1978;101:35–52.
84. Schwab RS, Chafetz ME, Walker S. Control of two simultaneous voluntary motor acts in normals and in parkinsonism. Arch Neurol Psychiat 1954;72:591–598.
85. Benecke R, Rothwell JC, Dick JPR, Day BL, Marsden CD. Disturbance of sequential movements in Parkinson's disease. Brain 1987;110:361–379.
86. Wiesendanger R, Wiesendanger M. Cerebello-cortical linkage in the monkey as revealed by transcellular labeling with the lectin wheat germ agglutinin conjugated to the marker horseradish peroxidase. Exp Brain Res 1985;59:105–117.
87. Divac I, Rosvold HE, Swarcbart MK. Behavioral effects of selective ablation of the caudate nucleus. J Comp Physiol Psychology 1967;63:184–190.
88. Botez MI, Botez T, Elie R, Attig E. Role of the cerebellum in complex human behavior. Ital J Neurol Sci 1989;10:291–300.
89. Ivry RB, Keele SW. Timing functions of the cerebellum. J Cog Neurosci 1989;1:136–152.
90. Schmahmann JD. An emerging concept. The cerebellar contribution to higher function. Arch Neurol

1991;48:1178–1187.
91. Fiez JA, Petersen SE, Cheney MK, Raichle ME. Impaired non-motor learning and error detection associated with cerebellar damage. Brain 1992;115:155–178.
92. Cummings JL. Frontal-subcortical circuits and human behavior. Arch Neurol 1993;50:873–880.
93. Buerger AA, Gross CG, Rocha-Miranda CE. Effects of ventral putamen lesions on discrimination learning by monkeys. J Comp Physiol Psychol 1974;86:440–446.

1996 Elsevier Science B.V.
Integrative and Molecular Approach to Brain Function.
M. Ito and Y. Miyashita, editors.

Neuronal activity during associative motor learning

Steven P. Wise[1], Longtang L. Chen[2] and Peter F. Dominey[3]

[1]Laboratory of Neurophysiology, NIMH, Poolesville, Maryland, USA; [2]Laboratory for Neural Information Processing, RIKEN, Saitama, Japan; and [3]Vision et Motricité, Bron, France and Institut des Sciences Cognitives, Lyon, France

Abstract. Advanced mammals can learn to make any limb or eye movement in response to virtually any detectable stimulus. We summarize selected neurophysiological data and one neural network model pertinent to such behavioral flexibility. Activity levels of supplementary eye-field neurons and the network's simulated neurons evolve in a systematic manner as monkeys learn arbitrary stimulus-saccade associations.

Key words: computational model, eye fields, frontal cortex, plasticity, premotor cortex.

Introduction

Neurobiologists have developed sophisticated models of how our motor system, and that of our fellow primates, guides the hand and eye toward (or away from) a target. However, our behavior would be scarcely recognizable were actions constrained to acquisition or avoidance movements. Our motor system can, based on experience and environmental context, select virtually any response to any stimulus, and such flexibility surely represents one of the higher functions of the brain.

Our goal is neurophysiological study of higher brain functions in nonhuman primates. Unfortunately, because they lack a well-developed language capacity, the study of these higher functions in nonhuman primates faces serious obstacles. According to some authorities, the ability to store the location of a visual stimulus represents a higher brain function, even "cognition", in nonhuman primates. To others, the ability to discriminate acoustic tones represents a precursor to language. Semantic quibbling aside, it seems existentially unimpressive to remember a simple location or to choose a higher pitched tone when to do so leads to reinforcement.

The present chapter discusses conditional motor learning, which appears to have the flexibility and diversity characteristic of a higher brain function (see [1,2] for a literature review). It can be defined as learning to select a motor response based on an instruction arbitrarily associated with that action [1—9]. Instructions usually consist of sensory, often visual, stimuli. No mere laboratory curiosity, conditional motor behavior turns up repeatedly in everyday experience: Imagine being unable to learn that a red light means "stop" or which telephone tone signals permission to dial. Thus

---

Address for correspondence: Steven P. Wise, Laboratory of Neurophysiology, NIMH, P.O. Box 608, Poolesville, MD 20837, USA.

would be life without conditional motor learning. We present two studies relevant to this higher brain function: one neurophysiological, the other a neural network simulation.

Behavioral design for conditional motor learning

Chen and Wise [10] recorded the activity of supplementary eye-field (SEF) neurons during oculomotor learning. In their task, the monkeys began each trial by fixating a small spot at the center of a video display (Fig. 1). A visual instruction stimulus (IS) was presented at that spot, termed the fixation point (Fig. 1, left). Each IS consisted of a combination of symbols, annuli, and bars of various hues and sizes. None of the ISs had any inherent relationship to a saccadic eye movement or eye-movement target. Nevertheless, each IS instructed the monkeys to make a saccade to one and only one of four potential eye-movement targets on each trial: 7° left, right, up or down from the central fixation point. The monkeys could not execute the saccade immediately upon receipt of the IS. Instead, they had to maintain gaze on the fixation point for a randomly varied instructed delay period of 1.5–3.5 s (Fig. 1, middle), after which the fixation spot disappeared. The disappearance of the fixation spot served as the trigger or "go" stimulus for the saccade. Then, the monkeys had 550 ms to fixate the target for 600 ms (Fig. 1, right) in order to receive reinforce-

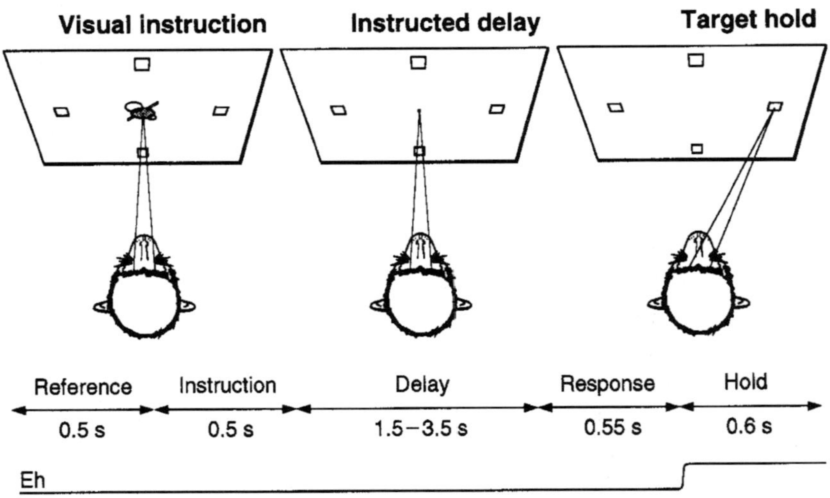

Fig. 1. Experimental design for the conditional oculomotor learning task [10,13]. Left: the monkey fixates the center of a video screen, where a complex visual stimulus appears. The stimuli each instruct a movement to one of the four targets. Middle: an instructed delay period ensues, in which the stimulus is replaced by a fixation spot for a randomly varied time. Right: the fixation spot disappears as the trigger or "go" stimulus. The monkey must then saccade to one of four targets (open squares) and hold fixation for 600 ms. Some stimuli were familiar to the monkey, others novel. Beneath the drawings of a monkey are the names and durations of the various task periods. Bottom: a schematic depiction of horizontal eye position (Eh) on the trial illustrated here, which requires a rightward saccade.

ment.

ISs were either familiar or novel. Most of the time, the IS on any given trial was selected randomly from a set of six to eight ISs, four of which were familiar and two to four of which were novel. Errors on familiar-IS trials were very rare (<2%) because the monkeys had extensive experience with those four ISs. By contrast, they had no experience with any of the novel ISs until the beginning of a learning block, and initially the monkeys made virtually random responses to those stimuli [11]. The monkeys had only reward or nonreward to guide subsequent response choices. After a brief trial-and-error phase, the monkeys learned to select the instructed saccade with a high and stable probability.

There are several important features of this experimental design and the subsequent data analysis. First, Chen and Wise focused their analysis exclusively on cell discharge rates during correctly executed trials. On correctly executed trials, the IS and the response were identical throughout learning. Thus, from the first correctly executed response (during the trial-and-error phase of learning) to the last (when learning had been consolidated) the stimuli and response were preternaturally equivalent. Second, the IS had to be attended because it instructed the saccade for each trial, on which any of several ISs might appear. And, third, the IS always occurred at a fixed (foveal) retinal location, which did not vary in terms of head- vs. eye-centered or viewer- vs. object-centered coordinates [12].

Neuronal activity during conditional motor learning

Chen and Wise chose to study SEF because it was adjacent to and connectionally similar with the dorsal premotor area, which is known to support conditional skeletomotor learning [1,2]. They categorized the observed single-neuron activity patterns into one of four classes: learning-dependent, learning-selective, learning-static, and unchanged [10]. Learning-dependent activity evolved during learning, i.e., it showed a significant time trend as the monkeys selected more "correct" responses to initially novel ISs. On familiar-IS trials, activity during the relevant task period significantly differed from that during a reference period. Learning-selective activity also evolved during learning, but on familiar-IS trials, when the monkey did not need to learn a new stimulus-response association, activity in the relevant task period failed to differ from the reference. In addition to learning-dependent and learning-selective activity, many cells showed no changes in task-related discharge during learning (unchanged), while others showed significant differences between familiar- and novel-IS trials without systematic time-trends during learning (learning-static activity).

Figures 2 and 3 illustrate learning-dependent activity in the SEF. Figure 2 shows a cell that had phasic, post-IS modulation for both downward and rightward saccades. The cell's modulation systematically increased as the monkey learned those two novel stimulus-response associations. The evolution of task-related activity was greater for downward than for rightward-saccade trials, but both were highly significant. Note that there was a fairly close correlation between the time course of performance improvements (open squares) and the learning-dependent increase in activity (filled

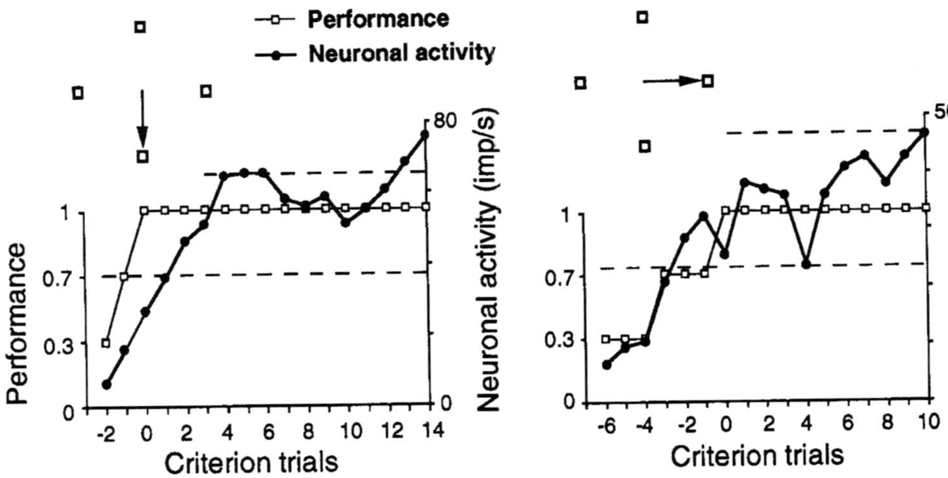

Fig. 2. Learning-dependent changes in activity in an SEF cell. Three-point moving averages of post-IS activity (circles) and performance (squares) for novel ISs instructing downward (left) and rightward (right) saccades. The first several correctly executed trials occur among several incorrect guesses. On those trials, the cell is relatively inactive. As the monkey learns the instructional significance of an initially novel IS, the cell discharges vigorously. The X-axis displays trials from the first instance of three consecutive correct choices in a three-point moving average (the criterion trial). Activity scales are in impulses/s. Performance scale shows proportion of correct responses per three trials.

circles). Because the visual stimulus and the motor response were constant throughout the learning block, Chen and Wise concluded that learning-dependent activity reflects the motor significance of the visual information.

Learning-dependent activity on novel-IS trials had two properties that helped rule out alternative interpretations: directional selectivity and convergence onto familiar-IS characteristics. Although not shown here, the cell illustrated in Fig. 2 had no significant modulation for the other two saccade directions, leftward and upward. The cell's directional selectivity rules out the possibility that learning-dependent activity reflected stimulus-reward association, expectation or anticipation of reward, or the affective valence of the various novel ISs. These related factors did not vary among saccade directions or ISs. Convergence of activity on novel- and familiar-IS trials, when they instruct identical saccades, resolved other interpretational issues. The dashed horizontal lines in Fig. 2 show the activity, plus or minus one standard deviation (SD), for the corresponding familiar ISs. Note that the familiar ISs were presented in trials that interleaved with those having novel ones. Modulation levels for the novel ISs converged on those seen for familiar ISs. This convergence argues against the possibility that the cell's activity primarily reflects stimulus features, such as would be expected in the inferotemporal cortex. As shown in Fig. 3, convergence was observed repeatedly for this SEF neuron. The top row of Fig. 3 shows three different blocks of trials, in which three different novel ISs instructed rightward

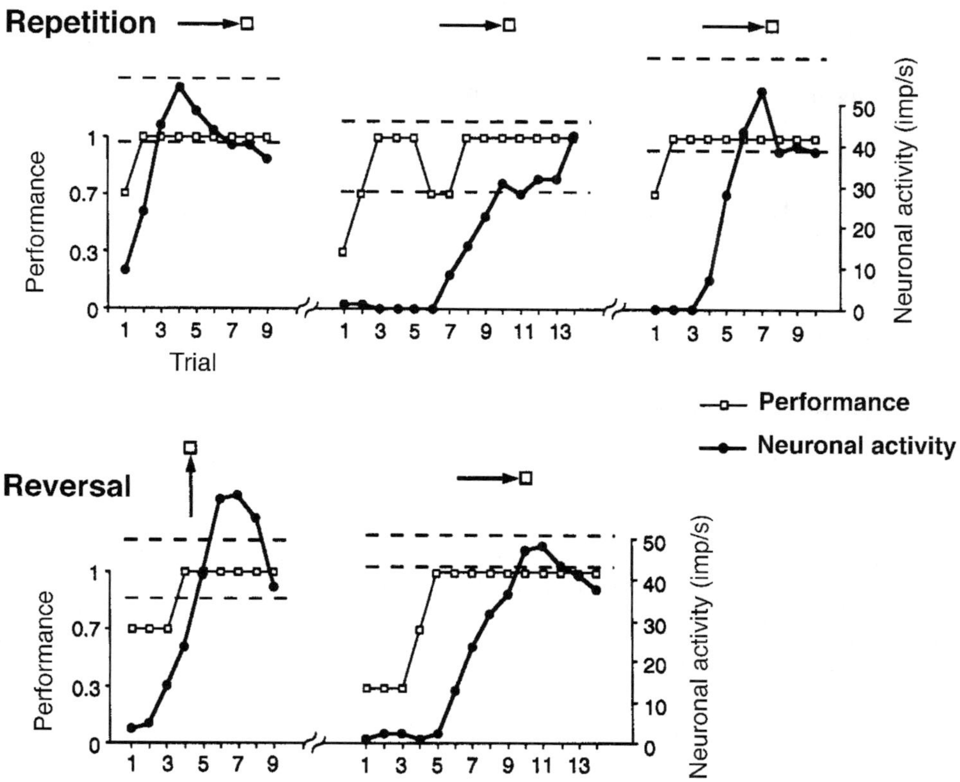

Fig. 3. Repeatability and reversibility of learning-dependent activity in an SEF cell. Top: repeated presentations of novel ISs. In separate blocks of trials, the monkey learned that three different, initially novel stimuli, instructed a rightward saccade. Bottom: reversal of the saccade direction instructed by a given IS. The saccade changed from instructing upward saccades in one block of learning trials to instructing rightward saccades in the subsequent block. Format as in Fig. 2.

saccades. Each time, activity rose into the SD range for the familiar IS. Figure 3 also shows the effects of reversing the instructional significance of an initially novel IS. After a block of trials in which the initially novel IS instructed upward saccades, it instructed rightward saccades. The cell's modulation decreased for the first correctly executed right-saccade trials (Fig. 3, bottom right), which indicates that its learning-dependent increase in modulation does not simply reflect sensitization to the IS.

Figure 4 shows a learning-selective SEF neuron. Learning-selective activity differed from learning-dependent activity in that this class lacked modulation (for the appropriate task period and saccade direction) for familiar ISs, i.e., on those trials when the monkeys did not need to learn a new conditional motor association. As the monkey learned the novel stimulus-response association, the cell's activity decreased until it eventually became completely silent.

Figure 5 shows the time course of learning-related changes in the SEF population,

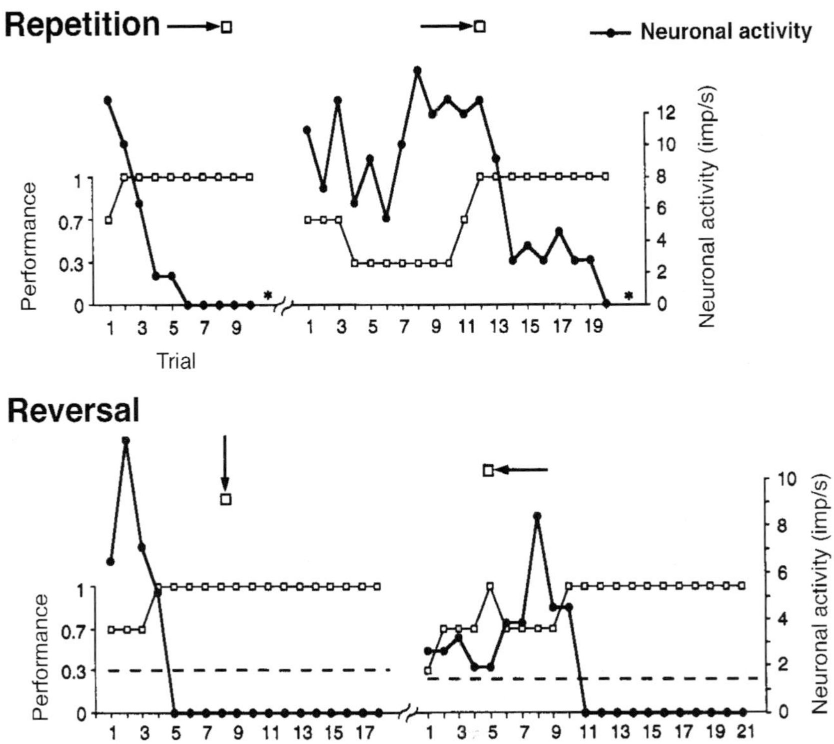

Fig. 4. Learning-selective changes in activity in an SEF cell, demonstrating repeatability and reversibility. Format as in Fig. 3. Asterisks on the top panels indicate zero activity for familiar-IS trials.

averaged over all saccade directions and all task periods. Learning-selective activity increased during the early phase of learning, reaching a peak just before the monkey reached the criterion trial (i.e., the first sequence of three consecutive, correct response choices, which is designated trial 0 in Fig. 5A). Learning-dependent activity — especially for increases during learning — predominated when the monkey reliably chose responses on the basis of learned visuomotor associations (Fig. 5A). However, the rigid distinction between learning-dependent and learning-selective neurons obscures some obvious similarities. Unlike the cells illustrated in Figs. 2 and 3, some activity classed as learning-dependent decreased during learning. As noted above, the learning-dependent classification depended on a statistical comparison with reference-period activity. Although the "decreasing" learning-dependent activity showed a significant task relationship for familiar-IS trials, its modulation level was often small. As a population, their activity levels closely resembled those for learning-selective activity (see Fig. 5B).

Together, the population curves shown in Fig. 5 provide strong evidence that much SEF activity reflects conditional motor learning. SEF is not the only such area: learning-dependent activity has also been reported for the dorsal premotor cortex

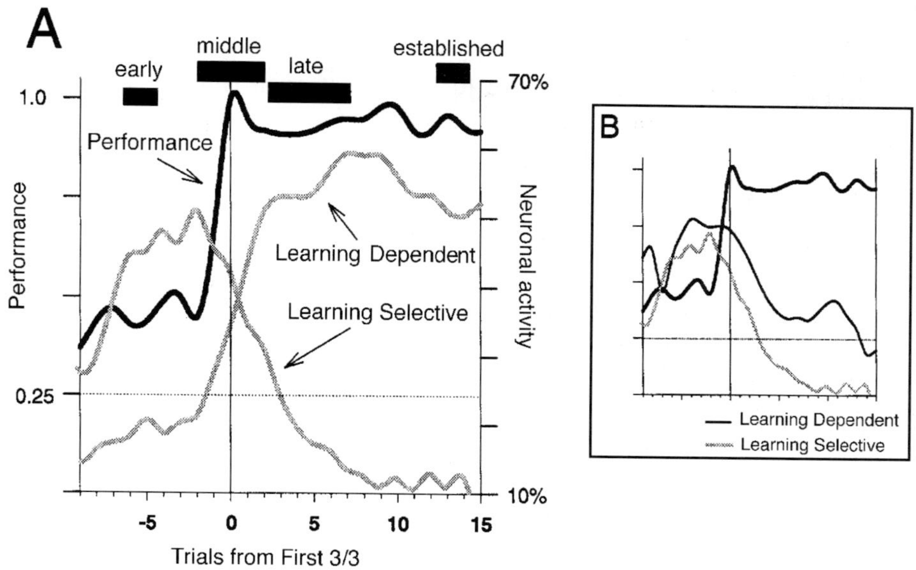

Fig. 5. Population curves from SEF, averaged over all directions and task periods, aligned on the criterion trial (see Fig. 2). **A**: Time course of activity changes in the subpopulation of learning-dependent cells with increasing activity during learning and in the population of learning-selective neurons (gray curves). The performance average for the same data set shown as the black curve. The approximate phases of learning are delimited at the top. **B**: Compares the subpopulation of learning-dependent cells with decreasing activity during learning with the learning-selective population and the time-course of learning. Axes and scales the same as in A.

during conditional skeletomotor learning [13,14]. But neurons displaying learning-related signals are not equally distributed among the motor, including oculomotor, areas of the cerebral cortex. Learning effects appear to be more common in SEF than in the "large-saccade" part of the frontal eye field [15] and in premotor cortex compared to the primary motor area [13].

Directional lability in SEF during learning and the population vector

Many SEF cells have different levels of activity associated with the various saccade directions. Chen and Wise, therefore, also estimated the preferred direction (PD) of SEF neurons during conditional motor learning [11]. They did this by computing the mean weighted circular average of unit vectors reflecting the cell's discharge rate for each of the four saccade directions. For example, the learning-dependent SEF cell illustrated in Fig. 2 had a PD mainly downward and somewhat to the right. PDs were apparent during all phases of learning, but they were rarely constant [11]. Only infrequently did a neuron's PD for novel ISs closely match that for familiar ISs throughout the learning process.

For familiar-IS trials, learning-dependent modulation might either have or lack a significant PD: termed directional and omnidirectional activity, respectively. For directional learning-dependent activity, the PD for novel-IS trials usually reoriented near the end of learning to resemble that for familiar-IS trials (Fig. 6). This property represents another aspect of the convergence onto familiar-IS-trial properties alluded to above. For omnidirectional activity, initially evident PDs dissipated with learning, even as some cells became more strongly modulated. Learning-selective activity typically showed a significant PD during the early phases of learning, but became unmodulated as learning progressed. Our findings show a pervasive lability in SEF PDs that may reflect a flexible and rapid remapping between inputs and responses within the premotor cortical network.

If, as described above, SEF directional neurons undergo dramatic changes in PD during learning, how could these neurons function as part of an oculomotor network? As Georgopoulos [16] and his colleagues have shown, representation of direction can be achieved by neuronal population, even in the face of considerable complexity of

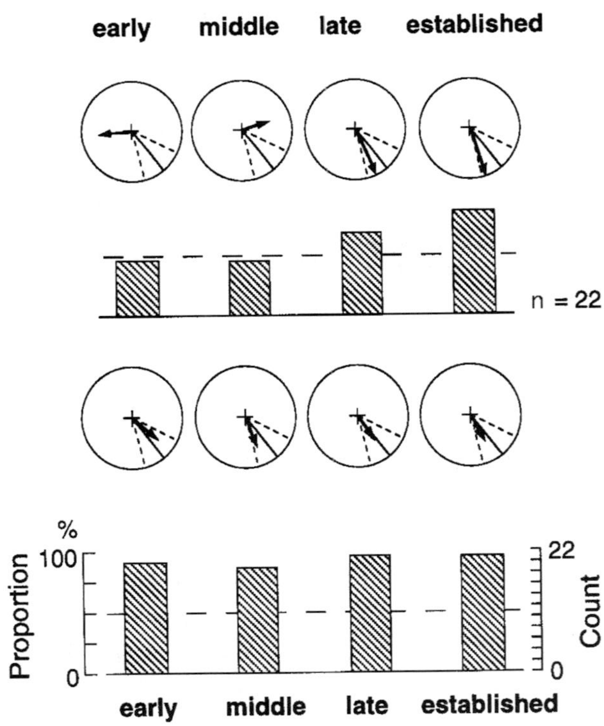

Fig. 6. Lability in preferred direction (PD) of an SEF cell during conditional motor learning. In each half (top and bottom) of the figure, the circles with solid and dashed radii show the PD for familiar-IS trials averaged over the entire learning block (solid) and its 99% confidence limits (dashed). The bar charts show the proportion of the learning-dependent population (n = 22) with PDs falling within 99% confidence limits of those for familiar-IS trials. Top: the PD during trials with novel ISs. Scale and axes the same as in the bottom half of the figure. Bottom: the PD during trials with familiar ISs, divided into the same learning phases as for the novel ISs. Note that the PD is relatively stable for the familiar stimuli, but labile for the novel ones, although both trial types for all four saccade directions are pseudorandomly interleaved.

individual cell activity [17]. Chen and Wise, accordingly, undertook a preliminary population vector analysis for learning-dependent activity. It was assumed that each neuron exerted an influence on the population in the direction of either: 1) its "instantaneous" directional vector, as computed from a circular mean weighted average for novel-IS trials during each learning phase; or 2) its PD for the familiar-IS trials averaged over the entire learning block. These assumptions were implemented in separated calculations. Each vector had a sign and magnitude, normalized to 1 in its maximal discharge direction, and –1 in its minimal discharge direction. As can be appreciated from Fig. 7, based on the second method mentioned above, the popula-

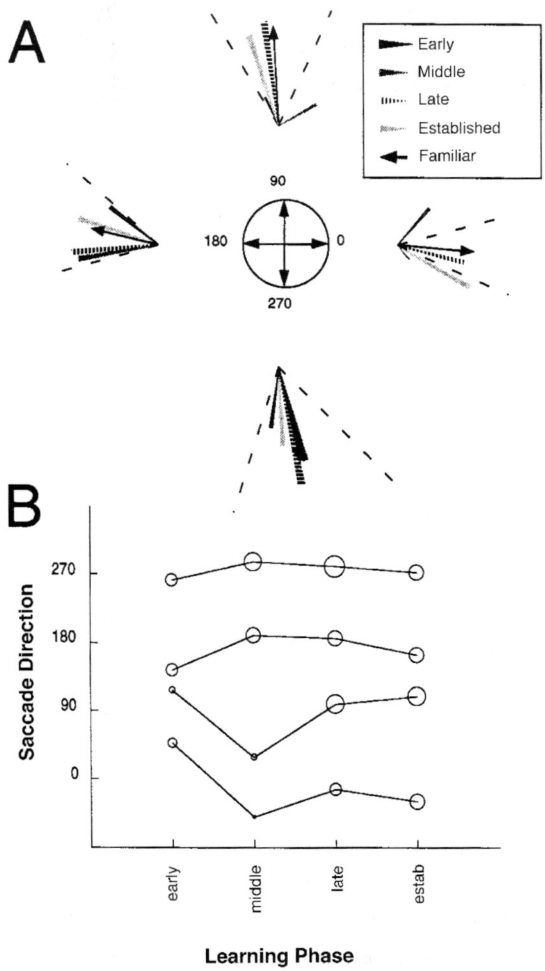

Fig. 7. Population vector for SEF during conditional motor learning, based on 22 learning-dependent cases (see Fig. 5). **A**: Population vector computed with respect to the circular mean weighted average for the familiar ISs in the entire learning block. Vectors with arrowheads and dashed lines indicate the population vector and 99% confidence limits for familiar-IS trials. **B**: Same data as in A, with the direction of the population vector plotted on the y-axis and the diameter of each circle proportional to the population vector's length.

tion vectors of familiar-IS trials pointed generally in the direction of the saccades. Moreover, for novel-IS trials the population vector for each learning phase usually fell within 99% confidence limits of population vectors for familiar-IS trials, especially toward the late and established phases of learning (Fig. 7A). Note that very rarely did individual PDs, among the 22 cases, point in exactly the saccadic directions, so this correspondence is not likely to result from the domination of the population by one or a few neurons. The correspondence between the population vector and the direction of movement is surprisingly close given the dramatic increases (Fig. 5A) and decreases (Fig. 5B) shown by the learning-dependent population, as well as the lability of their PDs. Further, because the calculations shown in Fig. 7 are based solely on learning-dependent activity, it appears that subpopulation alone sufficiently "represents" the saccade direction.

Conditional motor learning in anatomically structured models

From the foregoing, it seems that the PDs of individual SEF neurons are highly labile, but that the spatial information distributed over the SEF population as a whole has considerable stability. If that is so, what kind of mechanism could produce such properties? We are ignorant about the causal linkages, if any, between the activity Chen and Wise observed in SEF and the selection of saccadic eye movements. However, computational modeling can provide insight into how an information processing system could work, and the causal mechanisms of the model can be appreciated.

Dominey and his colleagues have recently developed a computational model of conditional motor learning based on the functional cortico-subcortical neuroanatomy of the oculomotor system [18]. In their model, which resembles one developed earlier by Fagg and Arbib [19], visual features of stimuli are represented in a simulated inferotemporal (IT) cortex, which projects to a region of premotor units. These premotor units have predetermined motor fields, and the model assumes that they participate directly in saccade generation. The strength, or weight, of the IT-to-premotor projection is labile. By reinforcement learning, synapses linking ensembles of IT neurons (which encode a given IS) to premotor neurons (which generate the appropriate saccade) become strengthened to yield error-free performance. Here, we present simulation results that provide insight into the learning-dependent, learning-selective and PD changes observed by Chen and Wise. Note that the allocation of the site(s) of synaptic plasticity to IT-projection targets does not imply that the projection must be monosynaptic or that the target necessarily corresponds to the SEF. In the original paper by Dominey et al., the target was posited to be the caudate nucleus, which is plausible, as well.

Figure 8 displays two simulated neurons, one with learning-dependent activity (gray circles) and the other with learning-selective activity (black circles). We have selected a learning-dependent neuron that contributes (as determined by the network architecture) to leftward saccades. The network operates in a winner-takes-all manner, hence the output of a premotor neuron can be considered a "vote" in its preferred

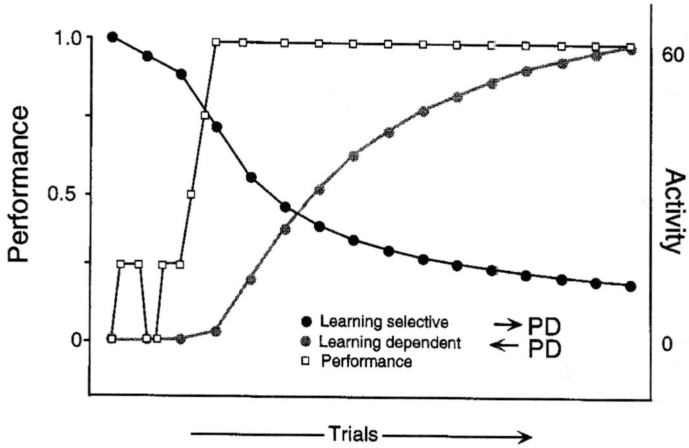

Fig. 8. Simulation of learning-dependent and learning-selective changes in a neural network model. PD represents the effective output of the simulated unit.

direction. The simulated learning-selective neuron contributes to rightward saccades in the same manner. In Fig. 8, the novel IS instructs a leftward saccade. Learning-dependent activity increases with learning, then stabilizes. Its activity increase results from augmentation of input from IT cells coding the novel IS, and it would foster the generation of a correct, leftward saccade. Learning-selective activity is essentially the inverse process. Initially strong IS activation of a premotor unit that does not participate in generating the correct saccade becomes weakened during learning as the invalid initial bias is removed. This decrementing synaptic weight results from a function built into the model, according to which each IT cell has a total synaptic output "weight" that is conserved. Thus, as synapses from an IT cell to some premotor cells are strengthened via the reinforcement for a correct response, other synapses from this same IT cell are weakened to conserve the total synaptic weight. Such a conservation function is necessary to prevent a long-term saturation of the network.

The model also sheds some light on the observation of PD lability (Fig. 6). As noted above, the simulated learning-selective cell illustrated in Fig. 8 contributes to rightward saccades (its PD construed as a motor field). The novel IS, which instructs leftward saccades, is initially biased to activate the learning-selective cell. During the early phase of learning, the learning-selective cell "responds" to the novel IS, and thus will contribute a rightward "vote" to the winner-takes-all output. Early in learning, this cell would contribute to "incorrect" saccades to the right. During the middle-to-late phases of learning, however, the initially novel IS will start to evoke "correct" saccades more frequently as the cells voting left begin to "outvote" those voting right. During this period, the learning-selective neuron continues to be activated by the novel IS, although to a lesser extent (Fig. 8). One consequence of this activation is the "incorrect" rightward vote noted above. Another is more subtle and more relevant to PD lability: activation of the cell during a "correctly" executed leftward-saccade trial will contribute a leftward component to the cell's PD, as it is usually defined. In the model, this PD component is due to the remaining influence of the novel IS. As this component is reduced, the cell's PD will change, depending on its activity for novel ISs instructing saccades in the other directions.

The model thus provides one explanation of PD lability. In this model, PD lability results from modifying sensory inputs to units with spatially fixed motor outputs, as opposed to modifying the effective output of these units. Of course, there are other possibilities, but this appears to be a plausible interpretation of the phenomena observed by Chen and Wise.

Conclusion

Most, if not all, vertebrates can move with respect to spatially localized stimuli. They can approach stimuli, especially if places or objects are associated with a positive affective valence, and they can avoid those with a negative valence. By contrast, conditional motor learning represents the kind of flexible behavior that is characteristic of advanced animals, including ourselves. We now know that a vast network of cells, including at least the striatum [20], the dorsal premotor cortex [13,14], and the supplementary eye fields [10,11,15], show discharge modulation that evolves during conditional motor learning. Many of these cells have activity that correlates very closely with the monkey's acquisition of stimulus-response association, whereas others show an initial increase followed by dramatic suppression as learning becomes consolidated. The evolving elements of this neuronal network are likely to play a central role in learning responses based on arbitrary associations, which, in turn, may underlie the behavioral flexibility that epitomizes the interaction of advanced animals with a rapidly changing, but partially predictable, environment.

Acknowledgements

PFD was supported by a postdoctoral fellowship from the Fyssen Foundation, Paris, France.

References

1. Passingham RE. The Frontal Lobes and Voluntary Action. Oxford: Oxford University Press, 1993.
2. Wise SP. Evolution of neuronal activity during conditional motor learning. In: Bloedel JR, Ebner TJ, Wise SP (eds) Acquisition of Motor Behavior in Vertebrates. Cambridge: MIT Press, 1996;261–286.
3. Halsband U, Freund H-J. Premotor cortex and conditional motor learning in man. Brain 1990;113: 207–222.
4. Halsband U, Passingham RE. The role of premotor and parietal cortex in the direction of action. Brain Res 1982;240:368–372.
5. Petrides M. Motor conditional associative-learning after selective prefrontal lesions in the monkey. Behav Brain Res 1982;5:407–413.
6. Passingham RE. Premotor cortex and preparation for movement. Exp Brain Res 1988;70:590–596.
7. Passingham RE. Premotor cortex and the retrieval of movement. Brain Behav Evol 1989;33:189–192.

8. Petrides M. Conditional learning and the primate frontal cortex. In: Perecman E (ed) The Frontal Lobes Revisited. New York: IRBN Press, 1987;91–108.
9. Kurata K, Hoffman DS. Differential effects of muscimol microinjection into dorsal and ventral aspects of the premotor cortex of monkeys. J Neurophysiol 1994;71:1151–1164.
10. Chen LL, Wise SP. Neuronal activity in the supplementary eye field during acquisition of conditional oculomotor associations. J Neurophysiol 1995;73:1101–1121.
11. Chen LL, Wise SP. Evolution of directional preferences in the supplementary eye field during acquisition of conditional oculomotor associations. J Neurosci 1996;16:(In press).
12. Olson CR, Gettner SN. Object-centered direction selectivity in the macaque supplementary eye field. Science 1995;269:985–988.
13. Germain L, Lamarre Y. Neuronal activity in the motor and premotor cortices before and after learning the associations between auditory stimuli and motor responses. Brain Res 1993;611:175–179.
14. Mitz AR, Godschalk M, Wise SP. Learning-dependent neuronal activity in the premotor cortex of rhesus monkeys. J Neurosci 1991;11:1855–1872.
15. Chen LL, Wise SP. Supplementary eye field contrasted with the frontal eye field during acquisition of conditional oculomotor associations. J Neurophysiol 1995;73:1122–1134.
16. Georgopoulos AP. Current issues in directional motor control. Trend Neurosci 1995;18:506–510.
17. Lurito JT, Georgakopoulos T, Georgopoulos AP. Cognitive spatial-motor processes. 7. The making of movements at an angle from a stimulus direction: Studies of motor cortical activity at the single cell and population levels. Exp Brain Res 1991;87:562–580.
18. Dominey P, Arbib M, Joseph JP. A model of corticostriatal plasticity for learning oculomotor associations and sequences. J Cognitive Neurosci 1995;7:311–336.
19. Fagg AH, Arbib MA. A model of primate visual-motor conditional learning. J Adapt Behav 1992; 1:3–37.
20. Tremblay L, Hollerman J, Schultz W. Neuronal activity in primate striatum neurons during learning. Soc Neurosci Abstr 1994;20:780.

1996 Elsevier Science B.V.
Integrative and Molecular Approach to Brain Function.
M. Ito and Y. Miyashita, editors.

Neural mechanisms of sequential procedural learning

O. Hikosaka[1], S. Miyachi[1], K. Miyashita[1], K. Sakai[1], S. Miyauchi[2], R. Takino[2], Y. Sasaki[2] and B. Pütz[2]

[1]*Department of Physiology, Juntendo University School of Medicine; and* [2]*Communications Research Laboratory, Tokyo, Japan*

Abstract. Brain mechanisms underlying procedural learning and memory are still poorly understood. As a first step to answer these questions, we designed a new experimental paradigm (called "2 × 5 task") for monkeys and humans. This task enabled us to examine the process of learning of new procedures as well as the execution of well-learned procedures repeatedly.

Using the 2 × 5 task, we examined the functions of the basal ganglia and the medial frontal cortex in procedural learning. This was done by injecting muscimol to inactivate the neural functions locally and reversibly. We found that there was anterior-posterior functional differentiation both in the basal ganglia and the medial frontal cortex. In the basal ganglia, the middle-posterior putamen was selectively related to storage or retrieval of procedural memory, while the anterior caudate/putamen was preferentially related to learning of new procedures. In the medial frontal cortex, the supplementary motor area (SMA) was related to execution of learned procedures, while the pre-SMA (anterior to SMA) was selectively related to learning of new procedures. The findings obtained in the monkey medial frontal cortex was confirmed by human experiments using functional MRI.

Key words: basal ganglia, caudate putamen, fMRI, monkey, presupplementary motor area.

Two types of memory

Memory can be divided into two groups: memory of events or facts, and memory for skills and rules [1,2]. They are often called declarative memory and procedural memory. An accumulation of declarative memories will give rise to knowledge; an accumulation of procedural memories give rise to intelligence. The usage of language, e.g., may be the most advanced form of procedural memory.

This classification of memory is derived from the observation that amnesic patients, who cannot remember daily episodes, can learn new behavioral procedures [3]. Because we know that the medial temporal areas including the hippocampus are crucial for declarative memory, we naturally assume that neural systems for procedural memory should be somewhere outside the medial temporal areas [1,4].

However, we still know little about the neural mechanisms of procedural memory and learning. How are procedural memories acquired? Where are they stored? How are the stored memories retrieved? Are different neural systems involved in these processes? We have no definite answers to any of these questions. It is partly because

---

Address for correspondence: Okihide Hikosaka, Department of Physiology, Juntendo University, School of Medicine, 2-1-1 Hongo, Bunkyo-ku, Tokyo 113, Japan. Tel.: +81-3-5802-1026. Fax: +81-3-3813-4954. E-mail: hikosaka@med.juntendo.ac.jp procedural memory is difficult to examine experimentally. The acquisition of procedural memory cannot be done just by showing something to the subject; the subject has to perform the procedure repeatedly. The presence of procedural memory cannot be indicated just by yes/no answer; it can be expressed only by performing the procedure.

Our research on procedural learning and memory is an attempt to break through such difficulties. Using the following behavioral paradigm we have obtained some hints on the neural mechanisms of procedural memory.

Procedure of 2 × 5 task

In front of the monkey, is a panel on which 16 LED buttons are arranged in a 4 × 4 matrix (Fig. 1) [5]. Beneath the panel is another button called "home key". If the monkey presses the home key, two of the 16 LED buttons are illuminated simultaneously. The monkey has to press them in a correct (predetermined) order which he has to find out by trial-and-error. This is called "set". If successful, the LED buttons turned off sequentially and another pair of LEDs, a second set, is illuminated which the monkey has to press again in a correct order. A total of five sets is presented in a fixed order for completion of a trial, which we call "hyperset". If the monkey presses a wrong button, the trial is aborted and the monkey has to start again from the home key as a new trial.

A major advantage of this 2 × 5 task is that new hypersets can be generated practically as many as possible. With daily practice, each monkey became able to perform more than 20 hypersets almost perfectly. Particularly important here is that even at this advanced stage we can ask the monkey to learn totally new sequences.

Even if one stops riding a bicycle for a long time, one can do it again without much difficulty [6]. Two of our monkeys also had a chance to take a rest for 6

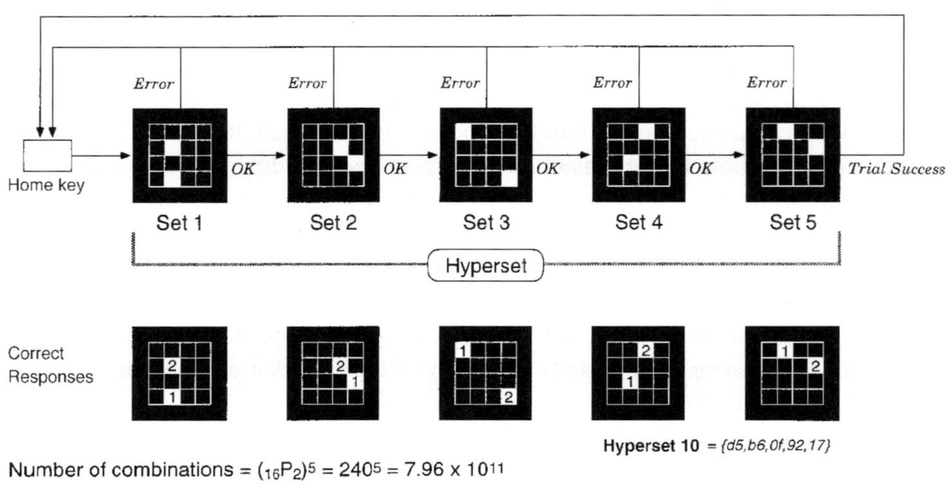

Fig. 1. Procedure of 2 × 5 task for monkeys.

months, forgetting about 2 × 5 task [7]. Now, would they remember any of the learned hypersets after the interruption? We compared the performances of the monkeys, namely the number of errors, on the 14 learned hypersets just before the interruption and the first time after the interruption. The median number of errors for these hypersets was just one before the interruption and after the interruption. Both were nearly complete performances, which contrasted with such frequent errors (more than 20) when the same hypersets were performed for the first time.

This observation prompted us to raise one question. In order to learn new things, the brain must be very plastic. In order to retain memories and skills, the brain must be solid. This seems a paradox. It is at least unlikely that the same neural mechanisms work both for learning new things and for storing long-term memories.

We thus hypothesized that different neural systems are involved in the two processes: one for learning of new procedures, and the other for execution of well-learned procedures (Table 1). Of course, there are other possibilities. But they can, in principle, be differentiated using the 2 × 5 task. Let me tentatively call these mechanisms "learning mechanism" and "memory mechanism".

If the learning mechanism does not work, new hypersets would not be learned while already learned hypersets would be performed normally. If, on the other hand, the memory mechanism does not work, learned hypersets would not be performed while new hypersets would be learned normally.

Here we see the advantage of the 2 × 5 task. It allowed us to examine the process of learning repeatedly using new hypersets while at the same time examine the retrieval of long-term memory using learned hypersets.

There are many brain areas that might be involved in procedural memory. Let us extend the idea of the procedural-declarative dichotomy of memory. If declarative memory is a kind of sensory memory, procedural memory is a kind of motor memory [8]. Consider piano playing and keyboard typing. Invariably critical for them are motor skills. Therefore, it seems natural to hypothesize that brain areas that are thought to be important for voluntary movements are also important for procedural learning and memory.

Heterogeneous functions of the basal ganglia in procedural learning

Some people have advocated the idea that the basal ganglia are particularly important for procedural memory [9–11], which seems analogous with the function of the hippocampus in declarative memory. To test the hypothesis especially in relation to the scheme in Table 1, we eliminated the activity of the basal ganglia locally and reversibly by injecting a small amount of muscimol, a GABA agonist [12].

Figure 2 shows the injection sites in the monkey striatum. The injections were classified into three groups based on the effects. Group 1: the anterior striatum, including the head of the caudate, nucleus accumbens, and the anterior part of the putamen. Group 2: the middle-posterior part of the putamen. Group 3: the middle-posterior part of the caudate. In most cases, the injections were bilateral, two to four locations. For control experiments, we injected saline into these areas. We performed

Table 1. Hypothetical mechanisms underlying procedural learning and memory. The mechanism for learning of new procedures (learning mechanism) and the mechanism for storage of long-term memories (memory mechanism) are assumed to be separate from each other. This hypothesis can be tested by the local functional blockade of each of the corresponding neural mechanisms, as indicated at top.

Dysfunction of:	Performance in:	
	Learned hypersets	New hypersets
Learning mechanism	No change	↓
Memory mechanism	↓	No change

essentially the same experiments in two monkeys.

In the control experiments in which saline was injected into the striatum, the mean number of errors before completing 10 successful trials was much smaller for learned hypersets than that for new hypersets (mean: 0.3 and 16.9 for monkey 1, 0.3 and 10.0 for monkey 2, respectively).

After muscimol injections into the anterior caudate and putamen, the number of errors increased for new hypersets (mean: 37.4 for monkey 1, $p < 0.05$ by two-tailed t test; 21.6 for monkey 2, $p < 0.001$). In contrast, the number of errors for learned hypersets did not increase significantly (mean: 2.5 for monkey 1, $p > 0.05$; 1.0 for monkey 2, $p > 0.05$). This result suggests that the anterior caudate/putamen contributes to the process of learning (learning mechanism) but not to the storage or retrieval of long-term memory (memory mechanism).

After the injections into the middle-posterior putamen, the number of errors increased for learned hypersets (mean: 3.0 for monkey 1, $p < 0.001$; 2.2 for monkey 2, $p < 0.001$), but not for new hypersets (mean: 20.9 for monkey 1, $p > 0.05$; 13.0 for monkey 2, $p > 0.05$). This suggests that the middle-posterior putamen is related to the memory mechanism, not learning mechanism.

The blockade of middle and posterior caudate produced no significant changes in number of errors for learned hypersets and new hypersets.

Of course, the basal ganglia would not work by themselves. Their intimate interconnections with the cerebral cortex should now be considered [13].

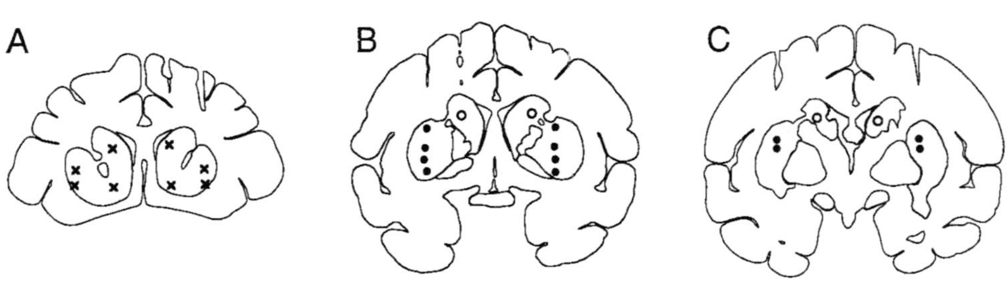

Fig. 2. Sites of muscimol injection in the monkey striatum. **A**: 6 mm anterior to the anterior commissure (AC); **B**: 2 mm anterior to the AC; **C**: 2 mm posterior to the AC. These injection sites were classified into three groups: anterior caudate and putamen (×); middle and posterior putamen (●); middle and posterior caudate (○).

Role of the monkey presupplementary motor area in learning of new sequences

There are many candidate areas in the cerebral cortex, but we so far have concentrated on the medial frontal cortex, especially the supplementary motor area (SMA). Our studies were motivated by many previous studies suggesting that the SMA is related to sequential procedures. Tanji and other people have proposed that the medial frontal cortex that was traditionally called the SMA should be divided into two regions: the SMA proper posteriorly and the pre-SMA anteriorly [14–18]. We in fact found that the SMA and pre-SMA are quite different in terms of procedural learning and memory.

Our immediate question was again based on the hypothesis for the separate mechanisms for learning and memory (Table 1): Is the SMA or pre-SMA related to the process of learning (learning mechanism) or related to the storage/retrieval of long-term memories (memory mechanism)?

We tried to answer this question by recording single-cell activities in these areas [19]. According to the hypothesis shown in Table 1, neurons belonging to the learning mechanism should be active during learning of new procedures, while neurons belonging to the memory mechanism should be active during execution of learned procedures.

Among 185 neurons we recorded, 35 were statistically more active in new hypersets than in learned hypersets; 21 were more active in learned hypersets than in new hypersets. We further found that neurons preferentially related to new hypersets tended to be localized in the pre-SMA. Neurons preferentially related to learned hypersets were distributed more widely, including the SMA and dorsal convexity of the cortex. There were some neurons that showed learning-dependent changes in spike activity more selectively depending on different sets or hypersets. They were distributed posteriorly, probably in the SMA.

The results suggest that pre-SMA neurons are preferentially related to the process of learning, while SMA proper neurons are related to the execution of sequential movements. Furthermore, we obtained essentially the same results in humans using functional MRI, as shown below.

Human fMRI study on procedural learning

The subject lay in the supine position in the MRI scanner and held a plate with four button switches. Through a mirror, the subject saw four white rectangles on a screen in which two circles appeared in different colors. The subject had to press the two corresponding buttons to turn off the circles.

The task was the same as the one used for the monkeys with some modifications [20]. For control, we used the task that we call "pseudolearning" task. True learning task and the pseudolearning task were nearly identical in terms of sensory motor events. The only difference was who learned the sequence: in the true learning, the subject learned; in the pseudolearning, the computer learned the sequence. Therefore, what we obtained in this experiment was the brain areas specifically related to learning of sequences, not sensory motor activities.

We found that a localized area in what we regarded to be the human pre-SMA was consistently active. The data shown in Fig. 3 were obtained in a single subject, but in different experiments. We regarded the active focus as the pre-SMA, because it was located slightly more anteriorly to the anterior commissure along the axis connecting the anterior commissure and posterior commissure.

In the learning vs. pseudolearning paradigm, we did not observe any significant activation in the SMA proper (which we identified in the region posterior to the AC). This was probably because we extracted the learning-selective activation by comparing the test scans (true learning) and the control scans (pseudolearning). To test this idea, we used another paradigm ("perform vs. watch paradigm") to extract movement-related activities unrelated to learning. In the test scan using the "perform task", the subject was allowed to press the corresponding buttons in any order; in the control scan using the "watch task", the target circles turned on and off at the same rate as if someone else (or the computer) was performing the task. In the test and control scans, the subjects watched the same stimuli turning on and off. The only difference was who actually performed the task. Of course, no learning was involved.

The results of this experiment using the perform vs. watch paradigm were clearly different: the SMA proper was activated consistently, while the pre-SMA showed little activation. Such contrasting activations of the SMA vs. pre-SMA were consistently observed in three subjects who were studied repeatedly.

These results suggest, in humans, that the pre-SMA is preferentially related to

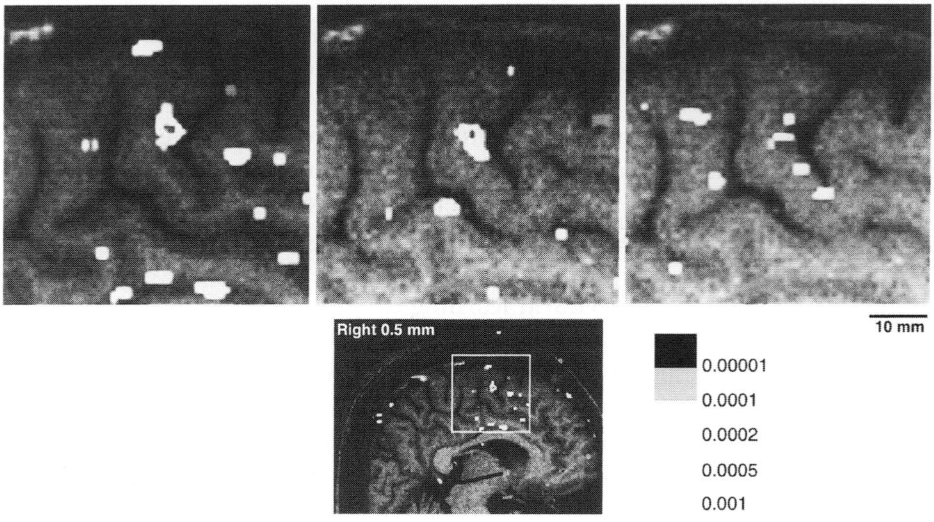

Fig. 3. Consistent activation of the pre-SMA during learning of sequential procedures in a single human subject. The SMA/pre-SMA region (white rectangle in the inset figure below) is enlarged for each of the three experiments. The subject experienced a new hyperset for each of the three experiments. Common to these data was a local activation that was slightly anterior to the anterior commissure (AC) relative to the axis connecting the AC and PC (posterior commissure), as shown in the inset figure below. The levels of significance for the t test images are color-coded according to the legend at the right-bottom.

learning of new sequences, not movements, while the SMA proper is preferentially related to movements, not learning. This suggestion is very similar to the suggestions we obtained in monkey experiments.

Conclusion

Our studies using trained monkeys, supported by human fMRI studies, suggested that there may indeed be separate mechanisms in the brain, one for learning of new procedures (learning mechanism) and the other for storage and retrieval of well-learned procedures (memory mechanism) (Fig. 4). However, each of these mechanisms may be distributed across anatomical boundaries. Thus, the pre-SMA in the cerebral cortex and the anterior striatum in the basal ganglia would constitute the learning mechanism. We would suggest that the middle-posterior part of the putamen, perhaps together with other brain areas, would constitute the memory mechanism.

However, the role of the SMA is, at present, still obscure. Furthermore, we have left out two important structures: primary motor cortex and cerebellum, which we will try to examine in the future studies.

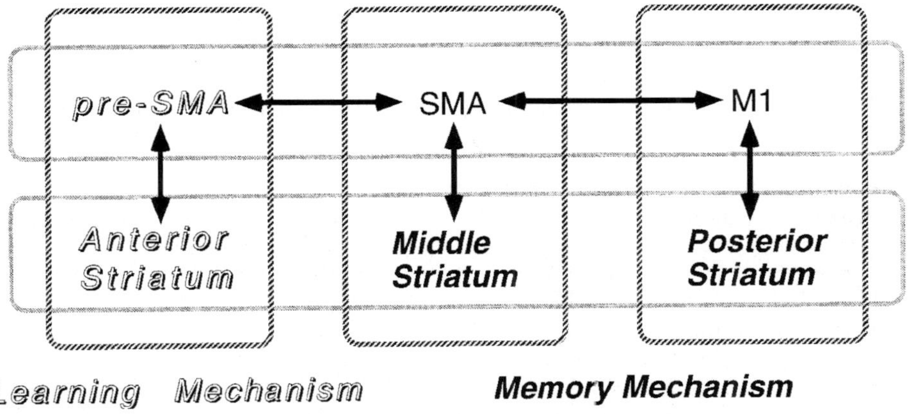

Fig. 4. Summary scheme showing the functional differentiation (learning mechanism and memory mechanism) in the neural systems consisting of the medial frontal cortex and the basal ganglia.

Acknowledgements

This research was supported by Uehara Memorial Foundation.

References

1. Squire LR. Mechanisms of memory. Science 1986;232:1612–1619.
2. Tulving E. How many memory systems are there? Am Psychol 1985;40:385–398.
3. Cohen NJ, Squire LR. Preserved learning and retention of pattern-analyzing skill in amnesia: Dissociation of knowing how and knowing that. Science 1980;210:207–210.

4. Mishkin M, Appenzeller T. The anatomy of memory. Sci Am 1987;256:62–71.
5. Hikosaka O, Rand MK, Miyachi S, Miyashita K. Learning of sequential movements in the monkey – Process of learning and retention of memory. J Neurophysiol 1995;74:1652–1661.
6. Kolers PA. Reading a year later. J Exp Psychol: Human Learn Mem 1976;2:554–565.
7. Hikosaka O, Miyachi S, Miyashita K, Rand MK. Learning of sequential procedures in monkeys. In: Bloedel JR, Ebner TJ, Wise S (eds) The Aquisition of Motor Behaviour in Vertebrates. Cambridge, MA: MIT Press, 1996;303–317.
8. Hikosaka O, Miyachi S, Miyashita K, Rand MK, Karadi Z. Procedural learning in monkeys – possible roles of the basal ganglia. In: Ono T, McNaughton BL, Molotchnikoff S, Rolls ET, Nishijo H (eds) Perception, Memory, and Emotion: Frontier in Neuroscience. Oxford, UK: Elsevier, 1996;403–420.
9. Knopman D, Nissen MJ. Procedural learning is impaired in Huntington's disease: evidence from the serial reaction time task. Neuropsychologia 1991;29:245–254.
10. Mishkin M, Malamut B, Bachevalier J. Memories and habits: Two neural systems. In: Lynch G, McGaugh JL, Weinberger NM (eds) Neurobiology of Human Learning and Memory. New York: The Guilford Press, 1984;65–77.
11. Saint-Cyr JA, Taylor AE, Lang AE. Procedural learning and neostriatal dysfunction in man. Brain 1988;111:941–959.
12. Miyachi S, Miyashita K, Karadi Z, Hikosaka O. Effects of the blockade of monkey basal ganglia on the procedural learning and memory. Soc Neurosci Abstr 1994;20:357.
13. Strick PL, Dum RP, Mushiake H. Basal ganglia "loops" with the cerebral cortex. In: Kimura M, Graybiel AM (eds) Functions of the Cortico-Basal Ganglia Loop. Tokyo: Springer, 1995;106–124.
14. Matsuzaka Y, Aizawa H, Tanji J. A motor area rostral to the supplementary motor area (presupplementary motor area) in the monkey: neuronal activity during a learned motor task. J Neurophysiol 1992;68:653–662.
15. Tanji J. The supplementary motor area in the cerebral cortex. Neurosci Res 1994;19:251–268.
16. Luppino G, Matelli M, Camarda RM, Gallese V et al. Multiple representations of body movements in mesial area 6 and the adjacent cingulate cortex: an intracortical microstimulation study in the macaque monkey. J Comp Neurol 1991;311:463–482.
17. Rizzolatti G, Gentilucci M, Camarda R, Gallese V et al. Neurons related to reaching-grasping arm movements in the rostal part of area 6 (area 6aβ). Exp Brain Res 1990;82:337–350.
18. Dum RP, Strick PL. The origin of corticospinal projections from the premotor areas in the frontal lobe. J Neurosci 1991;11:667–689.
19. Miyashita K, Hikosaka O, Lu X, Miyachi S. Neuronal activities in medial premotor area of monkey during learning of sequential movements. Soc Neurosci Abstr 1995;21:1928.
20. Hikosaka O, Sakai K, Miyauchi S, Takino R, Sasaki Y, Putz B. Activation of human presupplementary motor area in learning of sequential procedures: A functional MRI study. J Neurophysiol 1996;76:617–621.

Noninvasive measurements in humans

Integrative and Molecular Approach to Brain Function.
M. Ito and Y. Miyashita, editors.

Human cortical reactivity revealed by neuromagnetic recordings

Riitta Hari

Brain Research Unit, Low Temperature Laboratory, Helsinki University of Technology, Finland

Abstract. Neuromagnetic recordings reflect synchronous cortical neuronal activity with millisecond time resolution. With the state-of-the-art whole-scalp neuromagnetometers one can monitor temporal dynamics of human cortical functions during various complex tasks. Examples will be given about the reactivity of the human cerebral cortex reflected in both evoked and spontaneous neuromagnetic activity.

Key words: auditory cortex, brain rhythms, inverse problem, magnetoencephalography, somatosensory cortex.

Magnetoencephalography

In magnetoencephalography (MEG), one records weak magnetic fields outside the head [1,2]. MEG, although closely related to EEG, benefits from the skull's and scalp's transparency to magnetic fields: the MEG signals, which arise mainly from synaptic currents in fissural cortex can be picked up outside the head undisturbed by tissue inhomogeneities. MEG typically sees only a part of the whole cortical activity, which leads to more straightforward interpretation of the signal patterns than of the corresponding electric scalp potentials.

Compared with positron emission tomography (PET) and functional magnetic resonance imaging (fMRI), the advantage of MEG is its good temporal resolution which allows monitoring of cortical dynamics in millisecond scale. It is commonly suggested that combination of PET/fMRI with MEG should be performed by first defining active brain regions from PET/fMRI data and then using these sites as sources in the inverse solution of the MEG data to reveal the temporal behavior of the active brain areas. This approach will certainly work in many cases. It should be noted, however, that changes in the synchronization of a neuronal population, reflected in the MEG signals, may occur without significant changes in the mean neuronal firing rates, and thus without changes of blood flow and metabolism. Thus, modulation of neuronal synchrony, although potentially of high importance for local information processing, may remain undetected in PET and fMRI. Combination of structural information from MRI and functional information from MEG is at present in routine use in many laboratories.

The inverse problem, i.e., the deduction of neuronal currents from the measured external electric potential or magnetic field distribution, does not have a unique

Address for correspondence: Brain Research Unit, Low Temperature Laboratory, Helsinki University of Technology, 02150 Espoo, Finland. Fax: +358-0-4512969. E-mail: hari@neuro.hut.fi solution. Therefore, the generators of MEG signals are typically modelled with one or several current dipoles, each representing activation of a local cortical area. The safest way to interpret the complex cortical activation patterns is to build up the knowledge step-by-step, starting from brain areas with well-known activities.

Two main approaches exist to study reactivity of the human cerebral cortex by means of neuromagnetic recordings. One can either collect evoked responses triggered by various sensory stimuli or to record spontaneous activity during various tasks; interest in human brain rhythms is rapidly increasing with the emergence of neuromagnetometers with extended sensor arrays covering the whole scalp (for earlier multichannel devices, see Fig. 1). The following paragraphs will give examples of both of these approaches.

Evoked responses

Auditory system

The source locations of consecutive deflections of the auditory evoked magnetic fields vary slightly, implying that the center of gravity of activation changes as a function of time between different cytoarchitectonic areas of the auditory cortex. The most prominent signals peak about 100 ms after the sound onset or a change within a sound.

Several MEG recordings have documented the sensitivity of the auditory cortex to various changes in the sensory input [3]. One way to study these processes is to present the stimuli in an oddball manner: a monotonous sequence of "standard" sounds is interrupted from time to time by a physically deviant sound. The auditory cortex and also some other brain regions react strongly to the deviants, often with right-hemisphere dominance [4].

A random omission of some sounds of the sequence is a dramatic change of stimulation. When the omissions are attended by the subject, they produce strong

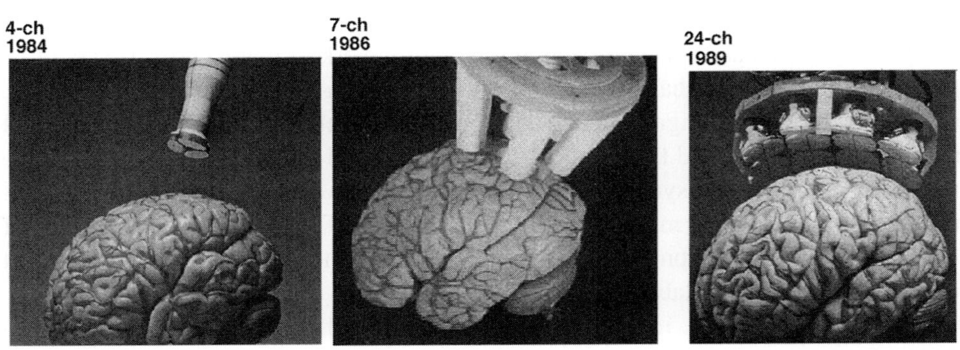

Fig. 1. Examples of multichannel neuromagnetometers constructed in the Low Temperature Laboratory before the era of whole-scalp coverage. The devices are at typical measurement distance above the brain model. Note that the instruments and the brain are presented without their normal surroundings for illustrative purposes only; the method is totally noninvasive!

responses which peak 145–195 ms after the predicted sound onset and are seen over both hemispheres for monaural stimulus sequences [5]. A major part of the signals appears to be generated in the supratemporal auditory cortex, suggesting that the sensory specific projection cortex keeps a track of quite complex changes in the stimulation environment. Here we may be evidencing effects of top-down processing at the level of the auditory cortex. The omissions also activate the superior temporal sulcus and frontal lobe, again more prominently in the right than in the left hemisphere [5].

Somatosensory system

Electric stimulation of median or ulnar nerve evokes a synchronous action potential burst which reaches the contralateral primary somatosensory cortex, SI, within 20 ms; the earliest cortical signal reflects activation of area 3b in the posterior wall of the rolandic fissure. Activation of some cytoarchitectonic areas (e.g., 1 and 2) of the hand SI cortex may remain undetected by MEG, because the net current flow is radial with respect to the skull. However, in the foot somatosensory region all walls of the cortex can be assumed to be tangential with respect to the surface (see Fig. 2). Consequently, activation of these areas is seen as a rotation of the field patterns and of the corresponding source currents [6]. It turns out that the activation sequence in the foot area cannot be explained by a sequence analogical to that seen in the hand area (Fig. 2). These functional differences, which are surprisingly strong compared with animal data, probably result from the human upright position which frees the hands for skilful work and leaves locomotion, position control and fight against gravity largely to lower limbs. Thus, the human cerebral cortex receives very specialized inputs from the upper and lower extremities.

The sites of earliest SI activity are somatotopically organized and the course of the central sulcus can be easily determined by stimulating of various body regions. Such information is useful for preoperative evaluation of patients when one wants to define the relative location of an epileptic focus or of some other abnormality with respect to functionally irretrievable areas.

Whole-scalp recordings after electric stimulation of upper limb nerves imply activation of several cortical regions. Signals from the contralateral SI cortex are followed by bilateral activation of SII cortices and by contralateral activation of the posterior parietal cortex. Recently, Forss et al. [7] observed additional signals in the mesial cortex, close to area 4, when the subject attended to infrequent, randomly presented ulnar and median nerve stimuli.

Spontaneous activity

Parieto-occipital α *rhythm*

Sources of the posterior 10 Hz α rhythm cluster predominantly in the parieto-occipital region and to a smaller extent in the occipital areas. The level of the α decreases

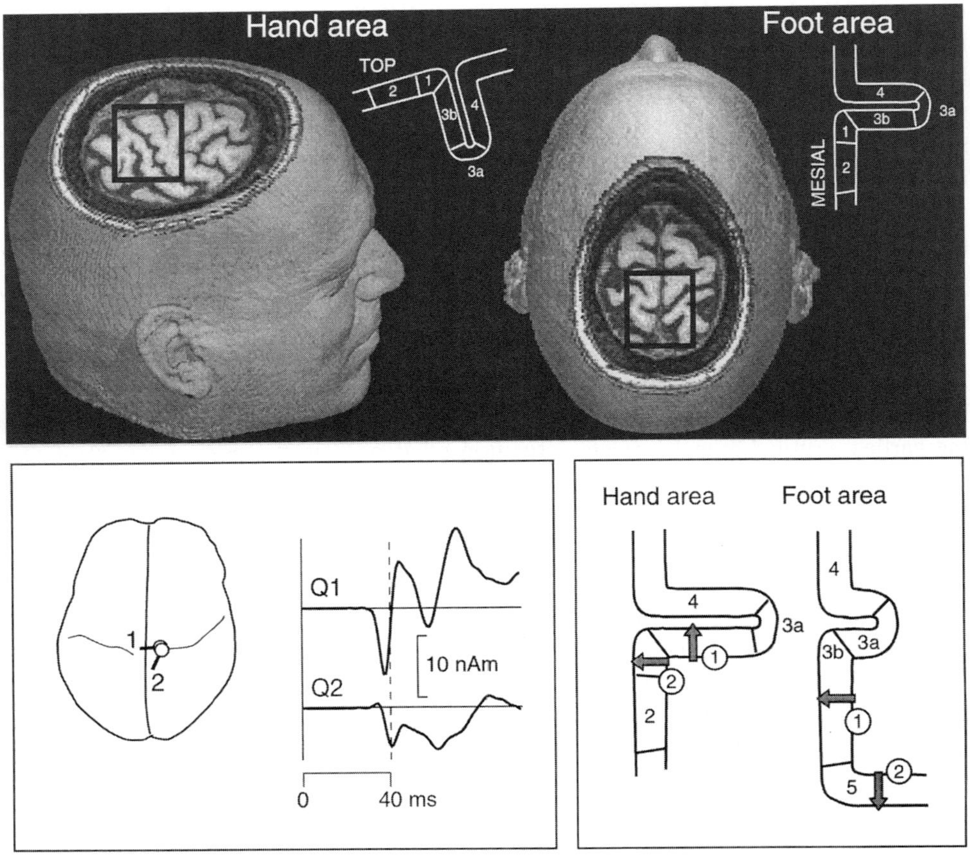

Fig. 2. Above: Schematic presentation of the cytoarchitectonic areas in hand and foot somatomotor area. Bottom left: Two dipoles (1 and 2) in the foot area used to model tibial nerve SEFs of one subject, and the corresponding source strengths (Q1 and Q2) as a function of time. These sources explained the measured responses extremely well during the first 100 ms. Bottom right: Models for sequential activation (circled 1 and 2) in the hand and foot somatomotor cortex after electric stimulation of median and tibial nerves, respectively. For detailed argumentation for the hypothesised structure and function of the foot area, see [6].

during visual stimulation. The suppression occurring during passive viewing of pictures is considerably intensified when the subject is asked to name the pictures, either silently or aloud [8].

Changes very similar to those elicited by visual stimuli may be produced also by visual imagery [9], suggesting involvement of nonprimary visual areas close to the parieto-occipital sulcus in generation and inspection of mental images. The same brain region was activated in a recent PET study in association with memory-related imagery [10].

Somatomotor rhythms

MEG signals over the somatomotor regions display dominant frequencies around 10 and 20 Hz. These frequency components can occur either separately or simultaneously. Both of them react with a rebound to the offset of a brisk movement but the rebound is slightly faster and much stronger for the 20-Hz than for the 10-Hz signals [11]. The signals originate close to the hand primary somatomotor cortex, with more anterior sources for the 20-Hz than the 10-Hz component [11,12]. Thus, the 20-Hz component has been assumed to receive contribution from the precentral motor cortex. This conjecture is supported by the observed reactivity of the rhythms after movements: Although the sources of the 10-Hz rhythms cluster close to the hand region, regardless of the body part moved, the 20-Hz rhythms follow the somatotopic organization of the body parts along the precentral gyrus, shifting from the most medial position for foot movements via the hand area to the most lateral location for mouth movements [13].

Strong rebounds of the 20-Hz activity also occur after electric stimulation of the median nerve at the wrist at intensities exceeding the motor threshold. The rebound is abolished by simultaneous exploration of objects with the fingers of the stimulated hand. Interestingly, the rebound is also strongly suppressed when the subject just imagines exploratory finger movements [14], indicating involvement of the primary motor cortex in motor imagery.

Recent data suggest that the somatomotor rhythms may have a role in the coordination of neural activity between the central and peripheral nervous systems. For example, Conway et al. [15] found a close coupling between rhythmic MEG signals from the motor cortex and the electromyogram from the interosseus muscle during sustained contraction. They hypothesized involvement of cortical neurons in the generation of motor-unit synchronization. Moreover, Salenius et al. [16] found an increase in the level of the 40-Hz rolandic MEG activity during and after slow index finger movements in a subject with unusually prominent 40-Hz EEG activity. The reactive 40-Hz rhythm originated in the motor cortex and showed tight phase coupling to the electromyogram from the moving muscle, with a time difference of 20 ms. It thus seems evident that, in some conditions, somatomotor cortical rhythms are closely related to the supraspinal control of the motoneurone pool. The frequency range of optimal control is likely to be highly individual.

One can speculate that the observed coherent rhythms in the motor cortex reflect association of different cortical units to functional assemblies which are necessary to control the relative contraction strengths of muscles involved in voluntary high-precision movements. Thus, one would expect changes in the coherent frequencies during varying movements performed with the same muscles.

Present trends

One trend in neuromagnetic studies of human cortical functions is to progress towards more real-life-like stimuli. For example, viewing pantomimic movements was recently shown to strongly activate the V5 complex, known to react to different moving visual stimuli [17]. Continuous monitoring of the movements of co-citizens is certainly of central importance for human social behavior and for understanding the intentions of other people.

Another trend is to narrow the unknown brain regions in terms of MEG signal generation. At present, best knowledge exists about the functions of the sensory projection cortices whereas, e.g., the frontal lobe has remained largely unexplored by means of MEG recordings. Functions of deep brain structures, which will always remain less accessible than the cortical functions to MEG, have been approached both by applying new current reconstruction algorithms [18] and by assuming sources in certain structures, like thalamus, and by comparing the resulting signal patterns with the measured ones [19].

Cerebellum, a dense package of neurons able to produce strong postsynaptic

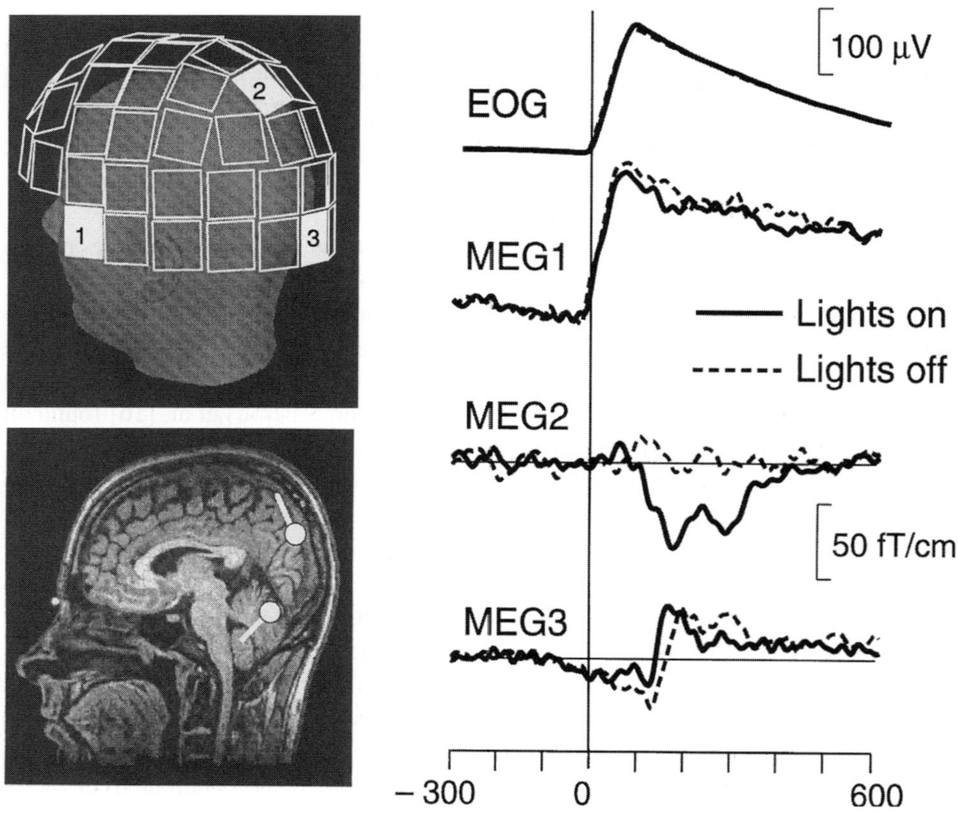

Fig. 3. Electro-oculographic (EOG) and MEG signals associated with leftward saccades; the MEG measurement locations are indicated on the sensor helmet on left. Signals were averaged with respect to saccade onsets indicated by the vertical line. Saccades performed with lights on and off in the experimental room are shown separately. On the left, source locations are projected to the subject's sagittal magnetic resonance image; the sites agree with activation of the posterior parietal cortex and the cerebellar vermis. Adapted from [21].

currents, has up to now remained out of the reach of noninvasive electrophysiological measurements. Magnetic signals of considerable amplitude have been recorded from an isolated turtle cerebellum [20]. Interestingly, Jousmäki et al. [21] recently recorded saccade-related signals which seem to arise from the human cerebellar vermis. The signals started, on average, 30 ms before the saccade onset and peaked about 170 ms after the onset (Fig. 3). During the peak activity in cerebellum, a clear signal was also seen in the posterior parietal cortex; this signal was more strongly suppressed than the cerebellar one by turning off the lights in the experimental room. The cerebellar signals probably originate from postsynaptic currents in Purkinje cells but the complex enfolding of the cerebellar cortex certainly results in considerable cancellation of signals generated in opposite walls of the folia. The possibility to noninvasively follow the temporal behavior of human cerebellar activation is of increasing interest for human neuroscience, since recent data suggest that the human cerebellum participates in functions outside motor control, like cognition [22] and sensory discrimination [23].

Conclusions

MEG provides a tool to characterize cortical reactivity under many conditions in conscious human beings. The long-term groundwork in building up the basic understanding of cortical dynamics during various "simple" stimulations and tasks now allows more real-life-like tasks to be applied and more complex source configurations to be properly interpreted.

Acknowledgements

This study was supported by the Academy of Finland and by Sigrid Jusélius Foundation. I thank Jyrki Mäkelä for comments on the manuscript and Mika Seppä for MRI surface renderings presented in the figures.

References

1. Hari R. Magnetoencephalography as a tool of clinical neurophysiology. In: Niedermeyer E, Lopes da Silva F (eds) Electroencephalography. Basic Principles, Clinical Applications and Related Fields, 3rd edn. Williams & Wilkins, 1993;1035–1061.
2. Hämäläinen M, Hari R, Ilmoniemi R, Knuutila J, Lounasmaa OV. Magnetoencephalography – theory, instrumentation, and applications to noninvasive studies of the working human brain. Rev Mod Physics 1993;65:413–497.
3. Hari R. The neuromagnetic method in the study of the human auditory cortex. In: Grandori F, Hoke M, Romani G (eds) Auditory Evoked Magnetic Fields and Potentials. Advances in Audiology, vol 6. Basel: Karger, 1990;222–282.
4. Levänen S, Ahonen A, Hari R, McEvoy L, Sams M. Deviant auditory stimuli activate human left and right auditory cortex differently. Cereb Cortex 1996;6:288–296.
5. Raij T, McEvoy L, Mäkelä J, Hari R. Human auditory cortex is activated by omissions of auditory stimuli. (Submitted).
6. Hari R, Nagamine T, Nishitani N, Mikuni N, Sato T, Tarkiainen A, Shibasaki H. Time-varying activation of different cytoarchitectonic areas of the human SI cortex after tibial nerve stimulation.

NeuroImage 1996;(In press).
7. Forss N, Merlet I, Vanni S, Hämäläinen M, Mauguiere F, Hari R. Activation of human mesial cortex during somatosensory attention task. Brain Res 1996;(In press).
8. Salmelin R, Hari R, Lounasmaa O, Sams M. Dynamics of brain activation during picture naming. Nature 1994;368:463–465.
9. Salenius S, Kajola M, Thompson WL, Kosslyn S, Hari R. Reactivity of magnetic parieto-occipital α rhythm during visual imagery. Electroenceph Clin Neurophysiol 1995;95:453–462.
10. Fletcher P, Frith C, Baker S, Shallice T, Frackowiak R, Dolan R. The mind's eye–precuneus activation in memory-related imagery. NeuroImage 1995;2:195–200.
11. Salmelin R, Hari R. Spatiotemporal characteristics of rhythmic neuromagnetic activity related to thumb movement. Neurosci 1994;60:537–550.
12. Tiihonen J, Kajola M, Hari R. Magnetic mu rhythm in man. Neurosci 1989;32:793–800.
13. Salmelin R, Hämäläinen M, Kajola M, Hari R. Human 20-Hz cortical oscillations reflect motorotopic organization. NeuroImage 1995;2:237–243.
14. Schnitzler A, Salenius S, Salmelin R, Jousmäki V, Hari R. Involvement of primary somatomotor cortex in motor imagery: a neuromagnetic study. Soc Neurosci Abstr 1995;21:518.
15. Conway B, Halliday D, Farmer S, Shahani U, Maas P, Weir A, Rosenberg J. Synchronization between motor cortex and spontal motoneuronal pool during the performance of a maintained motor task in man. J Physiol 1995;489:917–924.
16. Salenius S, Salmelin R, Neuper C, Pfurtscheller G, Hari R. Human cortical 40-Hz rhythm is closely related to EMG rhythmicity. Neurosci Lett 1996;(In press).
17. Levänen S, Salenius S, Uutela K, Hari R. Viewing pantomimic hand movements activates the human visual motion area. Human Brain Mapping Meeting, Boston 1996; Volume of abstracts.
18. Ribary U, Ioannides A, Singh K, Hasson R, Bolton J, Lado F, Mogilner A, Llinas R. Magnetic field tomography of coherent thalamocortical 40-Hz oscillation in humans. Proc Natl Acad Sci USA 1991;88:11037–11041.
19. Tesche C. Noninvasive imaging of neuronal population dynamics in human thalamus. Brain Res 1996;(In press).
20. Okada YC, Nicholson C. Magnetic evoked field associated with transcortical currents in turtle cerebellum. Biophys J 1988;53:723–731.
21. Jousmäki V, Hämäläinen M, Hari R. Dynamics of human cerebellar activation during visually-guided saccades. Report TKK F-A 756. ISBN 951-22-3126-3, ISSN 0355-7790, May 1996.
22. Fiez J. Cerebellar contributions to cognition. Neuron 1996;16:13–15.
23. Gao J-H, Parsons L, Bower J, Xiong J, Li J, Fox P. Cerebellum implicated in sensory acquisition and discrimination rather than motor control. Science 1996;272:545–547.

1996 Elsevier Science B.V.
Integrative and Molecular Approach to Brain Function.
M. Ito and Y. Miyashita, editors.

Integrative functions of the human frontal association cortex studied with MEG

Kazuo Sasaki[1,2], Toru Tsujimoto[1,2], Ryuichi Matsuzaki[1,2], Atsusi Nambu[1,2], Shin-ichi Kyuhou[1,2] and Satoru Nishikawa[1,2]

[1]*Kyoto University, Kyoto; and* [2]*National Institute for Physiological Sciences, Okazaki, Japan*

Abstract. We have studied on integrative functions of the human frontal association cortex with MEG. Among a few themes so far reported, the frontal mental θ wave is briefly described and discussed. Mental concentration on calculating, playing music and measuring time was found to be often associated with 5–7 Hz magnetic θ waves in the frontal areas when recorded with multichannel SQUID gradiometers. Electrical current dipoles responsible for the magnetic θ waves were estimated to distribute widely and successively in cortical areas in the frontal lobes in both the hemispheres. The frontal θ wave was also recorded during meditation of Zen (Buddhist) priests and yoga disciples. Dynamic activity with the θ waves appears to be a common feature of the human frontal association cortex to concentration on mental efforts with relatively abstract thinkings.

Key words: calculation, frontal mental θ wave, measuring time, meditation, playing music.

Background

We previously studied on integrative functions of the frontal association cortex in monkeys with chronically implanted recording electrodes in the cortex and found the no-go potential in the prefrontal area (in the principal sulcus) which is related to the decision not to move and subsequent suppression ("suppression will") [1,2]. The mode of development, disappearance and reappearance of the no-go potential according to initial learning, and confusion and relearning after reversal of go and no-go stimuli was so interesting [3] that we wanted to record, if any, the same potential in human subjects and to test the mode. EEG studies over the human scalp confirmed the no-go potential over the scalp with the same go/no-go task, but its localization in the brain could not be satisfactorily made with EEG analyses [4,5]. That was the moment on which we started to use MEG (magnetoencephalography) for better localization of the origin of cortical potential. Multichannel SQUID (Superconducting Quantum Interference Device) gradiometers were, in fact, effective for estimating the location of current dipoles responsible for the no-go potential, i.e., in and around the middle frontal gyrus [6,7].

With the MEG machine, we further studied integrative functions of the human

Address for correspondence: Prof Kazuo Sasaki, Department of Integrative Brain Science, Graduate School of Medicine, Kyoto University, 606-01 Kyoto, Japan. Tel.: +81-75-753-4480. Fax: +81-75-753-4486.

association cortex such as activity on a few kinds of thinking associated with frontal mental θ waves [8], motor speech centre [9], etc. In this report, we will deal mainly with the frontal mental θ wave. Before going into the subject, the basis of the MEG will be briefly mentioned; what are recorded with multichannel SQUID gradiometers?

Main sources of EEG and MEG recorded over the scalp are considered to be respectively extracellular and intracellular electrical currents due mainly to EPSPs (excitatory postsynaptic potentials) generated in either superficial or deep part of apical dendrites of pyramidal neurones in the cerebral cortex [7,10,11]. Extracellular currents are easily distorted, before reaching the surface of scalp, by several layers with different electrical impedances surrounding the cerebral hemispheres, especially due to shunting effect by the cerebrospinal fluid of low impedance as compared with the brain tissue and insulating effect by the skull of high impedance [5,7]. The shunting effect appears so irregular and complicated in three-dimensional configurations, with the fruid not only between the hemispheres but also in many sulci, that the distortion must be more or less difficult to be corrected in EEG analyses. MEG will be different in this respect from EEG, since magnetic fields due to the intracellular currents in the apical dendrites of pyramidal neurones (current dipoles) are directly recorded without such distortion as in the case of EEG due to the layers of different electrical impedances. The advantage in localizing the electrical current dipoles in the cerebral hemispheres with MEG was in fact demonstrated in several examples [3,6–9].

Methods

More than 50 adult volunteers were tested and, at least, six examinees for each theme. The 74-channel SQUID gradiometer system (dual 37 channels, Bti, Magnes 700; with detection coils of axial type) and the 122-channel SQUID gradiometer system (Neuromag 122; with detection coils of planar type) were used in magnetically shielded rooms.

Results and Discussion

Frontal mental θ wave on mental calculation and mental playing music

We found 5–7 Hz magnetic θ waves in the frontal association cortex of adult human subjects during concentration on mental calculation and mental playing music [8]. A subject was lying on a bed in a magnetically shielded room, and simultaneous recordings of the left and right hemispheres were performed with two sets of the 37-channel gradiometers. We recently used the Neuromag 122-channel gradiometers covering a whole head in sitting position [12]. MEG recordings were made continuously, usually for 2 min or sometimes 5 min, while a subject was calculating, e.g., successive subtraction of three from 100 in mind with the eyes closed (Figs. 1 and 2II). The subtraction was repeated from 100 until one, then from 101 until two, etc., continuously and as fast as possible for 2 min. Fast, accurate and continuous

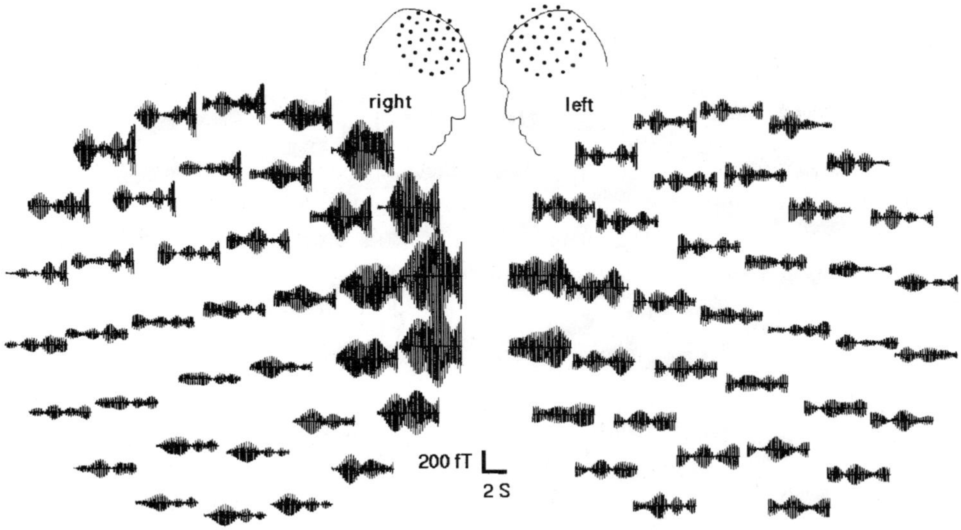

Fig. 1. Specimens of the frontal mental θ waves for about 5 s from a subject, recorded simultaneously with two sets of 37-channel gradiometers over the right and left frontal-parietal parts of the head. Thirty-seven dots (37 sites for detection coils of gradiometers) on the right and left sagittal views of the head correspond respectively to 37 traces of MEG-wave traces (right and left) (one trace among the recording positions on the left side was omitted because of noise). Raw MEG data were processed with sixth order digital band pass filter of 4.9–6.1 Hz. Two second time scale and 200 fT calibration for all MEG records. From [8].

calculation was essential in producing the marked frontal θ wave. The intensity and continuity of the mental effort for calculation appeared to be closely related to the amplitude and incidence of the θ wave. The same examinee was also asked to make several other kinds of mental work respectively for 2 min. Playing homophonic and/or polyphonic music in mind with its memorized score was effective in inducing the θ wave in three out of six subjects examined (Fig. 2III). Playing monophonic music in mind was ineffective in the three examinees. Imaging concrete objects, such as pictures, sculptures, routes for a walk (spatial task), etc., was not suitable for producing the θ wave. Increase of regional blood flow in the frontal cortex on mental calculation [13] and music performance [14] would be relevant findings to the present results.

Listening to the music of which the mental playing induced the frontal mental θ wave hardly elicited the frontal mental θ wave, even if the subject sympathized greatly with the music. Such great sympathy and also strong emotional excitation in feeling delight, anger, sorrow or pleasure, can induce another type of magnetic θ wave around the parietal part of the head in some subjects, and we call them "emotional θ waves" [15].

Electrical current dipoles estimated from the frontal θ waves on mental calculation are given for a subject in Fig. 3I. Specimen records from two different gradiometers

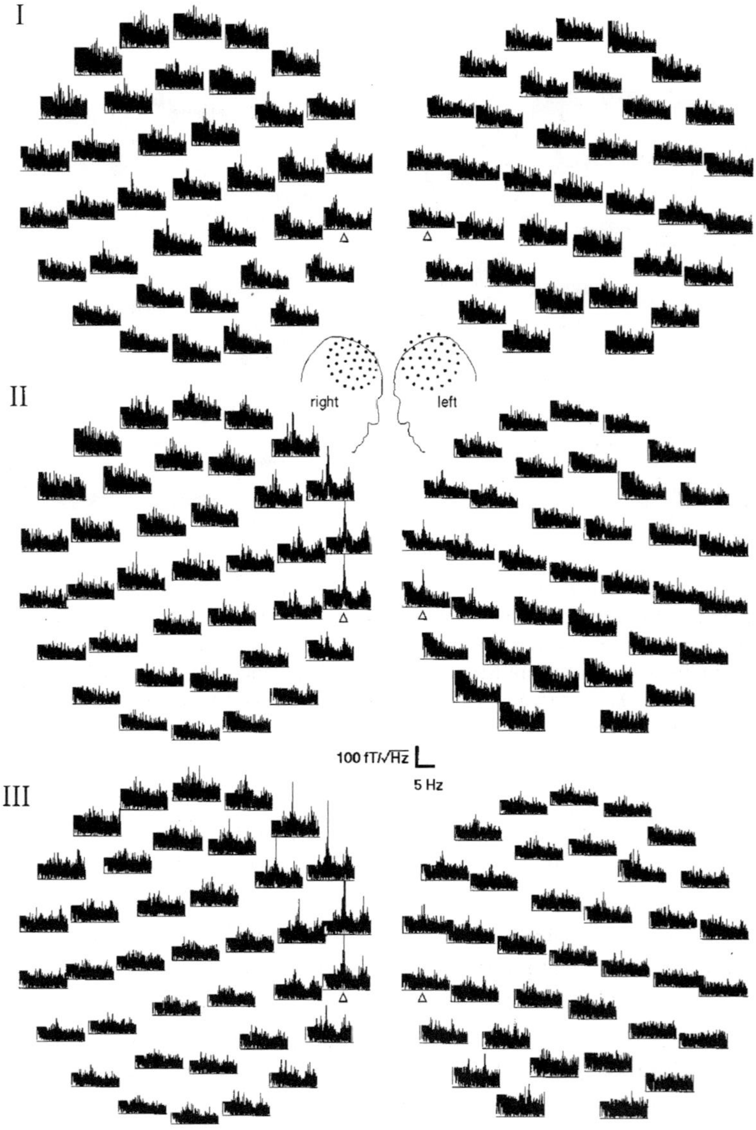

Fig. 2. FFT (Fast Fourier Transformation) frequency analysis of MEG waves for 2 min from the same subject as that in Fig. 1. **I** Resting. **II** Mental calculation. **III** Playing music in mind. Simultaneous recordings from the frontal-parietal parts of right and left hemispheres as illustrated by 37 dots on right and left sagittal views of the head in the inset figures (a record of a gradiometer among 37 on the left side was omitted because of noise). During calculation and playing music in mind, magnetic waves with peak frequency of about 5.5 Hz (marked by open triangles at two specimen records) occurred in anterior regions of the head much more frequently than on resting, especially on the right side (37-channel records in the left column). Frequency scale (abscissa), 5 Hz. Magnetic calibration (ordinate), 100 fT/√Hz. Low frequency signals below 2.5 Hz are cut. From [8].

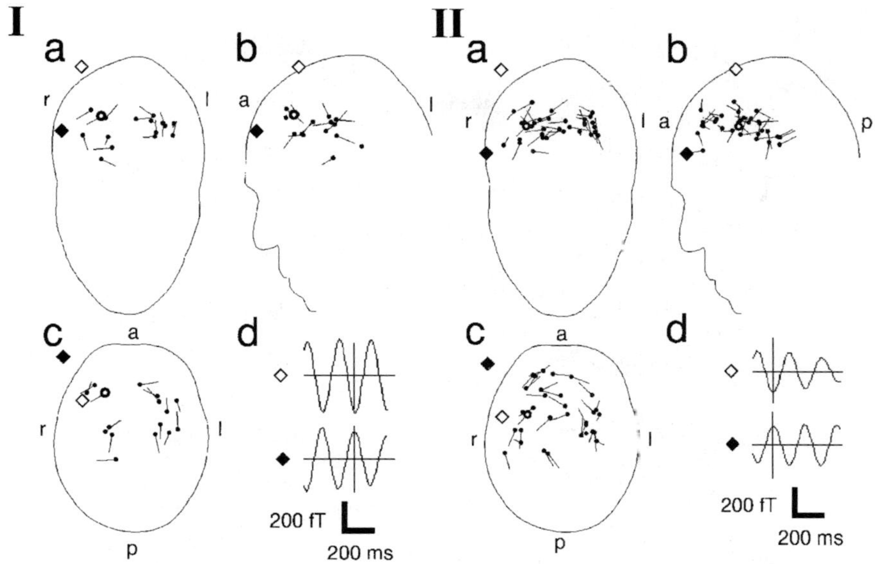

Fig. 3. Distribution of electrical current dipoles of the frontal mental θ waves estimated from MEG data on mental calculating (I) and playing music (II) with a subject. Position and direction but amplitude are shown. Current dipoles derived from magnetic θ waves with a phase difference of about 180° during 2 min of calculating and playing music are presented in coronal: a) sagittal, b) axial and c) views of the head. Open and filled squares in a–c indicate the sites of the two gradiometers which picked up the specimen records of θ waves in d. A current dipole marked by open circles in a–c was estimated at the time point of the vertical lines in d (I and II). The records in d show a difference of about 180° in phase. Two hundred milliseconds time scale and 200 fT calibration for records in d. Filtering conditions are described in Fig. 1.

show nearly 180° out of phase (Fig. 3Id). Only the peaks with 180° phase difference were picked up for estimating electrical current dipoles as shown in Fig. 3a–c. There must have virtually been much more dipoles during one session than shown in the figures. Figure 3I reveals that the electrical current dipoles for the θ waves appear repeatedly and widely in various parts of the frontal lobes in both the hemispheres during the 2 min of continuous and intensive calculating effort. Figure 3II presents the distribution of electrical current dipoles for the θ waves induced during 2 min of playing music in mind in the same subject as Fig. 3I. This distribution looks similar to that of the mental calculation shown in Fig. 3I. The data of Fig. 3II are superimposed on the six horizontal (axial) planes of MRI in the same subject in Fig. 4. They appear to be distributed in the frontal lobes and presumably in the prefrontal and premotor cortical areas.

Previous EEG studies reported that θ waves appeared over the frontal part of scalp during problem solving [16,17]. Because of their predominance in the midline frontal part, the θ waves were called "frontal midline θ rhythm" [18]. Such EEG findings can be explained by vector summation of the multiple electrical current dipoles

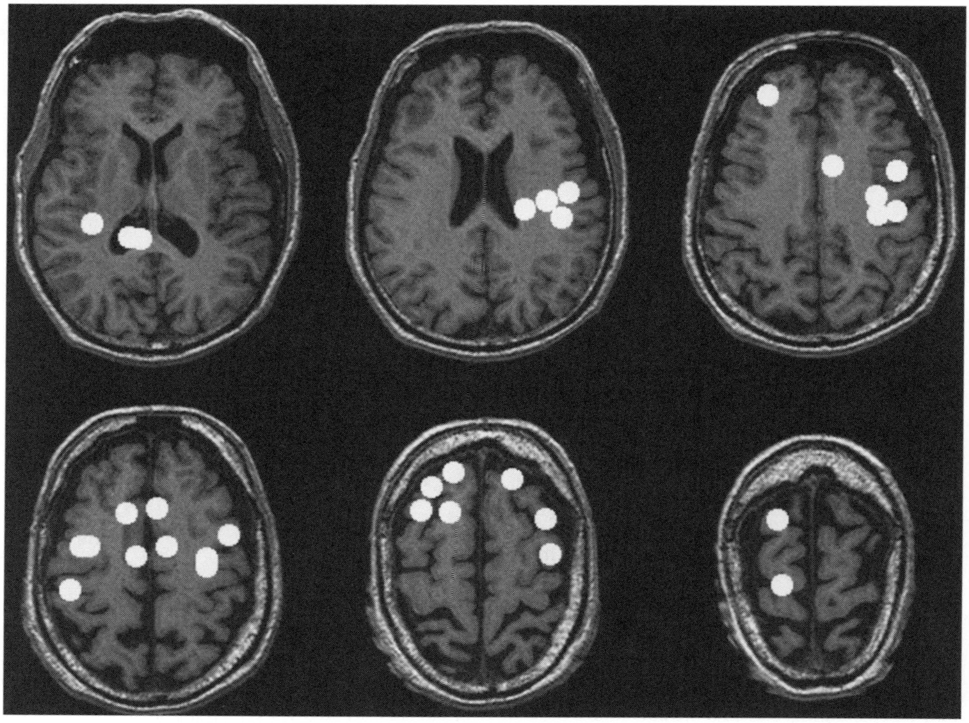

Fig. 4. Location of electrical current dipoles of the frontal mental θ waves on playing music for 2 min (Fig. 3II) is superimposed on six levels of MRI horizontal pictures of the brain (anterior, upward; left hemisphere, on the right side). Pictures from upper left to lower right correspond to horizontal sections from lower to upper levels. Only the positions of current dipoles are illustrated. They were distributed mostly in different loci of the cortical structures of the frontal lobes on both sides.

scattered in the left and right frontal lobes as described above. In fact, simultaneous recordings of EEG and MEG support the explanation [15]. It appears reasonable to call the frontal θ wave to be frontal "mental" θ wave [8].

Frontal mental θ wave on measuring time

As for the concept of metaphysical time, Immanuel Kant stated that: "die Zeit ist kein empirischer Begriff, der irgent von einer Erfahrung abgezogen worden "[19]. Henri Bergson stressed: "la durée réelle"[20]. Norbert Wiener discussed on "Newtonian and Bergsonian time" in the first chapter of his book "Cybernitics" [21]. It would be interesting to test our brain activities when we measure "time" subjectively or introspectively, as it seemed to be possibly one of the innate and abstract thinkings.

We tried to experiment on the measuring time by tasks modified from the stimulus paradigm for the contingent negative variation (CNV) [22]. A warning (W) light stimulus of one of two colours with duration of 100 ms was followed, after a fixed time of 1.5–6.0 s, by an imperative (I) light stimulus of the third colour with duration of 500 ms (see inset diagrams for stimulus paradigms in lower left and right in Fig. 5). In a task called go/no-go W-I, W stimulus of one colour indicates the subject to press a button by the thumb immediately on the onset of I stimulus (go trial) and W stimulus of another colour not to move (no-go trial). With the go trial, the examinee should try to move the finger earlier than the shortest reaction time in a simple reaction-time movement (less than 100–150 ms), i.e., it has to estimate the I stimulus to come, by "mental measuring time". The go and no-go trials were usually presented 100 times respectively at random time intervals and in irregular orders. In another task called "on/off W-I", W stimulus of one colour is followed by I stimulus and the subject presses the button after having seen the I stimulus (on trial), whereas W stimulus of another colour is not followed by I stimulus at all, so that the examinee should truly guess the W-I duration and push the button during the estimated (imaginary) time of I stimulus (off trial). The on/off trials are also delivered at random time intervals and in irregular order 100 times respectively, and the on and off responses are to be analyzed later separately.

Both the go trial in the go/no-go task and the off trial in the on/off task, in which

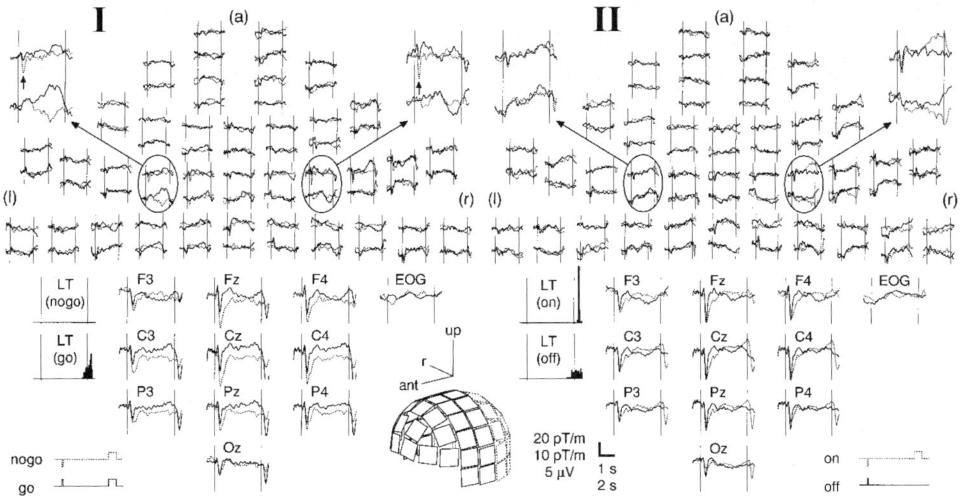

Fig. 5. MEG and EEG records with go/no-go warning-imperative (W-I) stimulus (I) and on/off W-I stimulus (II) tasks. Respective stimulus paradigms are schematically shown by solid and broken lines in lowermost left (go/no-go) and right (on/off) diagrams with solid (go and off) and broken (no-go and on) lines. Two different colour W stimuli by upward and downward brief pulses and the third colour I stimulus by upward square pulse. Records with 70-channel gradiometers (35 pairs of the longitude and latitude detection coils) over the anterior-central part of the head (pair of the coils in each square drawn by solid line in the inset helmet model) and records with 10 plate electrodes (F3-Oz) are aligned 100 times by the onset pulse of W stimulus (left vertical line). The onset of I stimulus (right vertical line) is 3.0 s after the onset of W. Latent times of movement onset are given by histograms of 50-ms bins (LT). Data of go (I) and off (II) with solid lines and those of no-go (I) and on (II) with broken lines. Two pairs of MEG records in I and II are enlarged. 1 s, 10 pT/m for MEG. 2 s, 20 pT/m for the enlarged MEG. 2 s, 5 µV for EEG.

the measuring time was required, revealed the frontal mental θ activity in much the same mode, although more predominantly with the off than the go trial. Figures 5–7 present examples from the results by the go/no-go and on/off tasks. Figure 5 gives MEG (70 channels in the anterior-central part out of the whole-head 122-channel gradiometers) and EEG (10 electrode positions) records on the go/no-go (I) and on/off (II) tasks. During 3 s of the W-I interval, magnetic and electrical changes occurred differently in different recording sites. In EEG records, go trials show "contingent negative variations (CNV)" [22] in contrast to those of contingent positive variations (CPV) on no-go trials, CPV being named and interpreted to be due to dysfacilitation of background and tonic EPSP currents in superficial parts of apical dendrites of cortical pyramidal neurones [23]. Also relatively sharp magnetic changes in the lateral-anterior part are noted immediately after the no-go W stimulus (about 200 ms peak latency) (marked by arrows in enlarged specimens), demonstrating the magnetic substrata for the no-go potential in the go/no-go W-I task [6]. Magnetic fields of the readiness potential are seen over the central part of the head on both sides (around the enlarged specimens), on the go and the on trials. In the off trial, latent times of movements (LT) were so varied that the readiness fields were not well synchronized to be visualized by this alignment of records. The 52 gradiometers in the posterior part omitted from the records in Fig. 5 showed only short-latency, transient activities evoked by the visual stimulations of W and I.

Frequency analyses of rhythmical magnetic fields around the W-I duration with Fast Fourier Transformation (FFT) are presented for the go/no-go (I) and the on/off (II) (III, EEG) tasks in Fig. 6. In this subject, a θ wave component of 5.5 Hz at the peak frequency is observed in the anterior part of the head on the go trial of the go/no-go task and the off trial of the on/off task. It is also noted that the θ component is more predominant on the right side than the left in this examinee. In the four examinees with the frontal θ wave on the measuring time, two were predominant on the left side, whereas the remaining two on the right sides. In all the four cases, the off trial was more marked than the go trial in inducing the θ activity as exemplified in Fig. 6. The SQUID gradiometers with the detection coil of planar type are much advantageous in localizing an activity in the cerebral cortical area by its spatially differentiating nature of the coil [12].

Time courses of the manifestation of the frontal θ wave are presented in Fig. 7, the case of the on/off task with 3 s W-I interval as presented in Figs. 5 and 6 being in II. The MEG records of the whole session with respective 100 trials of the on and off stimuli were filtered with a band pass of 5–6 Hz, rectified and then averaged separately for on and off trials by the onset pulses of on and off W stimuli (right half of Fig. 7II). The θ wave component is noted to develop gradually after the W stimulus and to cease after the imaginary I stimulus in several sites of the detection coils, i.e., the anterior part of the head. The same FFT analyses (left column) and the time courses (right column) are given for the W-I intervals of 1.5 s (I) and 6.0 s (III) with the same examinee. Although the θ component was much smaller and less developed with the 1.5-s interval, all these W-I intervals showed the same tendencies in the sites and time courses of the θ component. Even for the 6.0-s duration, the

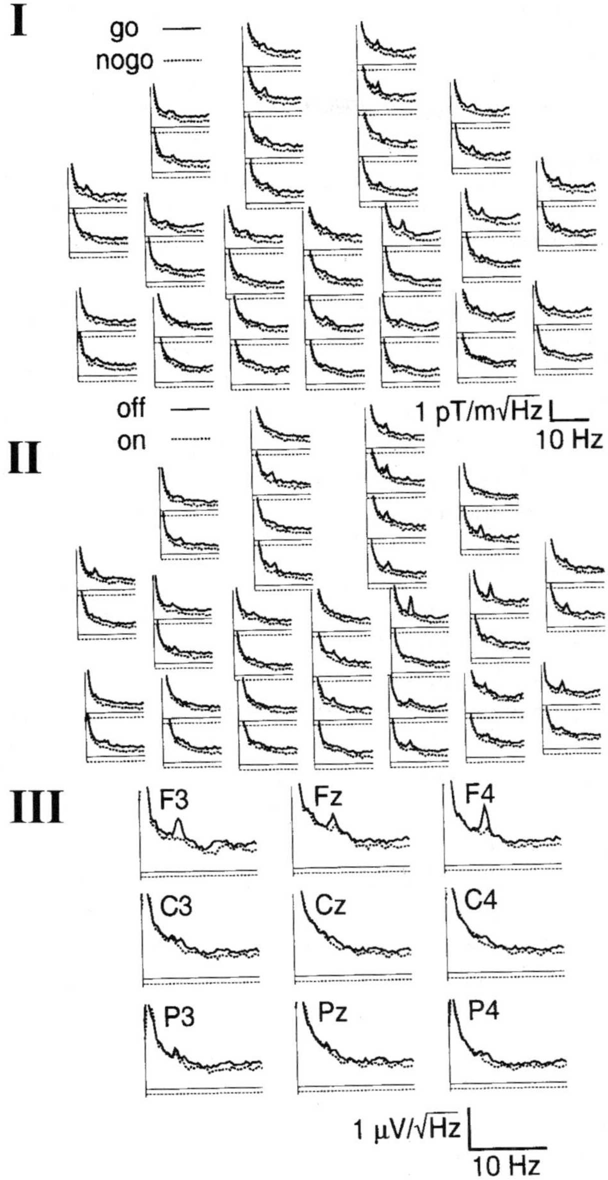

Fig. 6. FFT frequency analyses of MEG (I, II) and EEG (III) records on measuring time of W-I interval (3.0 s). 100 records for 3.4 s including the 3.0 s of W-I interval were analyzed. **I** go/no-go W-I trials. **II, III** on/off W-I trials. Solid lines for go and off. Broken lines for no-go and on. A slight shift in ordinates was given to distinguish the difference between the solid and broken lines. Records with 40 gradiometers in the anterior and medial part over the head are presented. Lateral two lines respectively on the left and right sides and posterior one line of the squares (containing paired detection coils) are omitted from those in Fig. 5.

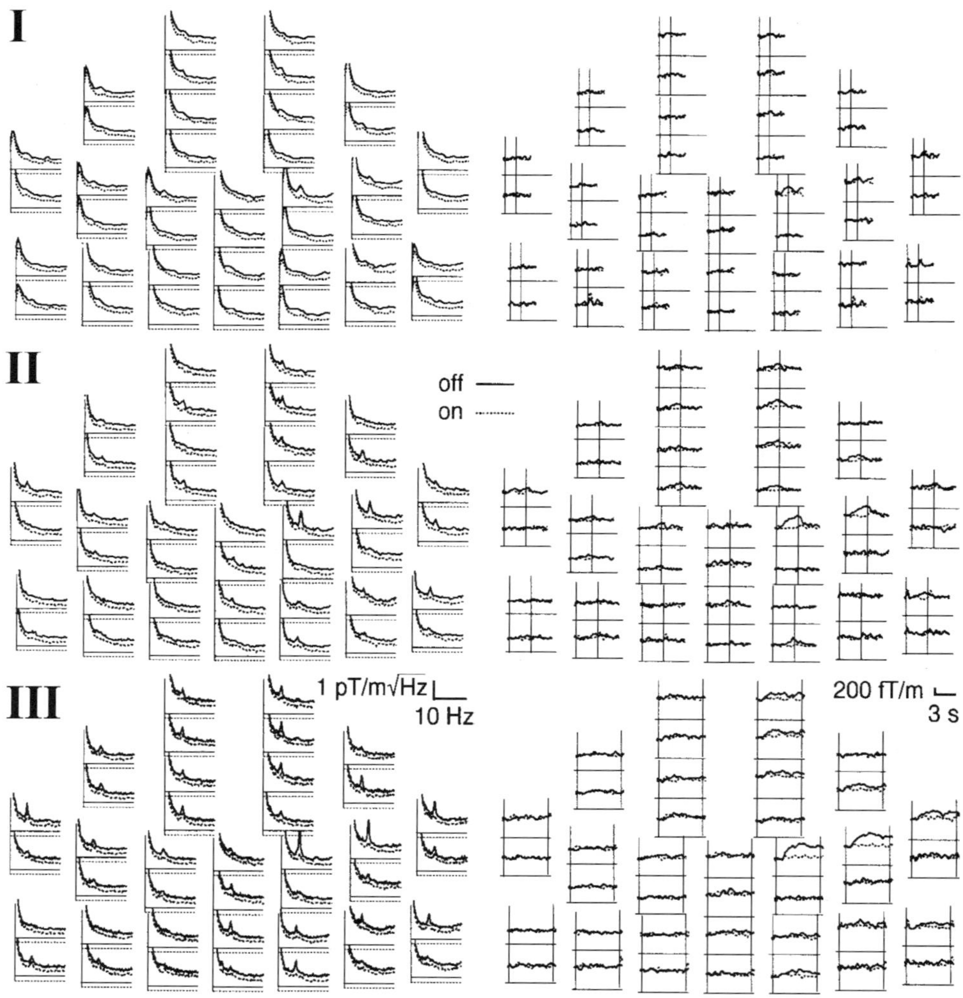

Fig. 7. Left diagrams: FFT frequency analyses of MEG records on measuring time of W-I intervals of 1.5 s (I), 3.0 s (II) (the same as Fig. 6II) and 6.0 s (III) with on/off W-I tasks in the same examinee. Right diagrams: Time courses of development of the frontal θ wave along W-I duration. θ waves (5–6 Hz) rectified, and aligned and accumulated 100 times by the W onset (left vertical line). The imaginary I onset (not given to the examinee) is indicated by the right vertical line. Abscissae in a common scale of 3 s. Records with the 40 gradiometers in the anterior and medial part as in Fig. 6 are presented.

frontal θ wave increased after the W stimulus, lasted for the duration and fell about 6 s later, i.e., around a time immediately after the estimated onset of the I stimulus (the falling phase not shown in Fig. 7III).

The concept of "time" has been one of the most difficult problem in metaphysics and psychological sciences [24]. In measuring duration of several seconds, the frontal lobes in both the hemispheres induced the frontal mental θ wave in much the same cortical areas with that in concentrating on mental calculation and imaginary music playing. The time course of the θ activity was found closely related to the duration of the measuring time, being correlated to the mental (subjective) effort in measuring time. The finding suggests the real contribution of the frontal association cortex to the mental measurement of time, due to so good temporal as well as spatial resolutions by use of MEG. The difference or the correlation between the subjective (mental) and objective (physical) time should be considered further [21].

Zen contemplation and yoga meditation

We found that the frontal mental θ waves can be recorded also on meditation of Zen (Buddhist) priests ("Zazen") and of yoga disciples (Sasaki et al. in preparation) [7]. EEG studies were previously made on Zazen [25] and yoga [26,27], but localization of θ waves, if any, was not clearly shown. The meditations in Zen and yoga may have a common feature in brain activities to the mental concentrations as mentioned above.

Acknowledgements

This study was supported by Grant-in-Aid for New Program (No. 01-05NP010, Leader Prof S. Ebashi) from the Ministry of Education, Science and Culture of Japan and by Grant of Uehara Memorial Foundation. We are indebted to expert technical help by Mr O. Nagata, Mr Y. Takeshima and Ms K. Matsuzawa in the National Institute for Physiological Sciences. We thank Prof H. Shibasaki and Dr T. Nagamine in Department of Brain Pathophysiology, Graduate School of Medicine, Kyoto University for help and advice in using Neuromag-122.

References

1. Sasaki K, Gemba H. Electrical activity in the prefrontal cortex specific to no-go reaction of conditioned hand movement with colour discrimination in the monkey. Exp Brain Res 1986;64:603–606.
2. Sasaki K, Gemba H, Tsujimoto T. Suppression of visually initiated hand movement by stimulation of the prefrontal cortex. Brain Res 1989;495:100–107.
3. Sasaki K, Gemba H, Nambu A, Matsuzaki R. Activity of the prefrontal cortex on no-go decision and motor suppression. In: Thierry A-M et al. (eds) Motor and cognitive functions of the prefrontal cortex. Berlin Heidelberg: Springer-Verlag, 1994;139–159.
4. Gemba H, Sasaki K. Potential related to no-go reaction of go/no-go hand movement task with colour discrimination in human. Neurosci Lett 1989;101:263–268.
5. Sasaki K, Gemba H, Nambu A, Jinnai K, Yamamoto T, Llinas R. Cortical activity specific to no-go reaction in go/no-go reaction time hand movement with colour discrimination in monkeys and human subjects. Biomed Res 1992;13(Suppl 1):5–9.
6. Sasaki K, Gemba H, Nambu A, Matsuzaki R. No-go activity in the frontal association cortex of human subjects. Neurosci Res 1994;18:249–252.
7. Sasaki K, Gemba H, Nambu A, Kyuhou S, Matsuzaki R, Tsujimoto T. Studies on integrative functions of the human frontal association cortex by use of MEG. Electroenceph Clin Neurophysiol 1996;(Suppl 47):181–190.

8. Sasaki K, Tsujimoto T, Nambu A, Matsuzaki R, Kyuhou S. Dynamic activities of the frontal association cortex in calculating and thinking. Neurosci Res 1994;19:229–233.
9. Sasaki K, Kyohou S, Nambu A, Matsuzaki R, Tsujimoto T, Gemba H. Motor speech centres in the frontal cortex. Neurosci Res 1995;22:245–248.
10. Sasaki K, Staunton HP, Dieckmann G. Characteristic features of augmenting and recruiting responses in the cerebral cortex. Exp Neurol 1970;26:369–392.
11. Sasaki K. Cerebro-cerebellar interconnections in cats and monkeys. In: Massion J, Sasaki K (eds) Cerebro-cerebellar interactions. Amsterdam: Elsevier/North-Holland, 1979;105–124.
12. Hämäläinen M, Hari R, Ilmoniemi RJ, Knuutila J, Lounasmaa OV. Magnetoencephalography – theory, instrumentation, and applications to noninvasive studies of the working human brain. Rev Modern Physics 1993;65:413–497.
13. Roland PE, Friberg L. Localization of cortical areas activated by thinking. J Neurophysiol 1985;53:1219–1243.
14. Sergent J, Zuck E, Terriah S, MacDonald B. Distributed neuronal network underlying musical sight-reading and keyboard performance. Science 1992;257:106–109.
15. Sasaki K, Tsujimoto T, Mastuzaki R, Nambu A, Kyuhou S. MEG study on cortical theta waves induced during concentrated mental work and emotional excitation in human subjects. In: Ono T et al. (eds) Perception, Memory, and Emotion. Elsevier Science Ltd, 1996;453–464.
16. Brazier MA, Casby JU. Crosscorrelation and autocorrelation studies of electroencepgalographic potentials. Electroenceph Clin Neurophysiol 1952;4:201–211.
17. Ishihara T, Yoshii N. Multivariate analytic study of EEG and mental activity in juvenile delinquents. Electroenceph Clin Neurophysiol 1972;33:71–80.
18. Yamaguchi Y. Frontal midline theta activity. In: Yamaguchi N, Fujisawa K (eds) Recent Advances in EEG and EMG Data Processing. Amsterdam, Elsevier/North-Holland, 1981;391–396.
19. Kant I. Kritik der reinen Vernunft. Stuttgart: Philipp Reclam Jun, 1966.
20. Bergson H. Essai sur les donntée immédiates de la conscience. Paris: Libtairie Felix Alcan, 1938.
21. Wiener N. Cybernetics. New York: John Wiley & Sons, 1947.
22. Walter WG, Cooper R, Aldrige VJ, McCallum WC, Winter AL. Contingent negative variation: An electric sign of sensori-motor association and expectancy in the human brain. Nature 1964;203:380–394.
23. Sasaki K, Gemba H, Tsujimoto T. Cortical field potentials elicited by go and no-go visual stimuli preceding visually-initiated hand movement in the monkey. Neurosci Res 1988;(Suppl 7):S97.
24. Pöppel E. Grenzen des Bewusstseins über Wirklichkeit und Welterfahrung. Stuttgart: Deutsche Verlags-Anstalt GmbH, 1985.
25. Kasamatsu A, Hirai T. An electroencephalographic study on the Zen meditation (Zazen). Folia Psychiat Neurolog Jpn 1966;20:315–336.
26. Banquet JP. Spectral analysis of the EEG in meditation. Electroenceph Clin Neurophysiol 1973;35:143–151.
27. Herbert R, Lehmann D. Theta bursts: An EEG pattern in normal subjects practising the transcendental meditation technique. Electroenceph Clin Neurophysiol 1977;42:397–405.

1996 Elsevier Science B.V.
Integrative and Molecular Approach to Brain Function.
M. Ito and Y. Miyashita, editors.

Central motor control and its disorders studied by multidisciplinary approach in humans

Hiroshi Shibasaki

Department of Brain Pathophysiology, Kyoto University School of Medicine, Shogoin, Sakyo, Kyoto, Japan

Introduction

Functional localization or specialization of motor control in the human brain used to be estimated based on the neurological deficits resulting from localized cerebral lesions or on the so-called positive phenomena such as epileptic seizures and involuntary movements. Recent advances in various kinds of high technologies, however, have enabled us to investigate the higher brain functions related to motor control in normal subjects. These techniques include: 1) electrophysiological studies such as movement-related cortical electrical potentials (MRCPs) or movement-related magnetic fields, and transcranial magnetic stimulation; 2) radionuclide techniques such as positron emission tomography (PET) and single photon emission computed tomography (SPECT); and 3) nuclear magnetic resonance (NMR) such as functional magnetic resonance imaging (fMRI) and magnetic resonance spectroscopy. Each of these techniques has its own characteristics in terms of spatial and temporal resolution. For example, activation studies with PET by using O-15 labelled water as a radiotracer provide a reasonably good spatial information, but chronologically it gives an overall activity throughout the period of approximately 2 min due to the half life of O-15. In contrast, electrophysiological techniques can provide us with a high-temporal resolution, actually in the order of ms. Electroencephalogram (EEG) recorded from the scalp electrodes, however, is strongly influenced by the different electrical conductivity of the tissues surrounding the cerebral cortex such as spinal fluid and skull, resulting in a widespread and distorted distribution of the potential over the scalp (shunt effect). Therefore, in order to identify the generator sources of the scalp-recorded EEG activities, special techniques for source localization are required. In contrast, the recently developed technique of the whole-head magnetoencephalography (MEG) can give us a reasonably precise information in terms of both space and time, although theoretically it can record only the electric current source which is tangentially oriented with respect to the head surface.

Whichever techniques are employed, the studies of voluntary movement either in

Address for correspondence: Hiroshi Shibasaki MD, PhD, Department of Brain Pathophysiology, Kyoto University School of Medicine, Shogoin, Sakyo, Kyoto 606-01, Japan. Tel.: +81-75-751-3601. Fax: +81-75-751-3202.

normal subjects or in patients with movement disorders are significantly influenced by various parameters of the movement itself such as the site, discreetness, rate, complexity and sequence of movement, intensity of volition, positive vs. negative movement, motor learning, and plasticity (motor recovery). In the noninvasive studies of higher brain functions related to movement, therefore, these parameters should be adequately controlled depending on the purpose of the study. Among these parameters, "complexity of movement" is difficult to be defined, because it is influenced by many other factors. In this article, the main focus is placed on two current topics in the field of human motor control: 1) bilateral activation of motor cortices associated with unilateral hand movement in physiological and pathological conditions; and 2) inhibitory mechanism of motor cortices.

Bilateral activation of motor cortices

In some patients with mirror movement, almost equal activation of bilateral primary motor cortices (M1) in association with unilateral hand movement has been demonstrated. By recording MRCPs, Shibasaki and Nagae [1] and Cohen et al. [2] demonstrated bilateral activation of M1 in the preparatory period of movement in patients with congenital mirror movement (Fig. 1). More recently, however, Mayer

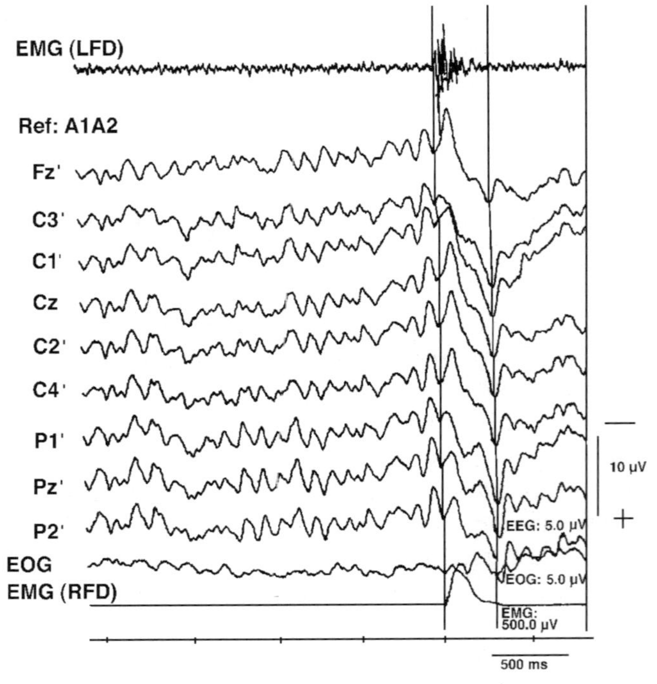

Fig. 1. MRCPs associated with voluntary grasping of the right hand in an 18-year-old male patient with mirror movement of the left hand following head trauma. The later slope of the premovement negative shift (NS′) is distributed symmetrically over the central region. LFD and RFD: left and right finger flexors, respectively.

et al. [3], also by recording MRCPs in patients with mirror movement, observed bilateral activation only during the execution period but not during the preparatory period of hand movement. By transcranial magnetic stimulation of the motor cortex in two patients with congenital mirror movement, Cohen et al. [2] demonstrated bilateral representation of hand movements in the motor cortex. The same authors demonstrated with PET an increased regional cerebral blood flow (rCBF) in the sensori-motor cortex (S1-M1) bilaterally in association with intended movement of unilateral hand [2]. We have made similar observations in a patient with acquired mirror movement (Fig. 2). These results are interpreted to suggest that those mirror movements are associated with the activation of M1 ipsilateral to the intended hand, probably due to either disinhibition of the M1 ipsilateral to the movement or prominent ipsilateral innervation of the corticospinal tracts. However, it is also possible that the generation of MRCPs and the increased blood flow in the S1-M1 ipsilateral to the intended hand movement might reflect the activation of inhibitory components of that area, because the patients attempt to suppress the mirror movement, as discussed later in this chapter.

During the recovery state from hemiparesis due to cerebral hemispheric lesion, participation of the ipsilateral hemisphere in the execution of voluntary movement of the recovering hand has been demonstrated by MRCP and rCBF studies. Kitamura et al. [4] demonstrated in two patients after recovery from hemiparesis a symmetric distribution of the later component of the negative slow shift called Negative Slope (NS′) preceding the movement of the affected upper limb while the movement of the intact hand generated asymmetric NS′ which was larger over the contralateral central region than the ipsilateral. Chollet et al. [5] studied rCBF by PET in six

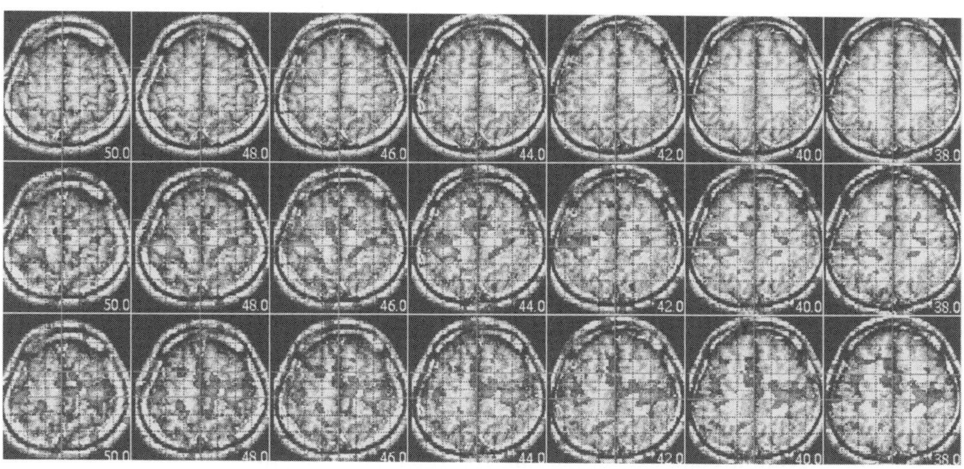

Fig. 2. Activated areas shown as an increase in rCBF by $H_2{}^{15}O$-PET during voluntary movement of each hand in the same patient with mirror movement as shown in Fig. 1, superimposed on his own MRI. Voluntary movement of the right hand (bottom row) is associated with the activation of not only the left S1-M1 (shown on the right side of the Fig.) but also of the right S1-M1, though to a lesser degree. Top row: rest for control, middle row: movement of the left (intact) hand.

patients after their recovery from a first hemiplegic stroke, and demonstrated the increase in rCBF in both the contralateral and ipsilateral S1-M1 and in both cerebellar hemispheres. Honda et al. [6] studied both MRCPs and rCBF in the same patients with recovering hemiparesis and demonstrated the results supporting those reports. Bilateral activation of the S1-M1 in these patients was interpreted to indicate the participation of the ipsilateral S1-M1 in the functional recovery of the paretic limb. It is important to note, however, that, even in normal subjects, complex or sequential movements of the unilateral hand are usually associated with activation of bilateral S1-M1, as described below.

Kitamura et al. [7] compared MRCPs associated with the voluntary, self-paced, sequential extensions of the middle followed by the index finger of the right hand with those associated with the simultaneous extensions of the two fingers in normal subjects. In that experiment, the motor tasks were designed so that the same muscles were involved in both tasks. As a result, a significantly larger NS′ was demonstrated at the midline vertex and over the bilateral central regions before the sequential movements as compared with the simultaneous ones. In view of the findings of subdural recording in patients with intractable epilepsy, which suggested the activation of the contralateral S1-M1 and bilateral supplementary motor areas (SMAs) within the 300-ms period preceding the movement onset [8,9], it was concluded that the S1-M1 was bilaterally activated immediately before the movement onset in the complex/sequential movement. Bilateral activation of S1-M1, in addition to that of the SMA, in the complex/sequential unilateral finger movements was also demonstrated by the PET-rCBF activation studies [10]. The activation of ipsilateral S1-M1 in the hand movements was also demonstrated by fMRI study, although it was to a much lesser degree as compared to that in the contralateral hemisphere [11].

The simultaneous recording of MEG and EEG during the motor paradigm often gives rise to discrepant results. The most frequent findings are the contralateral predominance of the magnetic fields preceding the movement onset and the almost equal distribution of electric fields over both hemispheres [12] (Fig. 3). In view of the fact that MEG can record only the current source tangentially oriented with respect to the head surface whereas EEG can record the current source regardless of its orientation, the above discrepancy may be explained by postulating that the crown of the precentral gyrus is activated bilaterally whereas the anterior bank of the central sulcus is activated mainly contralaterally. In addition, even in the subjects whose movement-related MEG shows activation of only the contralateral M1, desynchronization of the background activities time-locked to the movement onset, which can be investigated by the temporal spectral evolution, may be observed in the bilateral central regions [12,13]. Thus, since the unilateral finger movement can be associated with bilateral activation of the S1-M1 even in normal subjects, especially when the movement is complex and/or sequential, the bilateral activation of S1-M1 observed during the motor recovery in hemiparetic patients should be interpreted by taking a possibility into account that even a simple movement of the recovering hand may in fact serve as a relatively more complex movement for those patients due to the remaining difficulty.

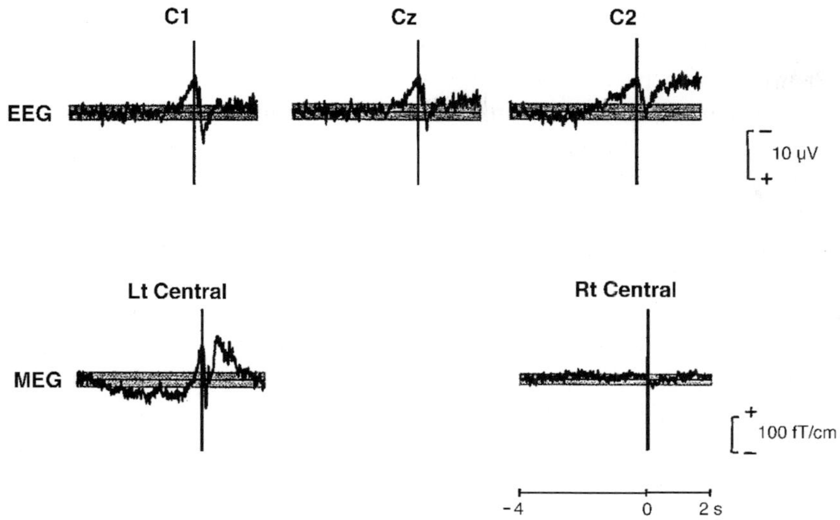

Fig. 3. Movement-related cortical potentials (EEG) and magnetic fields (MEG) simultaneously recorded in association with voluntary self-paced movement of the right index finger in a healthy subject. The late component of the premovement slow shift (NS′) is observed bilaterally on EEG, but almost exclusively at the contralateral (left) central region on MEG. (Recorded by Nagamine et al [12].)

Inhibitory mechanism of motor cortices

Recently, negative motor phenomena have drawn particular attention of many investigators who are involved in the field of motor control. The negative motor phenomena include various clinical conditions such as negative myoclonus or asterixis, atonic seizures or drop attacks, negative chorea, freezing phenomena seen in parkinsonian patients, and cataplexy and sleep paralysis seen in patients with narcoplesy, and physiologically muscle atonia observed during rapid eye movement (REM) sleep, and silent period of EMG discharge evoked by various stimuli (evoked silent period) [14].

Negative myoclonus is a myoclonic involuntary movement caused by abrupt interruption of muscle discharge or silent period. Negative myoclonus of cortical origin is of particular interest because it is often preceded by an EEG spike or spike-and-wave complex (epileptic negative myoclonus) [15–18]. Cortical negative myoclonus can be stimulus-sensitive and induced by somesthetic stimuli such as electric stimulation of the peripheral nerve. We have reported three cases of progressive myoclonus epilepsy in whom electric stimulation of the median nerve at wrist during sustained wrist extension elicited, following markedly enhanced cortical components of the somatosensory evoked potentials (SEPs), abrupt wrist drop associated with the silent period of 120–330 ms duration in the wrist extensor muscles [16]. The induced silent period was often either preceded or followed by an abrupt increase in the EMG discharge (cortical reflex positive myoclonus or C-reflex).

Since there was a positive correlation between the amplitude of SEP and the duration of the induced silent period, and furthermore, as the silent period was elicited also in the nonstimulated hand muscle only when the giant SEP was observed at the central region ipsilateral to the stimulus, it was interpreted to be a cortical reflex negative myoclonus, which is a negative counterpart of the conventional form of cortical reflex myoclonus.

Physiological mechanisms underlying the generation of cortical or cortical reflex negative myoclonus are still undetermined, but on the basis of the close similarity of its electrophysiological features to those of the cortical or cortical reflex positive myoclonus [19–21], it is plausible to speculate the abnormal hyperexcitability of the inhibitory components of S1-M1. With regard to the inhibitory M1 neurons, Schmidt and McIntosh [22] identified neurons in the monkey motor cortex which inhibited the muscle contraction when electrically stimulated (intracortical microstimulation). In humans, Wassermann et al. [23] observed inhibition of muscle contraction by transcranial magnetic stimulation of the motor cortex. Although the precise mechanism as to how these neurons inhibit the anterior horn cells remains to be elucidated, a strong input to those inhibitory motoneurons from the somatosensory cortex as the result of the hyperexcitability of the sensory neurons might explain, at least partially, the cortical reflex negative myoclonus.

In order to clarify the possible role of this inhibitory motor system in voluntary movement, we investigated the MRCP and PET-rCBF associated with voluntary relaxation of muscle or voluntary cessation of muscle contraction in normal subjects. The subjects were trained to drop the wrist by promptly terminating the contraction of the wrist extensor muscle without any accompanying contraction of other muscles, especially that of the wrist fexor. Before the voluntary contraction, both the early component of the premovement slow shift (Bereitschaftspotential; BP) and NS′ quite similar to those seen preceding the voluntary contraction of the wrist flexor muscle, which was adopted as a control task, were observed [24]. The topography of the BP preceding the voluntary muscle relaxation was only different from that preceding the contraction in that the former was more often associated with a positive slow shift over the frontopolar area as compared with the latter. Desynchronization of the background activity as studied by the temporal spectral evolution of the movement-related magnetic activities [13] demonstrated almost identical patterns in both movement tasks (Terada et al., unpublished data). Furthermore, the rCBF was almost equally increased in the contralateral S1-M1 and bilateral SMAs in both relaxation and contraction tasks (Terada et al., unpublished data). All these findings suggest the presence of an inhibitory system in the S1-M1.

Kitamura et al. [25] reported that the voluntary extension of either the middle or index finger alone was preceded by a larger NS′ than that preceding the simultaneous extension of the two fingers, suggesting an important role of S1-M1 in the discrete finger movement. Remy et al. [26] reported a greater increase in rCBF in the SMA and the contralateral S1-M1 in an isolated movement of a single finger than in the simultaneous movement of all five fingers. Since an isolated movement of a single finger is accompanied by intention, as well as effort to suppress the movement of other fingers, the larger increase in NS′ of the MRCP, as well as in rCBF in the discrete finger movement may be actually due, at least in part, to the activation of the inhibitory system of M1.

Another clinical condition in which the inhibitory motor system might be involved is dystonia. As the result of dysfunction of the basal ganglia input to the motor cortices, dystonia might be actually associated with a deficient inhibitory system of M1, causing excessive contraction of other muscles which otherwise do not participate in the intended movement. Deuschl et al. [27] reported diminished amplitude of NS′ over the contralateral central region before the movement of the affected hand in patients with focal dystonia or writer's cramp. We have observed diminished amplitude of NS′, but not of BP, over the central region contralateral to the affected hand in a patient with writer's cramp (Fig. 4). Furthermore, we found diminished amplitude of the late component of contingent negative variation (CNV) over the contralateral central region in patients with writer's cramp when the movement of the affected hand was employed as a response task to the imperative stimulus in auditory reaction time paradigm [28]. Ceballos-Baumann et al. [29], by studying six patients with idiopathic torsion dystonia with PET-rCBF, found, among other findings, significant underactivity in bilateral S1-M1 during paced joystick movements in freely selected directions, although the same authors [30] found rather overactivity on movement in bilateral S1-M1 in patients with acquired hemidystonia. The latter finding was interpreted to reflect the motor reorganization in the affected

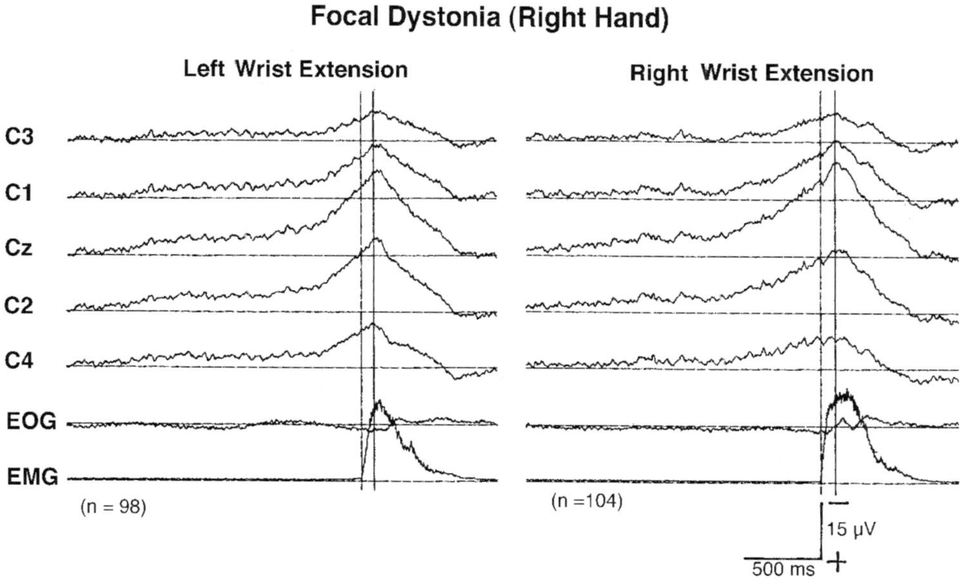

Fig. 4. MRCPs associated with voluntary, self-paced extensions of the left and right wrist in a right-handed patient with writer's cramp. With the movement of the right (affected) hand, NS′ is less asymmetric while in the movement of the left (intact) hand it is larger on the right hemisphere. (Recorded by Dr A. Ikeda.)

cortex. These findings may reflect impaired excitability of inhibitory motoneurons within M1, and support the above hypothesis.

Thus, if we postulate the existence of excitatory and inhibitory systems in the motor cortex, we can explain the positive and negative myoclonus as hyperexcitability of each system, respectively. In addition, as discussed above, dystonia might be associated with hypofunction of the inhibitory system.

Neural activation and noninvasive techniques

Several important factors should be taken into consideration when the results obtained by using different kinds of noninvasive techniques are interpreted. First, with regard to the orientation of neurons, particularly that of apical dendrites, within the activated brain area, EEG and MEG can record only the synchronous activation of a group of neurons which are oriented in the same direction, whereas theoretically the PET-rCBF study can detect the activity regardless of the orientation of neurons (Fig. 5). Secondly, direction of the overall current flow within a group of synchronously activated neurons with respect to the head surface is important for electrophysiological study, but not so much for the rCBF study with PET. EEG can detect summation of activities generated from both the tangentially and radially oriented current sources, whereas MEG can record only the tangentially oriented source. Thirdly, activation of the inhibitory neurons can be detected by all the techniques (EEG, MEG and PET-rCBF), but the target neurons which receive the inhibitory input from the inhibitory neurons cannot be detected by any of the techniques. Therefore, in addition to the characteristics of each technique employed as described in the Introduction, the biological characteristics of activated brain area should also be taken into account for the accurate interpretation of the data. Obviously this

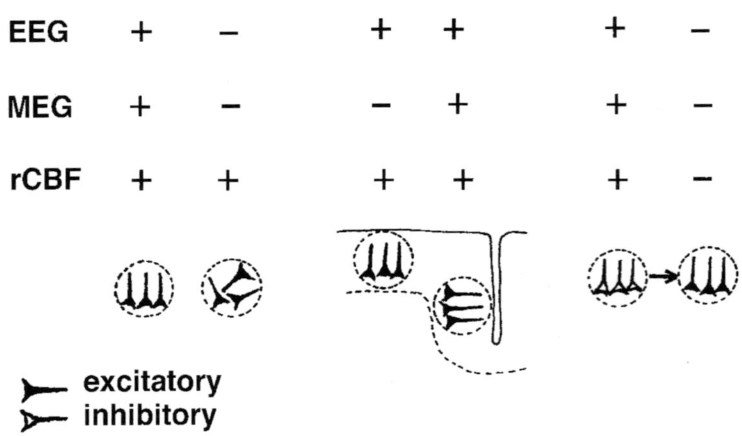

Fig. 5. Schematic illustration showing the different detectability of neural activation by EEG, MEG and PET-rCBF study, in terms of the orientation of neurons within the activated brain area (left column), the direction of the overall current flow in a group of activated neurons (middle column), and the activation of inhibitory neurons (right column). +: detectable, –: not detectable.

discussion holds true also for the brain functions other than those related to motor control.

References

1. Shibasaki H, Nagae K. Mirror movements: application of movement-related cortical potentials. Ann Neurol 1984;15:299–302.
2. Cohen LG, Meer J, Tarkka I, Bierner S, Leiderman DB, Dubinsky RM, Sanes JN, Jabbari B, Branscum B, Hallett M. Congenital mirror movements. Abnormal organization of motor pathways in two patients. Brain 1991;114:381–403.
3. Mayer M, Bötzel K, Paulus W, Plendl H, Prockl D, Danek A. Movement-related cortical potentials in persistent mirror movements. Electroenceph Clin Neurophysiol 1995;95:350–358.
4. Kitamura J, Shibasaki H, Takeuchi T. Cortical potentials preceding voluntary elbow movement in recovered hemiparesis. Electroenceph Clin Neurophysiol 1996;98:149–156.
5. Chollet F, DiPiero V, Wise RJS, Brooks DJ, Dolan RJ, Frackowiak RSJ. The functional anatomy of motor recovery after stroke in humans: A study with positron emission tomography. Ann Neurol 1991;29:63–71.
6. Honda M, Nagamine T, Fukuyama H, Yonekura Y, Kimura J, Shibasaki H. Movement-related cortical potentials and regional cerebral blood flow change in patients with stroke after motor recovery. J Neurol Sci 1996;(In press).
7. Kitamura J, Shibasaki H, Kondo T. A cortical slow potential is larger before an isolated movement of a single finger than simultaneous movement of two fingers. Electroenceph Clin Neurophysiol 1993;86:252–258.
8. Neshige R, Lüders H, Shibasaki H. Recording of movement-related potentials from scalp and cortex in man. Brain 1988;111:719–736.
9. Ikeda A, Lüders HO, Burgess RC, Shibasaki H. Movement-related potentials recorded from supplementary motor area and primary motor area. Role of supplementary motor area in voluntary movements. Brain 1992;115:1017–1043.
10. Shibasaki H, Sadato N, Lyshkow H, Yonekura Y, Honda M, Nagamine T, Suwazono S, Magata Y, Ikeda A, Miyazaki M, Fukuyama H, Asato R, Konishi J. Both primary motor cortex and supplementary motor area play an important role in complex finger movement. Brain 1993;116:1387–1398.
11. Kim S-G, Ashe J, Georgopoulos AP, Merkle H, Ellermann JM, Menon RS, Ogawa S, Ugurbil K. Functional imaging of human motor cortex at high magnetic field. J Neurophysiol 1993;69:297–302.
12. Nagamine T, Kajola M, Salmelin R, Shibasaki H, Hari R. Movement-related slow cortical magnetic fields and changes of spontaneous MEG- and EEG-brain rhythms. Electroenceph Clin Neurophysiol 1996;(In press).
13. Salmelin R, Forss N, Knuutila J, Hari R. Bilateral activation of the human somatomotor cortex by distal hand movements. Electroenceph Clin Neurophysiol 1995;95:444–452.
14. Fahn S, Hallett M, Lüders HO, Marsden CD (eds) Negative Motor Phenomena, Adv Neurol, vol 67. Philadelphia: Lippincott-Raven, 1995.
15. Lance JW, Adams RD. The syndrome of intention or action myoclonus as a sequel to hypoxic encephalopathy. Brain 1963;86:111–136.
16. Shibasaki H, Ikeda A, Nagamine T, Mima T, Terada K, Nishitani N, Kanda M, Takano S, Hanazono T, Kohara N, Kaji R, Kimura J. Cortical reflex negative myoclonus. Brain 1994;117:477–486.
17. Rubboli G, Parmeggiani L, Tassinari CA. Frontal inhibitory spike component associated with epileptic negative myoclonus. Electroenceph Clin Neurophysiol 1995;95:201–205.
18. Baumgartner C, Podreka I, Olbrich A, Novak K, Serles W, Aull S, Almer G, Lurger S, Pietrzyk U, Prayer D, Lindinger G. Epileptic negative myoclonus: An EEG-single-photon emission CT study indicating involvement of premotor cortex. Neurology 1996;46:753–758.
19. Shibasaki H, Yamashita Y, Kuroiwa Y. Electroencephalographic studies of myoclonus. Myoclonus-related cortical spikes and high amplitude somatosensory evoked potentials. Brain 1978;101:447–460.

20. Shibasaki H, Yamashita Y, Neshige R, Tobimatsu S, Fukui R. Pathogenesis of giant somatosensory evoked potentials in progressive myoclonic epilelpsy. Brain 1985;108:225–240.
21. Kakigi R, Shibasaki H. Generator mechanisms of giant somatosensory evoked potentials in cortical reflex myoclonus. Brain 1987;110:1359–1373.
22. Schmidt EM, McIntosh JS. Microstimulation mapping of precentral cortex during trained movements. J Neurophysiol 1990;64:1668–1682.
23. Wassermann EM, Pascual-Leone A, Valls-Sole J, Toro C, Cohen LG, Hallett M. Topography of the inhibitory and excitatory responses to transcranial magnetic stimulation in a hand muscle. Electroenceph Clin Neurophyisol 1993;89:424–433.
24. Terada K, Ikeda A, Nagamine T, Shibasaki H. Movement-related cortical potentials associated with voluntary muscle relaxation. Electroenceph Clin Neurophysiol 1995;95:335–345.
25. Kitamura J, Shibasaki H, Takagi A, Nabeshima H, Yamaguchi A. Enhanced negative slope of cortical potentials before sequential as compared with simultaneous extensions of two fingers. Electroenceph Clin Neurophysiol 1993;86:176–182.
26. Remy P, Zilbovicius M, Leroy-Willig A, Syrota A, Samson Y. Movement- and task-related activation of motor cortical areas: a positron emission tomographic study. Ann Neurol 1994;36:19–26.
27. Deuschl G, Toro C, Matsumoto J, Hallett M. Movement-related cortical potentials in writer's cramp. Ann Neurol 1995;38:862–868.
28. Ikeda A, Shibasaki H, Kaji R, Terada K, Nagamine T, Honda M, Hamano T, Kimura J. Abnormal sensorimotor integration in writer's cramp: Study of contingent negative variation (CNV). Mov Disord 1996;(In press).
29. Ceballos-Baumann AO, Passingham RE, Warner T, Playford ED, Marsden CD, Brooks DJ. Overactive prefrontal and underactive motor cortical areas in idiopathic dystonia. Ann Neurol 1995a;37:363–372.
30. Ceballos-Baumann AO, Passingham RE, Marsden CD, Brooks DJ. Motor reorganization in acquired hemidystonia. Ann Neurol 1995b;37:746–751.

Index of authors

Keyword index